The Magazine
in America
1741–1990

The Magazine in America

1741–1990

John Tebbel

Mary Ellen Zuckerman

New York   Oxford
OXFORD UNIVERSITY PRESS
1991

Oxford University Press

Oxford   New York   Toronto
Delhi   Bombay   Calcutta   Madras   Karachi
Petaling Jaya   Singapore   Hong Kong   Tokyo
Nairobi   Dar es Salaam   Cape Town
Melbourne   Auckland

and associated companies in
Berlin   Ibadan

Published by Oxford University Press, Inc.,
200 Madison Avenue, New York, New York 10016

Oxford is a registered trademark of Oxford University Press

Library of Congress Cataloging-in-Publication Data
Tebbel, John William, 1912–
The magazine in America, 1741–1990 /
John Tebbel, Mary Ellen Zuckerman.
p.   cm.   Includes bibliographical references (p.   ) and index.
ISBN 0-19-505127-0
1. American periodicals—History.
I. Zuckerman, Mary Ellen.   II. Title.
PN4832.T43   1991
051'.09—dc20   90-7874

2 4 6 8 9 7 5 3 1
Printed in the United States of America
on acid-free paper

Foreword

In 1991, the American magazine industry will celebrate, or at least acknowledge, 250 years of its existence. That is longer than the life of the Republic, and although books and newspapers are older still, they do not surpass periodicals in their influence on life in this country. Numerically, magazines outnumber all the other media, and in the tremendous breadth of their specialization, they reach into the smallest segments of American life.

As a historical and cultural phenomenon, periodicals have not enjoyed as much attention from scholars as newspapers and books in the literature of media. For the period from 1741 to 1905, the rock on which all research must rest is the four volumes of Frank Luther Mott's *A History of Magazines in America,* winner of a Pulitzer Prize and coming as close as humanly possible to meriting the word "definitive." Everyone who explores this period must begin with Dr. Mott, and we have done so, but we have also viewed his work through the eyes of later research and the perspectives provided by recent social and cultural historians.

The bulk of our book, however, deals with the period from 1918 to the present, and here we have had the advantage of both scholarly and popular examinations of the periodical press, which have proliferated remarkably in our time. It was, for instance, the recognition by one medium of another's importance that led the *New York Times* in the eighties to create a department in its Business section called "The Media Business," where developments in the magazine world, among others, have been given more and better coverage than ever before in the newspaper press. This coverage was of exceptional help to us in sorting out the intricacies of the periodical scene today. Other magazines, too, examine what is going on in their world, notably *New York,* and the business press has increased its coverage. *Folio,* the trade magazine created by Joseph Hanson, has been particularly helpful. In the past two decades or so, there have also been numerous books examining the more spectacular events of the periodical business.

It is not hard to understand why magazines have been given so much attention in our time. More and more, they are seen as living history, recording the incredibly complicated life of this century in a breadth of detail unapproached by any other medium. Looking back, we can see that this has been the case from the beginning. Most of the visual images we have of nineteenth-century life in America, for example, come from magazine pages that first displayed woodcuts and, in time, photography to illuminate the world.

The flow of life in this country, which the magazines have mirrored, is so immense that it nearly overwhelms the historian who would attempt to encompass 250 years of it. It took many thousands of magazines to make up the composite mirror described in these pages. Obviously, in describing such an immensity, there are bound to be omissions: of people, of events, of magazines themselves that become the victims of a necessary triage in the writing of a single-volume history. Another Frank Mott and three or four more volumes would be necessary to do justice to the period from 1905 to the present. That is unlikely to happen in the foreseeable future, and consequently the present volume, with its bibliographical references, must suffice.

We have attempted, perhaps optimistically, to include what is most significant and what most truly reflects the development of magazines in America and regret that it is inevitably not possible to cover everyone and everything.

We have had essential help from many people in writing this book. Sarah Miles Watts deserves first thanks for bringing us together as collaborators. The librarians at the State University of New York College at Geneseo, particularly Harriet Sleggs, helped to track down hard-to-locate items and submitted endless interlibrary loan requests. Library research work by three Geneseo students—Ron Brown, Sue Hesselbarth, and Jim Kelly—helped enormously. For aid in clarifying some of the intricacies of modern print technology, thanks must go to Mark Thalheimer at the Gannett Center for Media Studies, Columbia University.

In a more personal way, Mary Ellen Zuckerman wishes to thank, for their continuing encouragement, Roseann Bellanca, Ann and Nir Berzak, Audrey Broner, Mary Carsky, Liz Faue, Lana Flame, Joanne Gerr, Lynn Gordon, Miriam and Mitchell Silverman, Bob Jacobsen, Debbie Joseph, Carol and John Malach, Connie Waller, Ruth Waller, and Julia Walker. She also wishes to thank Alice Rubenstein for her insights and support during the past four years. Finally, she thanks her husband, Miron, for all his love.

Southbury, Conn.   J. T.
Geneseo, N.Y.   M. E. Z.
July 1990

Contents

Part IV Magazines Since the Second World War

I

The Creation of
Magazine Audiences

(1741–1865)

Publishing for Elites:
The First Magazines
and Their Audiences

One of the oldest axioms in the business world is that the way to succeed is to find or manufacture a need and fill it. We have built a vast industrial system on that idea, powerfully aided in this century by the art of advertising. Media history demonstrates clearly that the printed word has been subject to the same demand-and-supply theory. First, the demand was for books, which the Cambridge Press began to satisfy in 1639. Then, as the population grew, the need was for newspapers, the first of which was published in 1690. Magazines were the last to appear, half a century later.

There had been no apparent need for them until that late date. In America, as in England, magazine reading had been essentially a pleasure of the upper classes. Colonials interested in magazines imported them from England. It was not until the literary and practical arts in America had developed sufficiently that an audience was created large enough to justify publication of American periodicals.

There is some doubt about who originated the first magazine in America, in 1741. Credit is often given to the ubiquitous Benjamin Franklin, who is known to have planned in 1740 a magazine he called, in the usual grandiose style of the day, *The General Magazine, and Historical Chronicle, for All the British Plantations in America.* He did not pull it off his press, however, until February 16, 1741, three days after his rival Philadelphia printer Andrew Bradford had appeared with the similarly titled *American Magazine, or A Monthly View of the Political State of the British Colonies.*

Both these ambitious printers were taking a large risk, setting the tone for all subsequent magazine publishing. There was no pressing need for a new medium in America, and the waiting audience was difficult to measure. In all

3

the colonies that year, the total population amounted to not much more than a million people. In the next half-century it would grow to about 4.5 million, black and white, scattered along more than 1,200 miles of Eastern seaboard and extending westward at some points a thousand miles. The roads were so bad it took a stagecoach, which presumably would carry the magazines, from eight to ten days to reach remote settlements. The mails were a little faster, but it was a long time before either magazines or newspapers would be admitted to their custody. When Queen Anne's Act of 1710 established a general post office, it did not provide for periodicals.

With such immediate difficulties, and a host of others, it is surprising that even Franklin would attempt a magazine, yet between 1741 and 1794, forty-five brave printer-publishers made the attempt. They believed that a need for magazines existed, and they hoped to make money supplying it. Their premise was simple. If such British magazines as the *Gentlemen's* and the *London Magazine* could find profitable audiences, it was reasonable to suppose that there were enough of the same cultured, educated men now in America, many of them once English residents themselves, who would welcome a magazine like those London successes.

In fact, the early entrepreneurs had no intention of creating a unique American product. What they intended was to produce British magazines in America, faithfully copying their overseas counterparts and filling much of their pages with material clipped brazenly from the English publications. There was little, if anything, that was American about these early magazines, even for some time after the Revolution.

Yet a striking paradox exists. While they were faithfully copying and emulating, the editors and publishers (often the same) were motivated by a rising patriotism, an urge to show the English, who openly sneered at nearly everything American, that the colonies (and then the new nation) were worthy of respect.

The upstart Americans earnestly wanted to be admired, leading one of the first magazines, even in 1741, to assert that among the reasons it had been started was so "That the Parliament and People of Great Britain may be truly and clearly informed of the Constitution and Government in the Colonies, whose great Distance from the Mother-country seems in some sense to have placed them out of her View." This feeling evolved quickly into an urge to show the English that Americans were not uncivilized barbarians, but people capable of producing a cultural life as good as the mother country's. Eventually, after the Revolution, these feelings crystallized into a patriotic nationalism, which was frequently tempered by self-doubts.

Patriotism gave the early magazine publishers a reason to seek financial support, of which they were always in need. "Shall we not exert ourselves to appear as respectable abroad as we really are at home?" the *New York Magazine* asked more than rhetorically in 1796. But most of the periodicals produced in the eighteenth century were not likely to convince doubters abroad or at home. The first attempts in Philadelphia were brief: Bradford folded his magazine in three months; Franklin, in six. Boston printers issued three numbers of the *Boston Weekly* two years later, but they were feeble collections of English clippings, some mediocre verse, and a smattering of news.

It was not until the eve of the Revolution, however, that a magazine more truly American appeared. It was the *Royal American,* produced in Worcester by that master printer and publisher Isaiah Thomas, already a newspaper entrepreneur and book publisher. Involved as he was with the imminent revolution, Thomas kept it going only six months before he turned it over to another printer, but in that time and for nine months afterward, it printed a series of engravings by Paul Revere and articles about domestic events by some of the best writers and became the first periodical to use illustration liberally.

One magazine published in this early period may have been the best. H. E. Brackenridge's *United States Magazine,* begun in 1779, was the prototype American periodical—that is, it was distinctively American in character. It was given shape by a brilliant editor, and it dealt primarily with things American. Brackenridge, who became a Pennsylvania Supreme Court justice after the war, was an accomplished writer and contributed much of the magazine's material, giving it an individuality and a quality previously lacking.

One of Brackenridge's wartime targets was James "Jemmy" Rivington, the Tory editor who returned to New York after the British captured it and made life as miserable as he could for George Washington and his struggling army. Another Brackenridge object of scorn was General Charles Lee, the ambitious eccentric who hoped to put himself in Washington's place and was eventually cashiered. Brackenridge printed an indiscreet letter from Lee to a young lady, and in an odd preview of a similar incident involving Ethan Allen and Rivington, the general arrived at Brackenridge's door armed with a horsewhip. The publisher wisely did not respond to his knock but put his head out an upstairs window and inquired what was wanted.

"Come down and I'll give you as good a horse-whipping as any rascal ever received," Lee shouted up to him.

"Excuse me, general," Brackenridge replied, "but I will not come down for *two* such favors."

Obviously, a new medium was being born in America after the Revolution, but it had been a hard and exceedingly slow birth since 1741. From that date until 1794, there were never more than three magazines in the country at any one time, and half of those published in the period were issued during its last eight years. Nevertheless, the shape of things to come was emerging. Publishers were already finding specialized audiences within the larger one.

Eighteenth-century magazine publishing, the beginnings of today's industry, reached another high point in 1787 with the issuing in Philadelphia of the *Columbian Magazine* by an engraver named John Trenchard. The *Columbian* was the first periodical to provide a retrospective view of revolutionary history and the politics surrounding it. Two years later, Isaiah Thomas turned again briefly to periodicals with the *Massachusetts Magazine,* notable primarily for its excellent engravings and the variety of its contents, much of it original.

More than any other magazine of the period, *New York*, published in the young metropolis by Thomas and James Swords ("T. and J. Swords" was their logo) displayed a particular interest in the theater, the first to do so in an extensive fashion, making it a valuable source for theater historians. It also carried poetry and contributions from such writers as Charles Brockden Brown and Noah Webster. A "Monthly Register" of current events, set in small type, foreshadowed the present *New Yorker*'s listings. A significant departure from past formulas was the large amount of space given to extracts from legal, cultural, and medical books, often excerpted at length.

Indifference was the greatest obstacle magazines had to overcome in these early days. A vigorous public demand for their product would have enabled the publishers to overcome all their other problems much more easily. But Americans were too busy working to be reading magazines, as they did newspapers. With its last issue, *New York Magazine* uttered a despairing cry: "Shall every attempt of this nature desist in these States? Shall our country be stigmatized, odiously stigmatized, with want of taste for literature?"

In form, the first American magazines looked much like the British publications they copied, mostly octavo pages of five to six by eight to nine inches, ranging from thirty-four pages in the earliest issues to sixty-four near the end of the century. A few were even larger; for example, the *American Museum* had a hundred pages in every issue. The rag paper looked rough, but it was extremely durable and survives today in the collections of libraries. Covers were not always used but appeared in a lighter-weight colored stock when publishers thought it wise to be more expansive. Text was set in what we would call 6-point type today, although a few were in the larger 10-point.

The content of the early periodicals was a mix of comments on manners, fashion, social life, religion, morals, and politics—no long fiction, however. In 1782, the *Columbian Magazine* warned that, "Novels not only pollute the imaginations of young women, but also give them false ideas of life. . . . Good sentiments scattered in loose novels render them the more dangerous."

Instead of novels, or excerpts from them, editors gave their readers the kind of short piece that would later evolve into the short story. They were called "character sketches," stories out of someone's life, or "fragments," brief sentimental stories intended "to improve the heart," as one editor put it. They were written in the most stilted language—"The conscious blush crimsoned her countenance," one began—and if they "improved the heart," it was only in the sense of reinforcing the patriarchal moral values by which society was presumed to live. Young women who strayed from prescribed paths, these stories said, would either be punished by others or succumb to their own misery. These views would dominate fiction for more than a century, both in magazines and in books.

The line between magazines and newspapers was still blurred in the late eighteenth century in terms of content. After the Revolution, both continued to devote much space to politics, and both printed a good deal of literary material. Magazines even published a respectable quantity of news, particularly calendars of such current events as births, deaths, and marriages, as well as such newspaper staples as meteorological tables and current prices.

If there was one perennial subject in the journals, it was women, a topic explored from every conceivable angle, nearly always by men. Women were both deified and deplored, and that was to be the case well into the next century. But there were other controversial subjects. Dueling was attacked and defended; alcohol and its social consequences were debated; and already magazines were talking about the great controversy that would divide the nation— slavery. Education was discussed often, particularly in Noah Webster's *American Magazine*, in which he argued against use of the Bible in schools (although most other editors favored it) and advocated (along with his fellow editors) a strong emphasis on basics. Reading the *American* today on education is to experience a curious time warp.

What emerged from the clutter of verbiage in the eighteenth-century magazines was a composite picture of the nation's social and political life, in greater depth than the newspapers could provide, particularly when illustrations appeared. The writing was often atrocious, and much material was borrowed at first, but magazines from the beginning held up the mirror to national life, thus becoming a prime source for the social historian.

2

Rise of the General Magazines

As the nineteenth century began, there was a new energy in the business of making magazines. Old problems remained, but they were not as frustrating as they had been, at least for the better publications. Between 1800 and 1825, there was a surge in the number of new periodicals, the first warnings of a veritable magazine tsunami between 1825 and 1850, induced by the technological breakthrough in printing with the invention of the cylinder press and the rapid growth of a highly literate population that was eager for knowledge and entertainment.

As magazines found their audiences in this expanding national market, the broad pattern of the future industry began to be established. Specialized audiences developed quickly, particularly those for religious journals, but the major event after 1825 was the rise of the general magazine, which would dominate the consumer field until our own time. During the first quarter-century, too, more attention was paid to politics in the magazines than to literature, in spite of a rising demand for a literature that Americans could call their own.

The most important of these early nineteenth-century periodicals was Joseph Dennie's *Port Folio,* one of the numerous weeklies that dominated the business. As was the case for most of the earlier periodicals, the *Port Folio*'s success was due almost entirely to its publisher. Joseph Dennie announced in his first issue (from Philadelphia) on January 3, 1801, that his eight-page quarto was "not quite a Gazette, nor wholly a Magazine, with something of politics to interest Quidnuncs, and something of literature to engage Students." Dennie was shrewdly aware of how necessary it was to harness politics with literature in order to succeed. In spite of his disclaimer, however, the *Port Folio* was unquestionably a magazine, not a newspaper.

Dennie's pen name in this anonymous era was "Oliver Oldschool," although he was known also as the "Lay Preacher." He had learned his craft on the *Farmer's Museum*, in Walpole, New Hampshire, after graduating from Harvard (rusticated, however, in his senior year) and making a half-hearted attempt to study law. He became editor of the *Museum*, wherein he wrote the "Lay Sermons" for which he is best known, and surrounded himself with such congenial spirits as Royall Tyler, America's first playwright. Offered a government job in 1799 as secretary to Secretary of State Timothy Pickering, Dennie soon moved over to writing politics in a Philadelphia Federalist paper, *Gazette of the United States*, and two years later launched his magazine.

In spite of its strong Federalist leanings, the *Port Folio* was less devoted to politics than to literature, and Dennie's friends, who also became contributors, included Joseph Hopkinson, the composer of "Hail, Columbia"; Richard Rush, who became minister to England and France; John Quincy Adams; Charles Brockden Brown, sometimes called the first American novelist, though not without dispute; Alexander Wilson, the ornithologist; and Gouverneur Morris. Over the whole magazine, however, hovered the Spirit of Dennie himself, a writer of wit and incisiveness, a superb critic, and a man of taste—a charismatic figure.

For eight years, the magazine sparkled under its founder's direction, but then, like so may others, Dennie became ill and discouraged with the difficulties of keeping the *Port Folio* alive and sold it. Under others, it survived as a much less entertaining monthly until 1825, when it expired just before the great boom in magazines began.

Early nineteenth-century magazines proliferated until nearly every town of any consequence in America could boast a weekly literary miscellany of some kind, but nearly all of them were transitory, seldom lasting more than two or three years. Having no longer any need to steal entirely from British publications, they stole from each other.

There was also a continuing struggle with postal regulations. As a result of the Post Office Act of 1794, postmasters general were given the power to accept or refuse any magazine, with no reason required. These federal functionaries, in general, did not like magazines, arguing that they cluttered up mail sacks, which were already too large. If they had been able to get away with it, they would have given access only to religious publications. But some individual postmasters were much more lenient, and this enabled magazines to survive. More and more periodicals came to be accepted until, by 1825, most went through the mails unless there was a cheaper way. At the beginning of the century, there had been only a dozen American magazines; by 1825, there were nearly one hundred.

Among them were a few that stood out from the others. One was the *Saturday Evening Post*, evolved in a circuitous way from Franklin's old newspaper, the *Pennsylvania Gazette*, thus providing the *Post* of our time with the claim, however dubious, that it had begun with Ben and so was the oldest magazine in America of continuous publication. One of its rivals was the *New York Mirror*, founded in 1823 by two poets, George Pope Morris and Samuel Woodworth. During its two decades of life, it was a prime factor in creating and sustaining the Knickerbocker literary school. Still another was the *United States Literary Gazette,* a Boston semiweekly, whose contributors included William Cullen Bryant, Henry Wadsworth Longfellow (still a student at Bowdoin), and other New England writers.

The first attempts to reach women as a particular audience were made soon after the turn of the century, but major efforts had to wait until the century's second quarter. From the beginning, however, clothes and their possible effect on female virtue interested male editors most. When French fashions arrived in America soon after 1800, one outraged contributor to the *Monthly Anthology* in 1804 wrote: "We have imported the worst of French corruptions, the want of female delicacy. The fair and innocent have borrowed the lewd arts of seduction. . . . I will not charge them with the design of kindling a lawless flame. They will shudder at the suggestion. But I warn them."

Medical journals were beginning to proliferate, although few of them lasted more than a short time. From the beginning, both the scientific and general magazines regularly carried articles about agriculture, but in 1810 a periodical devoted solely to the subject appeared in Georgetown, the *Agricultural Museum,* forerunner of many others to be published before the century ended.

Among the special classes of magazines beginning to rise in the first quarter of the century were the theatrical reviews, testifying to the growth of American cultural interests. They were as ephemeral as the plays themselves, however, as were the comic periodicals, forerunners of the comic-book craze of our time as well as of general humor magazines. Many of them were highly political, and the humor was primitive.

Fiction in the magazines continued to be considered disreputable by many people, including editors. It had to struggle for general recognition in the magazine world, as it did with the public, because for so long it was lumped in with such amusements as theater, dancing, gambling, cockfighting, and horse racing. Nevertheless, fiction remained an irrepressible force.

An obstacle to the development of both fiction and articles was the paucity of Americans who thought of themselves as professional writers, at least for some time. The occupation not only was considered a little disreputable but was unprofitable as well. Often editors refused to pay for contributions, and they reserved the right to do anything they liked with manuscripts, even rewriting what did not please them. Payment, in fact, was rare in America until 1819. When the *Atlantic* at its founding in 1824 became the first to promise payment to contributors, its editors believed some might not accept it and urged them to take the money as an honorarium, "for the principle of the thing."

Editors fared little better. Often there were no salaries at all, not even an honorarium, and on other periodicals, editors were paid in proportion to the magazine's financial condition, which was usually uncertain. That made magazine editing a part-time job in most cases.

After 1825, however, the face of magazine publishing began to change rapidly with the dawning of what was termed first by Mott as the "Golden Age of Magazines." It was an age of upheaval everywhere in the world. American magazines recorded and discussed revolution in France, Belgian independence, the struggle of Spanish liberalism, the rise of Young Italy under Mazzini, the Metternich repression in Austria, and the emergence of Greece as a nation. At home, many Americans felt that even greater changes were taking place, and their thinking and feeling about what was happening were reflected in the great surge of new magazines, a development helped immeasurably by the arrival of the cylinder press, the first major change in the printing process since the fifteenth century. Obviously, an America was coming into being that required something more substantial than the kind of magazines known previously, with their many limitations. A new general audience was waiting for them.

To reach a new literate mass readership, an extraordinary explosion—no other word seems adequate—of magazines occurred between 1825 and 1850. Since then, there have been several other golden ages even more spectacular in some respects, but it remains true that this period represented a major turning point in the history of American periodical publishing.

Reliable statistics are difficult to come by, and even those we have must be suspect, but roughly, the number of magazines rose from about 100 in 1825 to about 600 in 1850. The figures do not reflect the number of starts made in the period, which Mott estimates at between 4,000 and 5,000—magazines whose average life was two years.

While weeklies had predominated before, the general monthly now took center stage, and several important magazines were created by a different kind of entrepreneur. One of them was George R. Graham, no part-time dilettante or follower of a particular pattern but a born magazine publisher, although he started out to be a cabinetmaker (meanwhile studying law to better himself). In 1839, Graham bought a dying magazine called the *Casket* (one of many by that name) and changed it from a cheap-looking anthology of trivia into a bright and attractive periodical. A year later he bought the *Gentleman's Magazine*, which had been edited by the noted actor William E. Burton, and combined the two under a new title, *Graham's Magazine*. It quickly became one of the nation's three or four most important periodicals, displaying "a brilliance which has seldom been matched in American magazine history," as Mott described it.

For a short period in 1841–1842, Poe was *Graham's* literary and contributing editor, and the list of those who wrote for the magazine then and later was like a roll call of the best American writers and poets, accounting for a liveliness not common in the new crop of publications. Graham was the first publisher to understand that if contributors were paid well, there would never be a short supply of excellent writers in a magazine. One of the most frequent to appear was Nathaniel Parker Willis, usually considered to be the first "magazinist"; that is, one who makes a living by writing for magazines.

A rival of *Graham's* but of a different kind, was the *Knickerbocker Magazine*. After a faltering start in 1833, it acquired a new editor, Lewis Gaylord Clark, who made it one of the best general magazines in the country for the next quarter-century; the magazine was affectionately known as "Old Knick." Most of those who wrote for *Graham's* also wrote for the *Knickerbocker*—Irving, Cooper, Bryant, Willis, Longfellow, Hawthorne, and Whittier, among others.

There were all kinds of anomalies in the burgeoning magazine world of this period. One was the prevalence of the literary weeklies, springing up all over the country, two or three in a week on occasion. It took little more than faith and a hundred dollars to start one, but few survived.

Two of the longest-lived magazines were born in the decades before the Civil War, lingering into our own time as spectres from that other and far different world of the nineteenth century. One was the *North American Review*, begun in May 1815 and continuing into the late 1930s. Among its long list of editors was Jared Sparks, the noted Harvard historian; Edward T. Channing, a young lawyer and William Ellery's brother; Edward Everett, famous as scholar, orator, and writer; John Gorham Palfrey, professor of sacred literature at Harvard; James Russell Lowell; Charles Eliot Norton, president of Harvard; Henry Adams, and several lesser lights. Inclined to be dull for its first sixty years in spite of its brilliant contributors, the cream of New England intellectuals, the *North American Review* veered from dullness to brilliance and back again as time went on, but it was never an inconsiderable publication. Before it died, the *Review* had been for more than a century and a quarter, in Mott's words, "a remarkable repository, unmatched by that of any other magazine of American thought. . . ."

The second was the *Youth's Companion*, which became a part of the national memory, loved by generations of boys and girls. Beginning in 1827 as an improvement over the standard diet for young people of primarily religious and moral instruction, the *Companion* hoped to entertain as well as instruct; and so it did under the editorship of Nathaniel Willis and his partner, Asa Rand. Willis sold the magazine in 1857 (he was seventy-eight and weary), and in other hands the *Companion* lost something of its paste-pot character, along with its anonymous contributors, and began to print stories by such popular writers of the day as Harriet Beecher Stowe.

When the premium craze gripped the magazine business in the late 1860s, the *Companion* led the way and built its circulation to new heights with this kind of promotion. To that sort of marketing could be added the other ingredients of its success: continued stories with special appeal to young people, and the shrewd cultivation of interests affecting the whole family. In the 1880s, the *Companion* introduced its long series of articles by celebrities about their work: Theodore Roosevelt, Grover Cleveland, Booker T. Washington, Henry M. Stanley, Robert Peary, Lillian Nordica, and Marcella Sembrich, among many others. After the first decade of the twentieth century, however, the magazine went into a slow decline, dropped to 300,000 in circulation, and was sold to the Atlantic Monthly Company in 1925. Four years later it was merged with another young people's favorite, the *American Boy*, yet leaves behind happy memories among former readers still alive.

All things considered, the magazine business had assumed by the 1850s much of the character it has today. Magazinists were now a distinct class, although many had other occupations as well, and there were a growing number of specialists writing for the fast-increasing number of specialized magazines. Among the new crop of contributors were women, writing under their own names and rapidly becoming great favorites with the mass-market audience through their paperback romances. The most prolific of them was Mrs. E.D.E.N. Southworth, who became a veritable fiction factory before the war. Her books, most of them serialized in magazines, remained in print well into this century, dreadful as they were by modern standards.

Surprisingly, considering their continuing numerous problems, magazines carried as much illustration as they could afford, even though it was expensive, large plates costing as much as $1,000. Engraving on copper and steel was the method employed, and such work was essential to women's magazines using fashion plates. Other periodicals used them to illustrate city life, to depict actors and actresses, or to illustrate science and nature pieces. Later, during the Civil War, magazines like the two periodicals published by Harper's used illustration to cover the battles and civilian life, much as the picture-text magazines of our time do.

The period from 1825 to the guns of Fort Sumter was a time of significant beginnings, but of no really remarkable achievements until the 1850s and 1860s. The great explosion produced not only an astonishing flood of periodicals but broadened the industry into all kinds of specialization as well as tapping the mass market. For most entrepreneurs, it was still not a profitable business. Nevertheless, in spite of all caveats, American magazines were offering readers a comprehensive view of national life in the 1850s—a mirror held up to an expanding, struggling, chaotic country that was on the verge of postwar greatness.

3

The Magazine as a Political and Cultural Influence

After 1835, with the advent of James Gordon Bennett and his New York *Herald*, newspapers devoted themselves more and more to coverage of the news. Before that, they had been little more than propaganda organs in the hands of political parties, on whose bounty they were largely dependent. At about the same time, magazines were becoming forums for public opinion in ways they had not been before, by their nature offering a more varied and more intellectual sounding board for political argument than the newspapers had been able to provide. There was some overlapping, of course; in certain cases it was hard to draw a line between what was newspapering and what was magazine making.

Reform of one kind or another (more often moral or religious in the early days) had always been a preoccupation of magazine editors, and they were not slow to take up the great question of the day—slavery. As early as 1819, the antislavery magazines began to bloom like spring flowers, particularly in the Northwest Territory. Quakers were the first builders of this political platform, soon to be a national stage. Charles Osborn began his *Manumission Intelligencer* in 1819 (it lasted only a year), and Elijah Embree launched his *Emancipator* in 1820, both of them in Jonesboro, Tennessee.

In 1821, the heroic figure of Benjamin Lundy emerged to start his *Genius of Universal Emancipation* in Mt. Pleasant, Ohio. Never was a periodical begun on such minimal terms. Lundy had no capital and initially only six subscriptions. Every month he took the copy for his magazine and walked twenty miles to Steubenville, where it was set and printed, after which he carried the copies to Mt. Pleasant on his back.

The next year he moved the magazine nearer to his Quaker friends in Jonesboro, settling in a northeastern Tennessee town called Greenville, but with an eye to better things, he decided two years later to move once again, this time

14

to Baltimore. It was a long walk, through forests and over mountains, but Lundy did it with his knapsack on his back. Once settled, he went on promoting the cause of gradual abolition and colonization. Lundy was a pioneer in the colonization movement; that is, freeing the slaves, then settling them somewhere else, not in the nation's backyard. This became a national crusade, once favored by Lincoln, and resulted in the establishment of Liberia.

Lundy did so well in Baltimore that he made the *Genius* a weekly in 1826 and might have become even more influential there if he had not been set upon and nearly killed by a slave dealer. Undeterred, he went on walking tours through the northern states, preaching abolition wherever he stopped and continuing to issue his magazine spasmodically. Lundy suspended the *Genius* temporarily in 1829 and walked to Bennington, Vermont, where he invited William Lloyd Garrison, whom he had met earlier in Boston, to become assistant editor of his periodical, even though he knew Garrison wanted no part of colonization but demanded immediate and complete abolition.

In spite of this major difference, Garrison agreed to join Lundy. After working with him for about a year, he was arrested on a criminal libel charge for a story he had written that annoyed the owner of a vessel whose cargo was a profitable link in the slave trade. It was after this incident that Garrison, released from jail, went back to Boston and started the abolitionist paper that made him famous, the *Liberator*. Lundy moved too, down to Washington, where he began to issue the *Genius* again as a monthly. But Lundy was a restless man and before long moved for the last time to Lowell, Illinois, where he died. The *Genius* was taken over by other Quaker sympathizers and persisted both in Lowell and in Chicago until 1855.

The antislavery magazines of these early days were more or less religious, written often in an overly sentimental style, and offering a platform for humanitarians of all shades. The problem was that they were single issue periodicals and so tended to be overshadowed by more popular publications that dealt with the other seething controversies in antebellum America: states rights, so intimately involved with the slavery issue; squatter sovereignty, another ingredient in the slavery stew; the tariff, a perennial subject of debate in American politics; and what was later called "nativism," the beginnings of "America for Americans" as a political test of patriotism in the eyes of its adherents.

By 1850, this cauldron was bubbling furiously, and it was stirred vigorously by a variety of political papers and periodicals. Looking back over the decade of the fifties, the superintendent of the eighth census, in 1862, reported that such magazines had greatly increased. In 1850, already 1,630 had fallen into that category, but by 1860 there were 3,242, nearly a 100 percent increase.

Large areas of the magazine business shunned politics, however, and did not open their pages to the great debates sweeping the nation. Such major general magazines as *Harper's* and *Graham's*, the women's magazines, and most of the religious quarterlies appeared to be operating on another planet, or at least well above the storm. This was all the more strange, since by the mid-fifties, the nation had turned into a tumultuous debating society, shot through with violence. On the eve of war, however, many of those magazines that had managed to stay aloof were embroiled in one way or another with the conflict that would be our greatest national tragedy.

On the highest level, the magazines became debating platforms for serious discussions of the issues—or partisan salvos, as the case might be. The newspapers were embroiled as well, of course, but by their nature they could not offer the same kind of broad-ranging, semi-national forums for political debate. It is in the magazines that we find not only discussion of slavery but also expression of personal and national feeling about the Compromise of 1850, the Kansas-Nebraska bill, the *Dred Scott* case—even John Brown's raid—not to mention the presidential campaigns themselves, which were fought out more thoroughly in the magazines than they were in the newspapers.

In light of present-day political controversies, it is worth noting that free trade was already established in the 1850s as an American shibboleth—what Francis Liebar called in a *DeBow's Magazine* article "one of the great subjects of national theology." A few other magazines were also aware of the economic causes of the national upheaval over slavery, as the *North American Review* and the *Atlantic* testified with their articles on "King Cotton."

Even the good gray poets of New England plunged into the emotional cauldron of the fifties. For the *National Era*, the Washington magazine in which *Uncle Tom's Cabin* first appeared as a serial, Whittier wrote his "Ichabod," a blistering attack on Daniel Webster for his part in the passage of the Compromise. Webster was later defended by the *North American Review*.

The differences between magazines and newspapers narrowed in the increasing violence of the slavery debate. The major papers were fiercely partisan on one side or the other, and they were more than equaled sometimes by equally partisan magazines. The difference was that few newspapers of any consequence kept apart from the struggle, while many specialized periodicals treated the issue tangentially, if at all. They were compensated for, however, by the virulence of *Frederick Douglass' Paper*, the personal organ of the great black activist, and Garrison's *Liberator*, while on the other side, *DeBow's Magazine*, the best in the South, was more dignified but no less fervent in arguing the slavery cause. Even such a relatively staid literary publication as *Putnam's*, a subsidiary venture of George Palmer Putnam's book publishing house, lent its voice to criticizing the Supreme Court after the *Dred Scott* decision in 1857.

In the Midwest, where the slavery issue was increasingly more a matter of bullets than words, there was a rash of purely political magazines, such as the *Squatter Sovereign*, founded at Atchison, Kansas, in 1855, and the *Herald of Freedom*, operating in Lawrence for two years until its enemies destroyed it, after which the proprietor donated his scattered type to be made into cannon balls (which were called "new editions" of the *Herald* when they were fired in the struggle for Kansas).

In 1856, for the first time, many magazines took an active part in the campaign that elected James Buchanan as president. *Putnam's* gave its readers an excellent report of a torchlight parade in New York and later covered the election itself as well as any newspaper. The new *Harper's Weekly* was also beginning its career as a reporter of national events, and the equally new *Atlantic Monthly* spoke on the same subjects with a New England accent. In these and other magazines, the debate on the slavery question boiled and bubbled during all of this fateful decade, as it had for so long. One of the solutions suggested was to annex Cuba, then beginning its long struggle against Spain, and in the Southern view, make the island an extension of the American slave system to compensate the South for the growing number of free-soil states. In its first number, and in its first article, *Putnam's* surveyed the pros and cons of this plan.

There was now an entire category of antislavery magazines, much of its existing among the North's church periodicals. Congregationalists and Methodists in New York led the way, followed by the Quakers. These magazines were not entirely devoted to the question, by any means, but there were others that were, including Garrison's *Liberator*, Gamaliel Bailey's *National Era*, and *Frederick Douglass' Paper*, as well as the *National Anti-Slavery Standard*, published by the American Anti-Slavery Society. In Pittsburgh, there was the *Saturday Visiter* [sic], published by Mrs. Jan Grey Swisshelm, who had learned to hate slavery during a stay in Kentucky.

Among several others in the abolitionist chorus were those devoted to the notion of solving the problem by colonizing the slaves elsewhere. The Washington D.C., *African Repository* was the oldest of these periodicals, and there were others issued by the state organizations of the American Colonization Society. Aside from Frederick Douglass' passionate journal, the chief black voice in this chorus was New York's *Anglo-African Magazine*, edited by Thomas Hamilton, but its pages were more literary then political.

The only interruption in the gathering storm was the Panic of 1857, whose economic implications modern scholarship has identified more and more with the outbreak of war. Banks failed, stores went under, business in general sank into steep decline. This deep depression, which one journal, the *National Magazine*, believed had no precedent, persisted for at least two years. Its severe impact on American life can be studied in human terms today in the pages of *Frank Leslie's Illustrated Newspaper*, really a weekly magazine despite its name. *Harper's* and other weeklies also covered these events.

Frank Leslie had introduced something new into the magazine business with his various publications. His real name was Henry Carter. He had used the pseudonym to protect himself from a father who sternly opposed this son who wanted to be an artist, signing "Frank Leslie" to the drawings he submitted surreptitiously to the *Illustrated News* of London. He later became foreman of that magazine's engraving room. Coming to America in 1848, Leslie worked for other periodicals before he started his own in 1854, *Frank Leslie's Ladies' Gazette of Fashion and Fancy Needlework*. It did so well that he launched the *New York Journal of Romance* a year later, both of these preliminary to fulfilling his dream of years, an American version of the London *Illustrated News. Frank Leslie's Illustrated Newspaper*, later to be renamed *Leslie's Weekly*, met the criterion.

It began in 1855 with only sixteen pages, selling for ten cents and containing the seeds not only of Leslie's success but of modern picture-text magazines and news weeklies. Americans had never seen anything like this periodical. In it, they could find news stories illustrated graphically with engravings in a way the newspapers could not match. The pictures were large and striking, and the stories were seldom more than two weeks after the fact. But the little weekly did not confine itself to the news. It covered music, the theater, art, horse racing, and sports; reviewed new books; and published fiction serials. Fashions were added during the second year, and Leslie would have covered religious news, too, if fiction had not seemed more profitable and crowded it out. Leslie was no lover of organized religion.

Engravings often illustrated these other features in the magazine, although sometimes they were simply pictures of natural, historical, or home scenes. One at least still remains in American memory—"The Monarch of the Glen," showing a magnificent elk sniffing the air in a wild setting. It was reproduced thousands of times, well into this century.

The news pictures for Leslie's periodical were frankly on the sensational side. Sometimes the news itself was sensational enough—William Walker's filibustering war in Nicaragua; the violent events in "bloody Kansas"—but often the stories and pictures covered titillating murders or sex scandals.

Leslie's struggled at first, and the publisher was almost ready to give up, but by its third year, it had more than 100,000 readers. The chief obstacle to even greater success was the arrival of *Harper's Weekly* in 1857, an event bitterly resented by Leslie. He countered the competition by reducing *Leslie's* price to six cents, two dollars a year, and this sent its circulation up to 164,000 before the war.

From that point, Leslie could not be stopped. He became the first magazine entrepreneur to establish a publishing empire: *Frank Leslie's New Family Magazine* (1857); *Frank Leslie's Budget of Fun* (1858), with others to be added after the war, when his career—a career nearly as sensational as the magazine—began to take off.

Both in its earlier and later versions, from *Frank Leslie's Illustrated Newspaper* to *Leslie's Weekly*, this magazine introduced readers to such world events as Commodore Perry's historic visit to Japan, David Livingstone's explorations in Africa, the Crimean War, and the Austrian War of 1859. Nationally, it showed Americans marvelous pictures of the Great West now opening up and other scenes of a rapidly developing, strife-torn nation. But the magazine was at its best, or at least its most interesting, in its coverage of New York, the home base. *Leslie's* examined, for example, the fragrant career of Fernando Wood, the mayor who would lead the Copperhead opposition to Lincoln and the war. It provided vivid pictures of such events as Election Day 1850.

Remarkably foreshadowing the magazines of our time, *Leslie's* was present each week with pictures and text to tell readers about the laying of the Atlantic cable, early labor strife, the principals in the *Dred Scott* case, the Panic of 1857, and (with telling snapshots, so to speak) the multiplying incidents leading up to war. At the same time, the magazine covered many sports and was the first to feature a billiards department.

Leslie's from the beginning was also a crusading journal. It delivered a broadside attack against what it called the "swill milk" business, describing in a series the shockingly unsanitary conditions of the dairies supplying New York with milk and following a trail of corruption that led through the distilleries supplying the cows with feed, straight to the doors of New York's politicians, deeply involved with profits from the whole operation. *Leslie's* attacked head on and alone for some time, until other magazines took up the fight.

Here the idea of the picture magazine, used as a political weapon, came to flower for the first time. The publisher's hard-hitting editorials were supplemented by the far more effective pictures of filthy dairies, diseased and dying cows, and wagons carrying uncovered milk into the city. Municipal officials were forced to make an inquiry, but it was the old story of the criminals investigating the crime. *Leslie's*, even under threat of a criminal libel action, kept up the pressure, hiring private detectives to unearth new evidence.

As the magazine's all-out assault on the entire city administration came to a climax, the mayor was forced to appoint a committee from the New York Academy of Medicine to conduct an impartial investigation. The resulting report confirmed everything *Leslie's* had been saying, and even though it took two more years to clean up the mess, the magazine enjoyed not only a moral triumph but a sharply increased circulation. Frank Leslie himself was presented with a gold watch and chain, suitably inscribed by grateful citizens.

With that kind of crusading, people could easily forgive the way in which *Leslie's* frequently sensationalized crime news in words and pictures. Jealously criticized by the newspaper press, the magazine found it easy to sustain any attacks as its circulation reached 200,000 for a time.

In his politics and conceptions of morality, Frank Leslie was ambivalent and could sometimes be found on both sides, as in the case of prizefighting, illegal nearly everywhere, which his paper covered assiduously but was just as likely to condemn the next day. The publisher's enterprise was evident in his transatlantic coverage of a bout near London (also illegal) between the American hero "Benicia Boy" Heenan and a British fighter. A writer and an artist were sent to cover the bout, arrangements were made for engraving, and twenty-four hours after the fight, *Leslie's* was on the London streets with an illustrated extra, after which the plates and 20,000 copies were hurried to a New York–bound ship. The result was a clean beat on all of *Leslie's* competitors, both newspaper and magazine. The edition sold 347,000 copies.

Fervently against abolitionism, *Leslie's* approached the major events leading up to the war in the same enterprising spirit with which it approached prizefighting. Although it called John Brown a "maniac," the magazine covered the famous raid and the subsequent hanging with sensational engravings and text. In fact, its graphic picture of Brown at the end of a rope did much to stir up further an already inflamed public.

In the end, Frank Leslie decided it was better business to appear nonpartisan, but he found that objectivity did not pay. *Leslie's* came under heavy fire from critical guns in both the South and the North. After Fort Sumter, Leslie saw that impartiality was not a feasible idea, and by mid-1861, the magazine was a strong supporter of the Union. He had come to this conclusion, perhaps, when he offered payment to soldiers on either side after Sumter was fired upon who could provide sketches from which the magazine's artists might make drawings. This offer was not well received by a war-frenzied readership, even though the result was a handsome four-page foldout depicting the bombardment, published only a week after the event.

Leslie represented a new breed of editor, considerably ahead of his time in some repects. The newspaper business was beginning to be dominated by the so-called "giants of journalism," strong-minded men like Bennett, Greeley, and Charles Anderson Dana, who were personalities in their own right, known to a national audience. Magazines had had no such figures until Leslie; George Graham, Louis Godey, and Mrs. Hale were only notable editors, not personalities in the popular sense. Leslie, at a peak in the 1850s, was a striking figure. Short and broad but handsome, his heavy black beard thrusting out below a pair of penetrating eyes, he virtually radiated energy; he was a man who had to be noticed. Later, after the war, he became better known yet as a result of his scandalous divorce and remarriage to one of his editors, a woman who was an even more vivid personality. Ironically, it was Mrs. Leslie who has become the legendary figure, not her pioneering husband.

Leslie built his successful magazine and his small publishing empire on a simple precept: "Never shoot over the heads of the people." That meant he was a mass-magazine publisher at heart and in practice. At a time when books (particularly paperbacks) and newspapers were just beginning to stretch out into that vast market, he was the first of the magazine entrepreneurs to do so in any comprehensive way, foreshadowing the future with his innovative ideas and techniques.

Yet he was not a national political influence because, in the end, he did far more to amuse and entertain than to inform and instruct. He presented a vivid picture of American life to his fellow Americans, but he did little to inspire them to think about it and to help shape national life. That was why two magazines of far different character, begun in the same decade and directed to relatively small audiences, not only had a great deal more influence, then and later, but laid a groundwork so firm that they survive today.

The first of these to appear was *Harper's New Monthly,* launched in June 1850 with a first printing of 7,500 copies, each one consisting of 144 two-column octavo pages, selling for twenty-five cents, three dollars a year. Its first editor was Henry J. Raymond, who would found a more important national institution, the *New York Times,* two years later. In the first issue were the first installments of two serials: *Maurice Tiernay: Solider of Fortune,* by

Charles Lever; and *Lettice Arnold,* by Mrs. Anne Marsh. There were also three short stories, two of them by Dickens; a few biographical sketches, including one on Longfellow, who had been the first writer asked to contribute, during the previous October; a department called "The Monthly Record of Current Events," written by the editor; a column reviewing new books, called "Literary Nuances"; and a section on women's fashions. There were a few pictures—of William H. Prescott, Archibald Alison (another lesser-known historian), and Thomas B. Macauley—along with woodcuts to illustrate fashions.

This magazine was the rather offhand inspiration of the five Harper brothers, proprietors of what was already the leading publishing house in New York. Speaking of their magazine venture years later, Fletcher Harper remarked: "If we were asked why we first started a monthly magazine, we would have to say that it was as a tender to our business, though it has grown into something quite beyond that." By that he meant the magazine would draw freely on the firm's authors and books for its contents, and the traffic would be two-way—a formula successfully adopted later by the Appleton family, the Scribners, the Putnams, and other publishing houses.

Some of this plan could be seen in the initial announcement that the new magazine would "transfer to its pages as rapidly as they may be issued" the tales of Dickens, Bulwer, Lever, and a list of other writers then contributing to British periodicals. "The design," said the opening announcement, "is to place within reach of the great mass of the American people the unbounded treasures of the Periodical Literature of the present day." No mention was made that this literature was to be pirated; the magazine did not pay for reprints until the late 1850s.

The Harper brothers intended their magazine to reach a specific audience—educated and upper class. This was the audience that had the money to buy Harper books and the leisure time to read them. It was also the audience that would later be called "opinion makers." Thus, the brothers were building not only a literary constituency that would be helpful to their publishing house but, less wittingly, a platform for the issues of the day to be discussed by serious people, readers (politicians, clergymen, civic leaders) who would have definite roles in shaping public opinion. It was, in short, a forum for the governing class—professionals, industrialists, those in public service—people who influenced decisions on issues and set standards. The brothers themselves belonged to this group. But they had also begun as printers, and the magazine was a way to keep their presses running when books were not on the cylinders.

The investment and the idea paid off immediately. Circulation increased to 50,000 in six months as readers discovered they could get in *Harper's* not only the latest fiction but news of developments in American culture and politics. Fiction may have been the biggest attraction. Where else could readers find in one magazine such writers as Thackeray, Trollope, Hardy, Twain, Emerson, Hawthorne and Melville, not to mention such anonymous tales as "Forty-three Days in an Open Boat" (an apocryphal story written by Twain). The illustrators were equally distinguished: Frederic Remington, Winslow Homer, C. S. Reinhart, and Edwin A. Abbey, among others.

Alfred H. Guernsey, a Greek and Hebrew scholar, succeeded Raymond as editor and guided the magazine through the Civil War. Guernsey was one of the great editors of his time, so technically skilled that he could extract the meat from a two-volume historical or biographical work and make an eight- or twelve-page article out of it.

In spite of its literary tone, *Harper's* was political from the first, since the brothers themselves were much interested in politics; one of them, James, was later to become mayor of New York. Raymond, a politician himself who later served in the House, edited his *New York Times* with one hand and *Harper's* with the other for five years. Thurlow Weed, later to become the political boss of New York State, had worked as an apprentice with James Harper on the same press and was a factor in the early life of the magazine. With such connections, it was not surprising that *Harper's* began to be a forum for public affairs early in its career, not only reflecting the growing conflict over slavery but also the growth of industry and the opening of the West.

No magazine of that time was so beloved by its readers. There were hundreds of subscribers who had their old copies bound up in leather and then took the volumes with them when they migrated across the plains as settlements spread westward. These collections were in the backpacks of hundreds of others who sailed around the dangerous Horn to the gold fields of California. From the Atlantic to the Pacific, thousands of families read their copies avidly by the uncertain light of whale-oil lamps. In a single decade, on the eve of the Civil War, *Harper's* had become the leading national monthly— more truly "a mirror of American life and ideas" than most other periodicals.

It was a diverse audience, even within its relatively limited socioeconomic boundaries. Some were poor scholars, some farmers with the advantage of an education, some rich and cultured people, but the one thing they all held in common was a passionate interest in culture, not only for themselves but for their children. *Harper's* made them feel that they were part of the growth of a great nation and inspired them to believe they could make it a model for the world if they took their responsibilities for it seriously. The magazine perfectly reflected their convictions about themselves and the country.

But the *Monthly* did not entirely satisfy Fletcher Harper, who was the heart if not the soul of his organization. The brothers had agreed that the magazine would be unavoidably political, but they did not want to emphasize it; their intent for the *Monthly* was primarily cultural. Fletcher, however, the youngest of the brothers (and the most attuned to the decade), understood that in the 1850s, with the strains between North and South increasing every day and strife in Kansas parading through the newspapers, it would be fatuous and unprofitable to ignore politics. In any case, he had become enamored of the magazine business and wanted a more direct voice in managing one. The launching of *Harper's Weekly* in 1857 was thus very much Fletcher's doing.

In some respects the *Weekly* was like its sister *Monthly*, but besides there being a greater accent on politics, there were more pictures; as in the

Monthly, excellent fiction and essays rounded out the package. As Mott and others discovered, its files from 1857 to 1916 provide researchers with an invaluable illustrated history of the intervening periods. The *Weekly* was Fletcher Harper's "pet enterprise," as his grandson J. Henry Harper was to write years later. Fletcher devoted most of his energies to it until he died, a factor that may have contributed to the bankruptcy of the publishing house in 1899.

The subtitle of the *Weekly* was *A Journal of Civilization,* and indeed it was. Fletcher borrowed a little from Leslie's pictorial techniques and at first advertised his product as a "family newspaper . . . before anything else, it is a first-class newspaper." But it was a magazine for all practical purposes, and while it competed with the newspapers in its Civil War coverage, most of its contents was magazine-like in character. The line between the two media was sometimes fuzzy, though. The *Weekly* began by running front-page editorials, as a newspaper might have done in those days, but these were soon switched to the second page to make room for the pictures that would be the magazine's chief attraction. It printed light essays, some innocuous departments, two pages of domestic and foreign news in small print, miscellaneous contributions in the fields of travel, biography, and general information, a little poetry, a page or two of advertising, and the inevitable serials, which were of exceptionally good quality: Dickens's *Tale of Two Cities* and Willkie Collins's *The Woman in White* among them.

It was the pictorial display that attracted the most attention, however. In an early issue was a full-page engraving of President Buchanan and his Cabinet; and the following year, a double-page picture of the warship *Leviathan.* Winslow Homer was among the first-rate artists who began their careers in these pages.

In the early years the editor and his magazine were wholeheartedly Democratic, supporting Buchanan with a fervor this weak reed did not deserve. The noble object of the *Weekly* was to "unite rather than to separate the views and feelings of the different sections of our common country," and it persisted in this optimistic endeavor nearly to the brink of war, in spite of the fierce sectional quarrels that the magazine covered so vividly and that were fast making armed conflict inevitable.

Such constant optimism brought sharp responses from jealous and more partisan rivals. *Putnam's Monthly* told its readers that "whoever believes in his country and its constant progress in developing human liberty will understand that he has no ally in *Harper's Weekly*," and Greeley's New York *Tribune* referred to the magazine sarcastically as the "Weakly Journal of Civilization."

These jibes failed to deter most readers, many of whom were weary of extreme views from either side. They supported the *Weekly* as they did Raymond's *New York Times*, which had promised that it would not "get into a passion" in those passionate times. The *Weekly* quickly had a circulation of 60,000 in 1857, which by November of the following year had risen to 75,000. On the eve of the war, which would mark its greatest early success, the figure was 90,000.

The Harper brothers did not have the market all to themselves, by any means. There were not only such competitors as *Putnam's* and *Peterson's*, but in 1857 there emerged from Boston a rival that would do battle with the *Monthly* from that day to this. The magazine was the *Atlantic Monthly*, and its first editor was James Russell Lowell, one of the most eminent of the New England sages.

Just as the Harper brothers' two magazines owed their existence and success to Fletcher, the *Atlantic* was forever indebted to Francis H. Underwood, whom Bliss Perry later called "the editor who was never the editor." Underwood conceived the idea of the *Atlantic* in 1853, declaring it frankly to be a magazine that would "bring the literary influence of New England to aid the antislavery cause." He began at once to look for capital and in 1854 thought he had the backing of John P. Jewett, the young Boston publisher who had issued *Uncle Tom's Cabin*. But Jewett went out of business in spite of his best-seller, and it was three years later before Underwood succeeded in getting enough seed money and a lineup of contributors to launch his magazine.

On two afternoons in the spring of 1857, May 5 and 6, Underwood met for dinner in Boston at the Parker House in company with Moses Dresser Phillips, a partner in the publishing house of Phillips, Sampson & Co. (of which Underwood was literary advisor) and with Emerson, Longfellow, Lowell, the historian John L. Motley, Oliver Wendell Holmes, and the architect James Elliot Cabot. It would be difficult to imagine a more distinguished set of sponsors and potential contributors.

Phillips wrote later: "We sat down at three p.m., and rose at eight. The time occupied was longer by about four hours and thirty minutes than I am in the habit of consuming in that kind of occupation, but it was the richest time intellectually by all odds that I have ever had."

Underwood had not placed all his eggs even in so remarkable a basket. He also had assurances of contributions from Hawthorne, Thoreau, Whittier, Thomas Wentworth Higginson, Prescott, James Freeman Clarke, and Harriet Beecher Stowe. No magazine to date could have boasted such a list. These were among the best writers in America, and they strengthened Underwood in his conviction that he could publish a magazine that would, at last, be truly American, with no vestiges of transatlantic culture clinging to its pages. With understandable modesty, Underwood also realized that it would be far better to have Lowell's name on the masthead as editor than his own; for himself he reserved the title of "office editor," meaning assistant editor.

Over two splendid dinners of oysters, steak, Burgundy, and brandy, the founders made their fundamental decisions: an American magazine to be called the *Atlantic Monthly* (Holmes suggested the title), with no contribution to be signed and even the editor to remain anonymous. "The names of the contributors," Emerson remarked dryly, "will be given out when the names are worth more than the articles." The magazine would devote itself to literature, art, and politics, but no one could believe the founder's promise that it would be the "organ of no party or clique." These men were all passionate abolitionists.

As the first number neared its final stage of makeup, Lowell apparently began to have second thoughts about his completely American magazine. Playing it safe, he sent a letter and bank draft to his friend and co-sponsor, Charles Eliot Norton, who was in London, asking him if he would shop around for an English serial and perhaps a few good short stories and poems that had not already been pirated.

This move had a tragicomic aftermath. Norton was successful and brought back with him several handwritten manuscripts in his trunk, which unfortunately disappeared on its way from the pier to his hotel in New York and was never recovered. The British authors involved were so angered by this loss, which was no fault of Norton's, that they refused to contribute again, and so the *Atlantic* appeared as pristinely American as the founders had planned originally.

It could not help being successful, given the quality of its writing. There were only 15,000 readers at the beginning and never more than 50,000 during the remainder of the century, but that was enough. Sheer quality carried it through the first year, when as its semi-rival *Harper's Weekly* wrote, "Not for many years—not in the lifetime of most men who read this paper—has there been so much grave and deep apprehension. In our own country there is universal commercial prostration and thousands of our poorest fellow-citizens are turned out against the approaching winter without employment. In France the political cauldron seethes and bubbles with uncertainty. Russia hangs like a cloud dark and silent upon the horizon of Europe; while all the energies, resources and influences of the British Empire are sorely tried . . . in coping with the vast and deadly Indian situation, and with disturbed relations in China."

Neither *Harper's* nor the *Atlantic* could elicit much concern from readers about events abroad, however. The deepening gulf between North and South was on the minds of everyone at home, and it was this conflict that the *Atlantic* felt compelled to deal with while it was dispensing good literature. As Edward Weeks, one of its great editors, wrote on the magazine's hundredth anniversary, under the force of the drive for abolition "even the most objective of them, men like Emerson and Lowell, wrote at white heat." The magazine, as Weeks observed, grew out of the aspirations and anxieties of the times, until eventually it came to contain "the heartbeat and anger, the tenderness and laughter, of Americans from all quarters."

So, with the introduction of these two extraordinary magazines, the periodical business approached the Civil War, that watershed in American history. Most of the earlier editors still alive were growing old, and their magazines were beginning to fail—men like George Graham, Lewis Gaylord Clark, and Nathaniel Willis. Obviously, a changing of the guard was taking place. Lowell, Leslie, Alfred Guernsey, and Robert Bonner (of the Philadelphia *Ledger*) were the new leaders.

The Panic of 1857 had also reduced the number of magazines—from about 685 in 1850 to about 575 in 1860. An increase did not occur again until 1863, when wartime demands for information and entertainment proved to be a boost for periodicals and newspapers alike, and continued when the war was over.

Graham's was the chief victim of the Panic, suspending publication after thirty-two years. The *Knickerbocker* almost disappeared but managed to survive. Those less lucky included the *Democratic Review,* once a home for the best writers, and the *New York Mirror.* The South, much less affected by the Panic, saw little change in its barely thriving magazine business, which would soon be devastated by the war. A symptom of things to come was a rising interest in the art of advertising, which produced in 1851 the first magazine devoted to that subject, *Pettengill's Reporter;* it lasted until 1859.

Frank Leslie took note of this development in 1857. "The art of advertising," he wrote, "is one of the arts most studied by our literary vendors of fancy soaps, philanthropic corn doctors, humanitarian pill-makers, and all the industrious professions which have an intense feeling for one's pockets. Every trick that can be resorted to for the purpose of inducing one to read an advertisement is practiced, and, it must be confessed, very often with complete success. How often have we been seduced into the reading of some witty or sentimental verse that finally led us, by slow degrees, to a knowledge that somebody sold cure-all pills or incomparable trousers."

In their defense, it must be said that advertisers had to use such subterfuges to get attention, even as they do today, but in the 1850s the reason was the laudable desire on the part of publishers to protect their small advertisers, consequently refusing to let the larger ones use display type—except for the monthlies, of course, which did not carry small advertisements because they were not profitable.

Magazines entered the war years stronger than they had ever been, becoming potent politically, better written and edited, assuming the shape of the future. The war would test them, as it did every other aspect of American life.

4

Magazines for Ladies

Among the special interest groups targeted by the new nation's publishers, women became the primary concern, at times approaching an obsession, of publishers and editors. Not only was there a rapidly growing number of magazines directed particularly to them, but, as noted earlier, they were endlessly discussed in the pages of more general periodicals.

Journals directed specifically to women began appearing in the 1790s, and within the space of a few decades, the female audience was able to choose from among numerous *Toilets, Miscellanies, Caskets,* and *Repositories* edited specially for them. Specialized though they might be, these magazines shared the same problems of the general publications. They had the same distribution difficulties, the same lack of a developed audience at first, print technology inadequate to their needs, and the inability to extract editorial material from contributors, who, in any case, wanted to be anonymous. They followed the general pattern of frequent failure and even more frequent optimistic start-ups. By 1830, forty-five women's magazines had appeared—and for the most part, disappeared. But with each new journal, more women acquired the reading habit.

When the great magazine explosion took place between 1825 and 1860, periodicals for women were active participants in this golden age. Their numbers multiplied and their circulations grew. On the eve of the Civil War, more than a hundred magazines for women had started, providing models for the later mass-market women's journals and serving a dual function. On one hand, they encouraged women to read more—not a difficult task since it was one of the relatively few occupations sanctioned by the patriarchal society—and on the other hand, they offered an outlet for the growing number of women writers who were beginning to dominate the expanding market for romantic fiction.

Among all the women's magazines in antebellum America, one stood out above the others, like a lighthouse rising from a sea of mediocrity. *Godey's*

Lady's Book was so popular that it was known simply as "The Book" to the thousands of women who read it. To understand its popularity, it is necessary to examine its predecessors briefly, since *Godey's* was what they might have been.

Early magazines had reached a wide variety of readers, but for some reason (perhaps it was merely coincidence) women had to wait until the colonies became a nation before their interests were specifically served. In 1787, even as the Constitution was being born, Noah Webster was assuring the "fair readers" of his *American Magazine* that "no inconsiderable pains will be taken to furnish them with entertainment." They were still being assured in the early 1790s, as the editor of the *Massachusetts Magazine* declared his intention to devote greater coverage to topics of interest to women. "The fair sex merits our highest attention," he added sententiously. By that time, editors of general interest magazines were already beginning to understand that the female audience was important because so many of them were literate and, in upper-class households at least, had the time to read.

What *would* interest women? the general magazine editors asked themselves and returned some obvious answers. Dress reform, of course; male editors were forever telling women what they should wear. (In the xenophobic climate of the new nation, they especially wanted them to become independent of European fashions.) The role of women in society, certainly. Early magazines recorded the beginning of a public debate continuing for two centuries and showing no sign of ending.

But women were a sideshow in the general magazines, and it was clear that a female audience existed capable of supporting its own periodicals. Prospective publishers were aware that such British publications of the 1770s and 1780s as the *Lady's Magazine* and the *New Lady's Magazine* had done well in the small American audience able to obtain and pay for them.

The time was ripe in 1792 for the first American magazine edited specifically for women, and it duly appeared: the *Lady's Magazine, and Repository of Entertaining Knowledge,* published in Philadelphia. Short-lived, it was the first in a long line of periodicals catering to American women and playing a significant role in their lives. It was primarily a literary magazine, containing no columns on household work, which would eventually be the mainstay of most women's journals. Books were reviewed; most notably, a nine-page tribute to Mary Wollstonecraft's *Vindication of the Rights of Women.*

The *Lady's Magazine* declared itself open to female writers, whose excellence, it said, "with the beams of intellectual light . . . illuminates the paths of literature." If pen names are any indication of sex, many of the contributions from females had already appeared in other magazines. There was competition for what was available, since the general journals were also looking specifically for women contributors. What they searched for, according to one of them, was "the elegant polish of the Female Pencil."

Of the several women's magazines begun in the early nineteenth century, most were weeklies, at least in the beginning. They lived precarious, short lives, beset by all the difficulties the other magazines were having. But the simple fact that they kept on arriving and struggling testified to the existence of an audience that was interested, and it was an audience obviously on the increase.

Editorial content in these early ventures ranged from fiction to death notices, from sheet music and theatrical reviews to international news. There were also essays, items collected under the heading "Scientific Miscellany," columns of advice on everything from love to hairdos, and articles designed simply to amuse. There was fashion and beauty information and, as Mott says speaking of one of them, "a good deal of *et cetera.*" While hot political debate went on in the other periodicals, women's magazines gave minimal attention to current events. There was no clear editorial pattern or philosophy. Only fiction was consistently present and prominent, and it came to characterize these women's journals for much of the nineteenth century and most of the twentieth.

Filling the pages was the most pressing job of these early nineteenth-century editors. They took whatever was available, as long as it fit their columns, whether stolen from foreign journals, contributed anonymously by readers, or written by their own hand. It would be some time before these magazines settled into a predictable, formulaic content. Most survived on pirated material. Editors pleaded constantly for original work, but it did not appear at once.

Illustrations were crude, as they were in other magazines, because technology had not yet caught up with editorial ambitions. Nevertheless, the *Lady's Magazine* printed two engravings "submitted with all deference to the fair daughters of Columbia." In 1827, the *Philadelphia Album and Ladies' Weekly Gazette* published the first full-page color fashion prints in this country, several years before *Godey's.*

The diversity of editorial content in these helter-skelter first women's magazines can be seen in the repetition of such title words as "Miscellany," "Repository," "Literary Cabinet," "Museum," "Casket," all signifying magazines (in any category) without an organized content. All that held them together was their intent to publish material that would interest female readers.

Fiction, as noted, was the cornerstone of this diversity. In the grandiloquent style of the period, the editor of the *Lady's Weekly Miscellany,* one of the most successful early ventures, announced, "Fictitious story being a species of literature which is pleasing to all, we shall resort to it as to a fount from which we may take large draughts without causing anxiety."

The cultural historian Russell Nye classifies the fiction published in America during the late eighteenth and early nineteenth centuries into four types: sentimental, satiric, gothic, and the historical romance. Most of what was in early women's magazines was frankly sentimental; it was a little later that women writers began to be masters of the historical romance, as they remain today. Short stories usually featured a young heroine faced with an obstacle to the consummation of her love. This was embellished with the struggle of the heroine to identify and reject evil and to make a choice of the heart that accorded with honorable and true principles.

These stories invariably contained an extractable moral of some kind, and there was a shrewd business reason why editors insisted on one in case it was forgotten. In antebellum America, as there would be again at the end of the century, there were legions of educators, religious leaders, and others we would now call opinion makers who protested strongly against the swelling tide of fiction, particularly stories and novels aimed at young women. Since the women's magazines were clearly on a high moral plane, however, their use of fiction was considered somewhat more acceptable.

Nevertheless, the editors of these periodicals thought it judicious to print warnings in their pages about the pernicious effects of fiction on the young. The *Lady's Magazine* pointed out, "Novels are a species of writing which can scarcely be spoken of without being condemned. . . . Be ever careful not to read those which inflame the passions or corrupt the heart."

Attacks against fiction increased proportionately with the floodtide of such stories during the prewar years, immensely stimulated by the paperback revolution. The fact that the most successful writers of fictional romance were women was especially galling to male critics. But editors knew their market, and they gave their female readers what they wanted. Some magazines, in an effort to appease the attackers, labeled fictional stories as "Founded on Truth," or "Founded on Fact." By passing off fiction as fact and making sure the tales had a moral ending, as they invariably did, magazines could claim to be avoiding the corruption of women while still catering to what they knew their female readers desired.

Since they existed chiefly on subscription revenues, there was little advertising in these journals. A book ad on the last page of the first issue of the *Lady's Magazine* was the only concession to commerce. Relying on subscribers, however, had its own major pitfall, as all the magazines knew. Many of them simply failed to pay up, eliciting piteous columns reminding readers that they were the only source of support not only for the magazine but also for the families of the entire staff.

Sometimes these appeals amounted to personal pleas for charity. In launching the *Ladies Literary Museum, or Weekly Repository*, in 1817, Henry C. Lewis explained to his readers that he had been out of work "during the last distressful winter, when the wolf often appeared at the threshold and howled at the ears" of his family. Remonstrating with laggard subscribers in later issues, he reminded them that revenues from the *Museum* were his family's only sustenance. Even Sarah Hale, that consummate editor who was to carry *Godey's* on to new heights, had no hesitation in sharing with her readers the economic ups and downs of her *Ladies' Magazine* and the impact on her family. But these public pleas had little effect. Even Mrs. Hale's journal had to be rescued eventually with Louis Godey's money.

Philadelphia had emerged as the magazine center by 1800, with New York a close second and closing in. But it was Boston, slipping out of its former leadership, that nurtured the most famous editor of women's magazines in the nineteenth century—Sarah Hale. In 1827, she first appeared on the horizon when her prize-winning poem appeared in the *Boston Spectator and Ladies Album*. It was the first time her real name had appeared in print; she had previously used the pseudonym "Cornelia."

In these early years of the century, women's magazines were beginning to spread away from the traditional centers. The first of the genre to be launched in the South was the weekly *Ladies Magazine,* issued in Savannah in 1819, and soon there were others in Washington, D.C., Baltimore, and Harper's Ferry. The first publication for women in the West, the monthly *Masonic Miscellany and Ladies Literary Cabinet,* was begun in 1821.

All these publications were short-lived, and all suffered from the problems of magazines everywhere: low salaries for staff and even lower payments for contributors, the difficulty of getting writers, and numerous technological problems. The best that could be said for them was that they did offer women an opportunity to exercise their writing talents and editorial skills. Most of the editorial material was written by women (often by the editors themselves), and many editors were women, although the publishers were likely to be men. The rise of such magazines made it possible for women like Mrs. Mary Clarke Carr to appear on the scene. In 1814 she started the *Intellectual Female, or Ladies Tea Tray* in Philadelphia. She edited this magazine with a more personal touch than any publication up to that time had seen, creating a style that would soon be taken over successfully by Louis Godey.

These struggling pioneers paved the way for the sudden success of women's journals in the 1830s and 1840s. They had piqued the interest and molded the reading habits of a growing women's audience. It was true that magazine entrepreneurs had a more difficult time of it in general than those who started newspapers and book publishing houses in the same period, but they persevered against formidable odds until the second quarter of the century witnessed their triumph. Between 1830 and 1860, at least sixty-four women's magazines appeared, and many of them no longer had to worry about subscribers. Some acquired a reader loyalty that would characterize this genre in later years, and they soon began to challenge the general magazines.

The chief challenger, of course, was *Godey's Lady's Book,* although *Graham's* and *Peterson's* (which began as general periodicals) were not far behind. These major journals led the entire magazine field in several areas, paying decent fees to authors, copyrighting material, and publishing high-quality engravings. They were among the first to permit, even encourage, authors to sign their real names, and George Graham went even further in his magazine by splashing these names on his cover and title pages for promotional purposes. Competitors, both among the general periodicals and among the rival women's magazines, had to copy these developments. What was happening to women's magazines eventually affected the entire industry.

Again, it was *Godey's,* "The Banquet of the Boudoir" as an envious rival called it, that led all the rest. The reason was simple. Louis Godey was the first great magazine editor, and Sarah Hale, who joined him later, was his equal.

They were an odd couple. Godey, the man who "brought unalloyed pleasure to the female mind," as one admirer (himself) put it, was a Dickensian figure—overweight, bouncy, benign, a man with a simple, uncomplicated heart and mind. The fact was that he loved women, in a completely nonsexual way. Born in New York in 1804 of poor French parents, he had little formal schooling and learned his trade from books and in print shops. His life was what has been popularly known as the Horatio Alger climb from rags to riches, although none of Alger's heroes did so. He died with a fortune of a million dollars in 1878. Along the way, he left a lasting impression on the magazine business.

Psychohistorians have tried to make something of Godey's close affinity with women, but aside from the devoted wife and five equally devoted children with whom he spent his free time, the remaining countervailing evidence suggests that he was simply a man who enjoyed the company of women and who understood them. He would have been at home anywhere with the ladies of his time, speaking their language easily. To him, they were always "fair ladies" or "fair readers," and he spoke to them in his magazine as one good friend to another. "We have received a note from some fair Lady, we presume," he would write, "requesting us to give another description of Love than that found in the February number. This shall be done, and another fair Lady has it now in charge." He always capitalized words like "Lady" and "Love."

Beneath this façade of a simple, chubby, benign soul, however, was a talented businessman. Knowing exactly what ladies wanted to read, he knew how to create a magazine that would satisfy them. Godey began his venture in partnership with Charles Alexander, a Philadelphia newspaperman who had given him his first job, on the *Daily Chronicle*. They called it simply the *Lady's Book*, but Alexander soon tired of the venture. Godey took it over and added his own name; he was not an entirely modest man. During his lifetime he was also involved with a variety of other successful magazine ventures, but the *Lady's Book* was his love.

Completely in keeping with the mawkish purity of the period, when concert pianos had to have their legs bound up in pantaloons so that the mere suggestion of "leg" would not offend lady patrons, Godey wrote, "Nothing having the slightest appearance of indelicacy shall ever be admitted to the *Lady's Book*." Thus any inadvertent reference to chicken or turkey breasts, for example, in a food article, would be changed to "bosom."

When Godey hired Mrs. Sarah Josepha Hale as his editor in 1837 by buying her troubled journal, the *Ladies' Magazine*, he acquired a woman of superior editorial talents who could be guaranteed to share his view of what should be in *Godey's*. Mrs. Hale's appearance in her later years was deceptive, too. Plump, like Godey, bespectacled, and looking like everybody's grandmother (she had been something of a beauty in her youth), she seemed like the apotheosis of respectable home-and-family life. Beneath that exterior, however, was a determined reformer. Her great cause was "female education." She had begun crusading for it in her own magazine, and now she carried that crusade over into her editorship of *Godey's*. What she did for women's causes has been unfortunately overshadowed by her reputation as the author of "Mary had a little lamb" and by her persuading President Lincoln to adopt her idea of declaring Thanksgiving a national holiday, but she did much more. Mrs. Hale was a pioneer and a prime force in getting women into the teaching and medical professions. An excellent writer herself, she produced a good and popular cookbook, along with several other volumes, and edited a number of anthologies and annuals—all of this while she was editing *Godey's*. Louis Godey's ego would not permit him to give her any better title than literary editor, but she was second in authority only to him, and, unquestionably, she ran the magazine.

"Literary editor" she might be, but *Godey's* was never a literary success; nor was it wholly devoted to Mrs. Hale's crusades. What it did was to assemble a distinguished list of contributors who wrote the kind of sentimental fiction and articles the readers wanted. They raised its circulation figure to an extraordinary (for the times) 150,000 just before the war. Without question, it was the most successful woman's magazine to appear in the first three-quarters of the nineteenth century, exerting a formidable amount of influence not only on its own readership but on other magazines as well when they copied the *Godey's* formula. It was also an important influence on the women's magazines begun later in the century, as they looked to this highly successful journal for ideas and guidance.

Godey made a major contribution to the business in demonstrating how to break away from the kind of miscellany that seemed to dominate the magazines of his earlier days. He retained the best, especially the full-page fashion prints, but he swept out much of the trivial and replaced it with material that reflected what his readers were feeling and thinking, encouraging them to contribute their own stories and poems. It was an admirable combination of Godey's feminine tastes with his businessman's ability to create and sell a product. He personalized his relationship with readers through a column he called "Arm Chair Chats," which appeared in the back pages. In it, he spoke directly to his readers, giving them a sense of the man who created the magazine they so much enjoyed. Although it would have been an excellent advertising medium, *Godey's* made no great effort to acquire sponsors; what few ads it carried were stowed away obscurely in the back pages.

In spite of his closeness to the readers and his ability to communicate with them, Godey understood (as Cyrus Curtis did later) that he was not an editor, and it was not until he acquired Mrs. Hale that the magazine really began to soar. She had no domestic scene of her own to distract her. Her husband had died in 1822, leaving her a widow with five children; she was compelled to go to work to support them. She had relied almost at once on her writing talents and soon won recognition with *Northwood*, a novel about slavery. That had led directly to the editorship of her first journal, the *Ladies' Magazine*. In this, she laid down the principles that led to the success of *Godey's* later on. Along with the sentiment, there was also good writing on moral and social issues of interest to women, all of which was guided by an editorial policy stated in the first issue: "The work will be national . . . American . . . a miscellany which, although devoted to general literature, is more expressly designed to mark the progress of female improvement." That was the policy she brought to *Godey's*, one that lifted it to unprecedented heights.

Deftly, she introduced the moral and social issues that occupied her mind into the layers of fiction and sentiment. Mrs. Hale believed strongly in the power of organized women to achieve goals, and Godey's publicized a variety of causes—among them, the need to raise funds for the erection of a monument at the Bunker Hill battle site, the collection of money for the dependents of lost seamen, and most of all, the improvement of female education. She campaigned persistently for the establishment of seminaries for women in the United States so that they could be trained for professions, thus giving them the opportunity to earn their own livings.

These appeals were sandwiched in with poetry, fiction, and the kind of articles common to other women's magazines; Mrs. Hale wrote many of these selections. She worked hard to include articles and stories by American authors, both in her own magazine and in *Godey's*. With all her editorial work, Sarah Hale was also a first-rate administrator. But when she merged her magazine with *Godey's* and made a smooth transition as editor, the publisher relieved her of other responsibilities, and that meant she could give her undivided attention to improving the literary quality and overall content of *Godey's*. She edited the magazine for forty years, from 1837 to 1877, a longer tenure than any other editor of a women's periodical could boast. She raised its literary quality and gave it direction on moral issues, although the Civil War, at Godey's insistence, was excluded from its pages.

Mrs. Hale's ideas about women and their role in society, expressed in *Godey's* and debated in others, reflected the divided attitudes of women themselves. Although she was the champion of education for women, she also strongly supported the notion that women should stay in their own clearly defined domestic realm. They must exert their force in society through influence, she believed, not by the use of direct power. In an 1846 editorial, she declared, "Remember that woman must *influence* while man *governs,* and that their duties, though equal in dignity and importance, can never be *identical.* Like the influence of the sun and air on the plant, both must unite in perfecting society; and which is of paramount value can never be settled."

With that essentially conservative view, Mrs. Hale understood that she must include not only popular women writers and promising female newcomers but also established literary names, and it was this mixture that helped *Godey's* reach the top. She assumed correctly that women, as much as men, would want to read the work of Emerson, Hawthorne, Longfellow, Poe, Simms, T. S. Arthur, and N. P. Willis. They rested comfortably in *Godey's* columns beside Harriet Beecher Stowe, Mrs. Ann Stephens, Mrs. Sigourney, and others. She salted down her literary stew with articles on history and travel, music and art, famous women, health, the care of children, and cooking. She advised her readers about books they might like to read and gave them information about colleges and schools for their children. In 1848, *Godey's* was the first to publish plans for a model house, and each issue that year carried a new plan. Pictures of the completed house were shown, with the list of materials needed and prices. Sixty years later, Edward Bok appropriated this feature for the *Ladies' Home Journal*.

In *Godey's* pages, in fact, could be found the foundation stones on which the great women's magazines of our time were built. Reports on the latest labor-saving device for housework were offered, be it Masser's Self-Acting Patent Ice-cream Freezer and Beater, a sewing machine, or one of the early models of washing machines. There were also quantities of consumer information, much as *Good Housekeeping* would provide in our time. Mrs. Hale did not institute a Seal of Approval, but she did advise her readers on the relative qualities of a host of competing products. The magazine also featured ready-made clothing for the first time, soon after it became available, and one of its later fashion plates showed a woman's hands actually touching the keys of the newly invented typewriter.

Mrs. Hale's formula was aimed, obviously, at the taste of the upper-middle-class women who comprised her audience. That was why literary writers appeared, but not too many of them. These women wanted culture, but they wanted it leavened with entertainment and sentimental stories, with predictable subjects and plots. There was a lack of intellectural depth even in the nonfiction. Significantly, such topics as female education were considered safe, fitting in with the idea of woman's proper realm. There was nothing in *Godey's* contents that presented any significant challenges to the status quo.

Some media historians believe it was *Godey's* fashion plates that were the real secret of its popularity. These hand-colored plates and other fashion illustrations, which many readers so looked forward to, were Godey's own idea. The magazine had published these plates from the time it began in 1830, showing off the latest Parisian styles and so providing a variety of models for women to copy. The plates themselves were hand-painted in Philadelphia, some at the magazine's plant and others by women artists working at home. These unsupervised women often colored the plates in differing shades, puzzling readers. Godey handled this situation skillfully, telling subscribers that the differences were intentional and that they could compare copies and decide which shades best suited their individual coloring and figures.

Mrs. Hale disapproved of the plates because she thought fashion was frivolous and because she thought that imitating European fashions would discourage women from focusing on native talent. Godey wisely insisted on retaining the plates; it was one of the magazine's most popular features.

It was surprising in the years before the war that *Godey's* was able to keep on increasing its circulation against such powerful competitors as *Graham's* or *Peterson's* when, unlike them, it refused to say a word about slavery and the political conflict that was rending the country. It was not a situation that could endure, and by the time the war began, *Peterson's* had topped *Godey's* remarkable circulation.

The war itself was a bizarre episode in the magazine's history; it actually developed a healthy circulation among soldiers. We can only speculate that they were escaping the somber realities about them by entering *Godey's* gen-teel world, just as some American soliders in the Great War would read the *Ladies' Home Journal*. Nevertheless, *Godey's* lost about a third of its circu-lation during the war because of its refusal to print any information on the conflict (which nearly all other magazines were covering) and also, of course, because it lost its Southern readers.

Both Godey and Mrs. Hale were aging by the time the war ended; they were in their seventies, long past the life expectancies of the time. Trying to rebuild circulation, it seemed that even Louis Godey had lost his famed inti-macy with his audience. He made another mistake by keeping the magazine's subscription price at three dollars, while *Peterson's* was selling for only two. Circulation began a slow decline.

There was a reorganization in 1877. Mrs. Hale resigned at last, ending one of the most remarkable editorial careers in the history of the industry. After a brief transitional period, a new editor took over, Mrs. D. G. "Jennie June" Croly, an experienced, astute journalist, but even she could not revive the ven-erable magazine as it struggled to adapt to a new era. She departed in 1888, and four years later the magazine was moved to New York, and its name was changed to *Godey's Magazine*. Six years later it was dead, swallowed up by the "Grand High Executioner of Journalism," Frank Munsey, who permitted it to disappear into his successful *Argosy*.

The influence of *Godey's* on other magazines in its heyday was remarkable. *Peterson's*, for example, was modeled explicitly on its rival, and the other formidable contender in the field, *Graham's*, was forced into its editorial direction by what *Godey's* was doing. These competitors used the same kind of fiction and fashion plates, both gained large circulations, and both led in innovative ideas. George Graham and Charles J. Peterson, his editor, were responsible for *Peterson's*. It was Graham's idea in 1842 to set up another magazine with the explicit intention of taking away readers from *Godey's*. They intended to do this by pricing their magazine lower and at the same time offering much the same kind of editorial material. Graham's own journal was more general in its appeal than *Godey's*, and he believed it would not be hurt by *Peterson's*. As matters worked out, however, Graham found himself with two rivals when only one had existed before and had to change his own editorial content accordingly.

It was *Peterson's* low price that eventually enabled it to outstrip both of the others, since their editorial contents were not that much different. At two dol-lars annually, it was a dollar less than either of the established journals, and that proved to be a significant difference, particularly in the hard times fol-lowing the Panic of 1857.

The wisdom of the Godey formula could be seen in the failure of those who failed to follow it. When Mrs. Mary Chase Barney started her *National Mag-azine, or Ladies Magazine* in Baltimore in 1830, her intent was to provide political information for women and to give them a forum to express their views on political topics. She refused to publish romantic fiction, telling those readers who complained, somewhat ambiguously, that they must "make love for themselves." It was a failure. In time, she dropped politics and emphasized literary contributions, along with travel, biography, and even fashion and etiquette.

The service and household departments upon which the women's magazines of our century were built did not exist for the most part in these earlier women's journals. *Godey's* and *Peterson's* printed information on domestic life, but many of the lesser-known periodicals did not. Only two made an attempt to emphasize practical homemaking. In Boston, Caroline Gilman issued her *Lady's Annual Register and Housewife's Memorandum Book* in 1838. In her skilled hands (she was the author of many stories and poems, as well as the successful editor of the *Southern Rose Bud*), the domestic topics neglected by others flourished in her journal, and it lasted eight years. In Philadelphia, an attempt to emulate her in 1843, *Miss Leslie's Magazine* (with a lengthy subtitle) failed after a year, in spite of its promise that Miss Leslie would give her readers "valuable information on household economy and domestic interests." Restored after failure, it was transformed into a woman's literary journal.

Not all American writers considered magazines for women a good idea, even though they contributed to them. Poe, for instance, who had given numerous stories to *Godey's* and *Graham's* and also served for a time as an editor of the latter, resigned from his job (which he needed, as always) and declared in his resignation letter, "My reason for resigning was disgust at the namby-pamby character of the magazine. . . . I allude to the contemptible pictures, fashion-plates, music, and love-tales." Hawthorne shared Poe's distaste for the "love-tales," and his blanket condemnation of those who wrote them as "a damned mob of scribbling women" has echoed from that day to this.

Some general magazines felt themselves compelled to hire a token female editor, or at least find a way of placing a woman's name on the masthead to attract female readers. *Peterson's*, for example, carried Mrs. Ann Stephens's name on its title page as editor for the first ten years, although Peterson himself actually edited the magazine. Mrs. Stephens did contribute many articles and columns, but Peterson had decided it was better to give readers the impression the magazine was edited by a woman. It was not until 1886 that the magazine acknowledged officially that Mrs. Stephens had never been the editor.

Ann Stephens, however, was one of several women who started and edited publications directed to their sex, a company including Caroline Gilman, Mary Barney, and Matilda Clarke. Women were coming into their own in the magazine world, not simply as readers and writers but also as influential editors.

Mrs. Hale, so vigorous on behalf of women in other directions, opposed woman's suffrage. As early as 1833, she wrote, "The term 'rights of women' is one to which I have an almost constitutional aversion." She feared that women would become less attractive to men if they demanded the vote. When the suffragists met at Seneca Falls in 1848, *Godey's* carried only a brief mention of the historic meeting in an editorial column.

But not all magazines dismissed the subject so easily. After the Seneca Falls convention, a number of journals devoted to the rights of women materialized. One of the most notable was the *Lily, A Ladies Journal Devoted to Temperance and Literature*. Founded by the feminist Amelia Bloomer (with the assistance of the Seneca Falls Women's Temperance Society), this journal advocated both dress reform and female enfranchisement. The *Lily* lasted for six years, from 1849 to 1856, moving west to Ohio with the Bloomers in 1853. Another noted feminist, Elizabeth Cady Stanton, published some of her early writing in this magazine.

Several other women's rights journals appeared during the antebellum period, but almost all had brief lives. They included the *Genius of Liberty* (1852–1854), published in Cincinnati; the *Una* (1853–1857), edited by Pauline Wright Davis in Provincetown first and then Boston, a relatively conservative journal; the *Pioneer and Women's Advocate* (1852), also published in Provincetown by Anna W. Spencer; the *Woman's Advocate* (1855–1860), published by Anna McDowell in Philadelphia, owned, edited, and printed by women and "designed to present the wrongs of women and to plead for their redress"; and the *Sibyl: A Review of the Tastes, Errors and Fashions of Society* (1856–1864), founded in Middletown, New York, by Drs. (for Doctoress) Lydia Sayers. The latter printed articles on a variety of reforms and was the official organ of the National Dress Reform Association. Unfortunately, several of these reform journals began fighting with each other in print about their devotion to the cause of women's rights.

It needs to be noted that while a rather wide range of people read the women's journals of the antebellum period, they were bought primarily by economic elites. Most women in that day did not have enough money of their own to buy magazines for themselves on a consistent basis. Men held the purse strings and most also held women accountable for the money they were given to spend. Nor did women below the upper classes have much time to spend reading periodicals; they were slaves to domesticity. It would be some time before the postwar technological revolution also revolutionized the housewife's work so that she would have some leisure time for reading. By that time, there were a substantial number of women with the education, the ability, and the means to buy and enjoy magazines—magazines of all types, "ladies'" or otherwise. That change in the sociology of magazine reading would come to affect the industry as a whole.

5

The Beginnings
of Black Magazines

One of the most neglected areas of magazine history is the rise and eventual success of black periodicals. Like everything else in the history of African-Americans, it was a victory a long time in coming. Not until the mid-twentieth century could it be said that there was a successful and well-established category of such magazines. Yet, almost unnoticed by white historians, a variety of periodicals written, edited, and owned by blacks had been available for most of the nineteenth century and waited only for the better opportunities of the twentieth to flourish.

The audience was obviously ready-made from the beginning. No other magazines printed anything about the concerns of black people. When slavery was discussed, however sympathetically, it was almost entirely from the white point of view. Lynching was seldom addressed at all, nor was segregation as it became entrenched at every level of society. Black magazines, struggling to survive under worse handicaps than any other category, provided a means of communication and information for black professionals and other interest groups in the community.

More than a hundred black publications were issued between 1838, when David Ruggles's *Mirror of Liberty* first appeared in New York, and the onset of the Great War. A few of these were general magazines, designed for a diverse audience, but most were highly specialized and targeted for such specific audiences as teachers, doctors, or musicians. Some were sponsored by institutions, often a church, while others were entirely private ventures. Surprisingly, they had a widespread geographical distribution. Although 90 percent of all blacks were still living in the South as late as 1900, their magazines of that century were published not only in that region but in the North and West as well.

Life was particularly hard for this rising new genre. Both before and after the Civil War, these new journals shared the production, distribution, and financial problems of white magazines. That was enough to cause a high rate of mortality among the others, but for black publishers, there were added obstacles. Before the Civil War, most blacks were slaves, and the others had scant funds to launch any such enterprise as a magazine. For the same reason, contributors were hard to come by, and, in any case, the number of blacks who could read both before and for long after the war was relatively low. As late as 1900, the U.S. Census reported that only 55.5 percent of blacks were literate, although these figures are dubious, considering the way they were collected. Another formidable difficulty was how to generate advertising when it became the mainstay of all magazines. It took black publishers a much longer time to attract major advertisers to their publications.

Summarizing all these difficulties, Charles Alexander, a black journalist writing in 1906, observed, "The task of publishing a Negro newspaper or magazine is one that requires at the outset, great versatility and talent, remarkable executive ability and exceptional courage and tenacity." If this was still the situation in 1906, one can only imagine what it must have been fifty years earlier.

Looking at the brighter side, there is no question that these early black magazines provided an opportunity for writers and editors they would not have otherwise had. Where else could they find an outlet for their work, except in special circumstances? Among those who benefited, for example, were Pauline Hopkins, Paul Laurence Dunbar, and William Stanley Braithwaite. These writers found a place to express their outrage, to talk about possible solutions to black problems, and to develop a sense of racial unity and pride. And for both writers and editors, black journals offered an opportunity to exert some influence of good among aspiring young black men and women.

In fact, as would hardly be surprising, it was the same social and political problems facing the black journals that prompted their formation. Historian Penelope Bullock, in her excellent and comprehensive view of black magazines through the nineteenth and early twentieth centuries, points out that in periods of hardship and declining status among blacks, more magazines appeared than were launched when conditions seemed relatively more favorable. Indeed, until the period of the Great War ended, one of the black media's most important functions was to protest inequity and injustices and to fight for improvement.

Before the Civil War, in spite of every obstacle, eleven magazines targeted to blacks were published. None survived the war itself, yet their very existence was a strong affirmation of the need for such journals. At the minimum, they provided a starting point and established a tradition.

In these prewar years, there were constant demographic and social changes affecting the content of these magazines. The audience was still pitifully small. In 1790, only about 8 percent of blacks were free and therefore able to constitute an audience. By 1860, this figure had increased to only 11 percent, counting both North and South; most of them were urban dwellers. Not only was the percentage still extremely small, but in those seventy years, free blacks had lost many of the few rights they possessed, having been disenfranchised in several Northern states (including New York and Pennsylvania) and legally discriminated against in all of them. As a result, blacks organized their own institutions—schools, churches, and magazines.

As one would expect, most of these prewar publications were issued in cities, and most were financed through other institutions, especially churches. There was little professionalism in the early magazines because nearly all the editors had regular jobs and practiced their craft on the side. Thus it happened that ministers, educators, and businessmen found themselves doing something outside their competence, but doing it surprisingly well in many cases.

Often these amateur editors had to write nearly all the copy for their magazines, relying on reprints from other sources to fill remaining holes. Those few who did contribute from outside—usually ministers, reformers, or aspiring writers—were not paid. As Thomas Hamilton, editor of the *Anglo-African Magazine*, observed, "The contributors to this magazine have performed a labor of love—the publisher has not yet been able to pay them—for which we present our loving thanks."

The scant advertising was mostly for local wares and services, and illustrations were too expensive to appear often, although Hamilton's journal and the *Repository of Religion and Literature* occasionally printed engravings created by John Sartain of famous black individuals. Sartain was one of the best illustrators of his day.

With so little advertising, black magazines had to rely more than the white publications on the always unpredictable subscription method of financing. The rates were low enough—usually a dollar a year—but even that was sometimes too much for those who thought they could afford it, and the delinquency rate was higher than it was for white publications.

With all these handicaps, it is remarkable—and a ringing testimonial to those who launched them—that four journals for blacks appeared between 1838 and 1848. They were slight and did not live long, but they nevertheless signaled the beginning of a black periodical press. There is some scholarly argument over which had the honor of being first, but it comes down to a matter of definition. As early as 1827, *Freedom's Journal* had appeared in New York City, but it was one of those cases (still with us) where the line between magazine and newspaper was blurry. A more solid claim could be made by two new journals that appeared in 1838, the *National Reformer*, issued in Philadelphia; and the *Mirror of Liberty*, in New York.

The *National Reformer* was the prototype antebellum black magazine. Begun in September 1838, it was the organ of the American Moral Reform Society, an interracial organization originally known as the American Society of Free Persons of Color, begun in 1835. At that time, the lumber merchant William Whipper called for the establishment of a magazine as soon as possible to convey the society's beliefs for the benefit of a larger audience than its members. That was the origin of the *Reformer*, which finally emerged three years later in Columbia, Pennsylvania, with Whipper himself as editor and the Board of Managers of the society in Philadelphia as publishers.

Whipper was already deeply involved in trying to improve conditions for blacks. As the son of a black servant and her white employer, he had a personal interest in reform. Fortunately, he had lived in the house of his father, who loved him enough to leave him his lumber business when he died. The grateful son made it even more successful, amassing a respectable fortune, which he devoted in large part to various antislavery causes. His house was well known (to blacks) as a stop on the Underground Railway.

A remarkable figure, Whipper was not only an astute businessman and a dedicated reformer, but he was also much interested in education and literature, joining Philadelphia literary and reading societies and founding a library for blacks. Under his editorship, the *Reformer* was a reflection of all his many interests, with both original and reprinted material. However, he could sustain it little more than a year; it expired in December 1839.

While it was still alive, the *Reformer* spoke out for a wide variety of reforms, including not only abolition but also equal rights for both blacks and women, temperance, and nonviolence. Whipper was particularly concerned about moral reform. He called it "the cornerstone of the temple of universal freedom and eternal justice."

Among the writers who appeared in the *Reformer* were Daniel Alexander Payne, William Watkins, and Henry Highland Garnet. Their articles and reprints from others were supplemented by information from the society and the Philadelphia Vigilance Committee. Bullock speculates that the magazine may have died from so mundane a cause as the distance between its editorial offices and the publishers, which in those days was a more formidable problem than it would be a few years later.

Beginning in July 1838, the *Reformer*'s rival, *Mirror of Liberty*, was longer-lived, lasting for three years under the editorship of David Ruggles, an abolitionist and civil-rights leader. Ruggles's full-time job was secretary of the New York Committee of Vigilance, an organization created to help those free blacks who were kidnapped and taken into slavery. He was also active in the Underground Railway, and had helped Frederick Douglass, among many others. As a third job, this busy man acted as corresponding secretary for the American Reform Board of Disfranchised Commissioners, a civil-rights group for free blacks.

In spite of these activities, Ruggles had found time for work that somewhat prepared him to be a magazine editor, working for a printer and bookseller. He had, in fact, operated a bookstore in New York, with a reading room for blacks, for which he charged a low yearly price. Ruggles had also learned both the printing and bookbinding trades in his active life. Most useful for his magazine venture, he had contributed articles to several antislavery newspapers, including the *Emancipator* and the *Liberator*, as well as producing pamphlets attacking the popular (white) idea of colonization.

As Whipper's magazine had done, the *Mirror of Liberty* reflected Ruggles's many interests as well as his beliefs. It contained, among other items, the reports and minutes of the New York Committee of Vigilance, and information for and about the American Reform Board of Disfranchised Commissioners. Ruggles provided a running account of state legislation affecting blacks and such items as a speech by William Lloyd Garrison celebrating the emancipation of slaves in the British West Indies. For a lighter touch, the *Mirror* printed a poem by John Greenleaf Whittier on slavery. There were few contributions from individual black writers, however, although black community leaders tried to support the journal in every other way. Still, it was a fiery magazine that might have survived even longer if it had not been for Ruggles's ill health and the old problem of insufficient funds to keep the publication going.

A third black journal began in 1841, this one sponsored by a church, which naturally titled it the *African Methodist Episcopal Church Magazine*, with the church's Book Concern as publisher. It was primarily a church organ, reporting the numerous activities of this organization. It was the church that had split away from a white Philadelphia congregation in 1787. Richard Allen, who had led the breakaway, consolidated his people with other similarly separated groups in 1816 to form the A.M.E. After Allen was made bishop, he started a publishing house, and the magazine was one of his creations. Ministers were the chief contributors, and they also acted as subscription agents. They worked hard, but it was not enough to overcome the usual lack of money and the difficulty of finding enough literate readers. It persisted, however, until 1848 before giving up.

Of an entirely different nature was the fourth of these early black magazines. It was *L'Album littéraire*, published in Louisiana and devoted to literature. This journal opened its columns to creative writing of all kinds in French by blacks, some of whom spoke and wrote the language in that part of the country.

The turbulent years between 1854 and 1863 produced seven new periodicals for blacks—or at least seven were planned. Not all of them made it into print. The emphasis in these journals was on the mounting conflict itself, and new questions were raised in their pages. Was it a good idea to emigrate in case the South should win? What were the practical problems involved in liberation? Was there any hope of overcoming racial prejudice? What kind of education should blacks have, and could they obtain it? All these matters were debated in the new magazines.

The prize for grandiose titles, even in that day of crowded mastheads, would have had to go to the *Repository of Religion and Literature and of Science and Art*, a second attempt by the A.M.E. to sponsor a magazine. This one was meant for both black and white audiences. As the editors declared, "It improves the minds of our people, as well as it encourages those of the white people, who are subscribers and well wishers to the colored people."

Proposed in 1858, the *Repository* appeared in the following year, backed by the church's literary societies. It was the first truly general magazine to be published by blacks, containing articles on literature, the fine arts, music, science, morals, and history. The magazine encouraged contributions by unknown writers.

This diversity could be attributed to the many interests of the editorial board, particularly of Bishop Daniel A. Payne, the chief editor and principal contributor. A son of free Southern blacks, Payne had operated a school in Charleston, South Carolina, until the state passed a law prohibiting the education of blacks. Moving up North, Payne studied in a seminary and eventually joined the A.M.E. church, which he served as a bishop for forty-one years, a position that enabled him to carry on his unremitting campaign for black education.

With the demands made on him by his church position, Payne could not give the new magazine his full attention, but his editorial board was constantly at his side, designating one of their number to act as executive editor, a rotating assignment, during the magazine's five years of life.

Probably the most influential black magazine to appear was begun in 1858 by Frederick Douglass, who had previously founded the newspaper *North Star*, later to be known simply as *Frederick Douglass' Paper*. His magazine was similarly titled, *Douglass' Monthly*. It focused entirely on political matters, reporting on legislation, speeches, and the activities of abolitionist societies and individuals.

This former slave, who was at once abolitionist, lecturer, and journalist, had strong opinions and the eloquence to convey them in the pages of his magazine as well as elsewhere. His work predominated in the *Monthly*, where readers could follow his progress from a belief in moral persuasion to the advocacy of military action against slavery. Other black writers contributed, too: James McCune Smith, Daniel A. Payne, Martin Delaney, and Henry Highland Barnet, to name a few.

Douglass, however, also gave space to such white writers as Gerrit Smith, Charles Sumner, Wendell Phillips, and Lewis Tappan. His magazine was the most solid black publication yet to appear, and no doubt would have gone on if Douglass had not in 1863 decided to devote all his time to recruiting black soldiers for the Union Army, an offer the army refused. Douglass then turned to full-time lecturing instead.

In leaving the field to others, Douglass had a few thoughts about how coverage of blacks in the media had changed. "I have lived," he wrote, "to see the leading presses of the country, willing and ready to publish any argument or appeal in behalf of my race, I am able to make. So that while speaking and writing are still needful, the necessity for a special organ for my views and opinions on slavery no longer exists. To this extent at least, my paper has accomplished the object of its existence. It has done something towards battering down that dark and frowning wall of partition between the working minds of two races, hitherto thought impregnable."

Another distinguished black journal of this period, although its life soon ended, was the *Anglo-African Magazine*. Of this journal, a media scholar, Charles Johnson, wrote, "The *Anglo-African* showed a kinship to none of the Negro publications that preceded it and to the few that followed. Its standards were clear-cut and high, its articles scholarly and superior."

This journal, edited and published by Thomas Hamilton, lasted from January 1859 through March 1860. Hamilton came from a family of journalists and reformers and had worked as a newspaperman, bookseller, and book publisher. William Hamilton, his father, had been active in the freed slave community of New York, not only opposing slavery but also supporting several benevolent causes.

Starting as a New York newsboy, Hamilton had distributed the *Colored American* as early as 1837, and he witnessed the birth of the *Mirror of Liberty* and the A.M.E.'s magazine, both of which he read. Going into the business himself at the first opportunity, he worked on such white antislavery publications as the *Evangelist* and the *National Anti-Slavery Standard*.

In 1859, Hamilton decided to go into the business for himself and brought out the first issue of the *Anglo-American*, meanwhile simultaneously starting a newspaper, the *Weekly Anglo-African*, with the help of his brother Robert. The newspaper lasted through the entire war, but the magazine survived for only fifteen issues.

While it lasted, however, the *Anglo-American* led the field. Its articles, on a wide variety of topics, were written by the best black writers of the time, and the magazine also printed biography, featured essays on the economic and social status of blacks, provided information on legal and education developments affecting them, and reported the news of the day involving blacks, John Brown's raid most notably. Readers were appreciative. One sent in ten dollars to start a fund that would place the magazine in public libraries.

One of the *Anglo-African*'s most prolific and admired contributors was Martin Delaney, who also published in it parts of his novel about slavery, *Blake, or, the Huts of America: A Tale of the Mississippi Valley, the Southern United States, and Cuba*. About a third of this work appeared in the magazine and the remainder in Hamilton's newspaper.

Among the other contributors were Payne, Barnet, Douglass, Smith, James Theodore Holly, Frances Ellen Watkins Harper, and James W. C. Pennington. Satire was contributed by William J. Wilson, including a biting piece, "What Shall We Do With the White People?" which Wilson signed "Ethiop." In it he wrote, "This people, the white, must be saved; quiet and harmony must be restored. Plans for removal of these white people, as all such schemes are— such for example as these people have themselves laid for the removal of others out of their midst—would be wrong in conception, and prove abortive in attempt. . . . We give them also high credit for their material progress. Who knows, but that some day, when, after they shall have fulfilled their mission, carried arts and sciences to their highest point, they will make way for a milder, more genial race, or become so blended in it, as to lose their own peculiar and objectionable characteristics? In any case, in view of the existing state of things around us, let our constant thought be, 'What for the best good of all shall we do with the White people?'"

Not all blacks opposed emigration, a consistently controversial issue. In 1856, the *New Republic and Liberian Missionary Journal* appeared to support emigration to Liberia, the aim of the mostly white American Colonization Society, which eventually did establish a colony of American blacks there. The editor of this magazine was John Wolff. A different journal was issued by the National Emigration Convention of Colored People, a group supporting the idea of encouraging blacks to emigrate to other countries in the Western Hemisphere, especially the West Indies, Canada, and Latin America. The idea of an international black literary journal had been proposed at their 1854 annual meeting. A plan was drawn up two years later for such a magazine, to be called *Afro-American Repository*. Editors were appointed and a publishing company formed, but for reasons unknown the project never materialized.

During 1863 and 1864, the *Students' Repository*, issued at the Union Literary Institute in Indiana and edited by its principal, Samuel Smothers, had a brief run. The articles were devoted almost entirely to education and self-improvement, and the overall idea was to promote black education generally as well as the institute, which provided manual training. When Smothers joined the Union Army in 1864, the *Repository* suspended publication.

Finally, in San Francisco in 1862, there appeared the first black magazine west of the Mississippi. The West had been something of a magnet for blacks as well as whites before the war, at least for those who could get there as free men. Many settled in San Francisco, creating an audience for the oddly named *Lunar Visitor*, edited by A.M.E. minister John Jamison Moore. This journal carried some local news, but it devoted most of its space to large issues—black education, legislation affecting blacks, and the need for unity among them.

Taken together, the antebellum black journals represented an exceptionally brave and reasonably promising start, considering the formidable obstacles. They laid a solid foundation for what was to come.

6

Magazine Publishing
in the Civil War

All the print media suffered from the dislocations of war after the great conflict began, but none were in a worse position than the magazines. Newspapers were in demand as never before to bring news from Washington and the battlefields, so their financial problems were considerably helped by dramatic increases in circulation. Magazines, by contrast, had difficulty holding their audiences together, except for those few competing with newspapers in covering the war, and were hard-pressed to meet the rising costs of paper and ink, as well as all the other necessities of production.

The results were inevitable suspensions of many periodicals and a general decrease in size. Most had to struggle to survive, and they were not helped by a federal tax on advertising in 1862, affecting the general periodicals more than the others. Fortunately, the government did not raise postal rates during the war. While some publishers tried to make up for their loss of advertising revenues by increasing subscription rates, many others did not. For those in the North, the reduction in circulation area was a problem impossible to overcome. Some of the leading magazines, like *Harper's* and *Godey's*, were popular in the South; so were the illustrated weeklies, particularly *Leslie's*, and the religious quarterlies. There was one compensation. Many of the Northern periodicals had a substantial circulation in the Union Army itself, even (remarkably) *Godey's*.

The South, of course, had a much worse time. All the print media suffered from the lack of supplies, and sometimes the need was so desperate that desperate measures had to be taken. Some books, for example, were printed on the backs of wallpaper. No doubt the most memorable example of Southern ingenuity was the wartime edition of *Les Misérables,* a best-seller in the North. Getting a copy for Southern printers to set from was a major problem in itself, but a daring Confederate sea captain managed to smuggle a copy from France through the Union Navy's blockade, and it was immediately translated. The book that resulted was a triumph of persistence. Bound in dirty Confederate gray, its type haphazard, its pages uneven, this volume became an object of scorn in the Northern publishing community, where it was known as "Lee's Misérables." Nevertheless, it was as widely read in the South as elsewhere, limited only by the number of copies available.

The problems Southern magazines shared with book and newspaper publishers were compounded by special difficulties. Coarse paper and inferior inks did not lend themselves to the kind of product the periodical publishers wanted to produce, and conscription made it extremely hard to get enough labor to put together whatever was available to make a magazine. Postage rates, too, were even higher than they were in the North. As a result of all these obstacles, some of the best magazines failed to survive the war; the *Southern Literary Messenger* gave up in 1864, and other prestigious journals, like *De Bow's*, had to suspend operations.

De Bow's, in fact, typified the struggle of periodicals in the South to stay alive. Begun by young James D. B. De Bow in 1846 as the *Commercial Review of the South and West*, it continued under various other names until, in 1855, it became *De Bow's Review and Industrial Resources, Statistics, etc.* and remained so until it had to be suspended in 1862, reappearing for a single issue in 1864, then returning briefly in 1866 as *De Bow's Review Devoted to the Restoration of the Southern States*, and finally, in other hands, disappearing entirely in 1880. Through all these transformations, the *Review* managed to convey large doses of politics, tempered with literary miscellany, to its shifting readership. It was a loyal interpreter of the antebellum South, particularly valuable for its articles on economics.

The casualties among Southern periodicals were numerous during the war. They included the two oldest journals of medicine and many religious magazines. Nevertheless, a few old leaders like the *Southern Presbyterian Review* managed to survive, and there were even a few new starts, although they did not last. Overall, the magazine business in the South was not significantly damaged, since there were no important publishing centers in any case.

In the North, wartime magazines merely intensified what they had already been doing. As a sounding board for the discussion of slavery, or the propagandizing of one point of view or another, these periodicals had been effective before Emancipation in debating freedom for the slaves and afterwards contributed to the new discussion about what to do with the freedom. Their tone, in this as in other respects, was political: state's rights, nativism, the tariff, and other issues of the day.

Naturally enough, the war dominated every issue of most magazines. Three out of five of the *North American Review*'s long articles dealt with the conflict, and the *National Quarterly* had about the same proportion. The general monthlies used even more war material, but it was balanced somewhat by fiction, poetry, and essays. They also carried contributions sent directly from the battlefields. The staid and respected *Knickerbocker Magazine*, nearing the end of its life, turned out to be, unexpectedly, a Copperhead journal in blatant alignment and sympathy with the South.

As the war began to wind down, however, the monthlies started to shift focus toward more usual patterns of literature. This was not so much true of the weeklies. They had been closer to the war than others, functioning almost like newspapers, with correspondents in the field and superb pictorial coverage as well. *Harper's Weekly* was ahead of all the others in the extent and quality of its war coverage, with the possible exception of *Leslie's*.

Magazines in general were as diverse in their approach to the war as their audiences. It was hardly surprising, for instance, to see the quarterlies reflecting the temperaments of their publishers in how much space they devoted to the war. The special trade and technical journals covered the conflict only when aspects of it involved their particular fields.

Strangely, in a time of such national turmoil, there were a few magazines that took a "plague on both your houses" attitude toward the war. At one of the darkest hours in 1864, the *Oneida Circular* wrote, "We had almost forgotten about the war. We have had so far, and are likely to have hereafter, but little to do with War. . . . Curing slavery by war is like curing fever by calomel; the remedy, though effectual, remains in the system as a cause of disease."

By contrast, there was one thriving class of magazines that devoted its entire attention to the conflict—the military journals. They flourished everywhere: state military journals, militia periodicals, and leading them all, the *United States Army and Navy Journal*. In the field, various units of the army had their own journals in the North, although they were inclined to be irregular. These transitory publications included the *Swamp Angel*, issued in 1864 from the Federal garrison on Morris Island in Charleston Harbor, and the *Red River Rover*, printed, as it said, "on board the steamer *Des Moines*, Uncle Samuel Publisher." The *Rover* was printed on ruled paper, the work of a printer from the Eighth Wisconsin.

Another of these colorful service papers was the *Camp Kettle*, which advertised that it was "published at every opportunity by the Field and Staff Officers of the Roundhead Regiment," by which was meant the 100th Pennsylvania. During the Vicksburg siege, soldiers were entertained with the *Yazoo Daily Yankee*, "published semi-occasionally by Mr. Mudsill, Mr. Small-Fisted Farmer, Mr. Greasy Mechanic & Co." On the other side of the lines at Vicksburg, the Confederate soldiers turned out the *Vicksburg Daily Citizen*, emulating publishers back home by using the only available material, wallpaper.

Besides the *Knickerbocker*, there were two prominent Copperhead magazines, the *Day Book*, published daily and weekly at various times, and the monthly *Old Guard*. The *Day Book* had begun life in 1849 in New York as a "saucy, racy, and spicy" weekly, but by 1861 it had become an organ titled provocatively the *Caucasian*, so virulent that it was barred from the mails. It resumed in 1863 under its old name. It died of its own venom, in which it was exceeded only by *Old Guard*, which, as Mott says, was "merely a curiosity."

The war brought out all kinds of writing in the magazines, some of it inspired, some ridiculous, and much of it hysterical. All this was embodied in the memorable paragraph by J. Quitman Moore from an article in the January 1861 issue of *De Bow's*. "With all its attendant evils," Moore wrote, "with all its tragic horrors—with all its mighty retinue of sorrows, sufferings, and disasters—war—civil war—war of kindred races—is not the greatest calamity that can befall a people. . . . There is in war a sublime and awful beauty—a fearful and terrible loveliness."

This was somewhat removed from the pure hatred of the *Southern Monthly*, which wrote of the Yankees, "They are a race too loathsome, too hateful, for us ever, under any circumstances, to be identified with them as one people."

The *Atlantic* and *Harper's Monthly*, those two distinguished literary journals, neither poetized nor propagandized the conflict. The *Atlantic*'s policy was that the war not be permitted to monopolize its pages. James Thomas Fields, it editor during those years, believed that the magazine should be a place where readers could retreat from the agonies of the day. He liked to print articles that were not merely reportorial, that looked toward the future. *Harper's* was being edited by Alfred H. Guernsey, a Greek and Hebrew scholar, who had become the magazine's second editor in 1856, a position he held until 1869. He succeeded in doing the nearly impossible: steering his magazine throughout the Civil War years on a nonpartisan course. He was rewarded by having *Harper's* called "more unexceptionable on the subject of slavery than any northern work of similar kind."

No doubt the most remarkable category of all the Northern magazines during the war was the one directed to women. These magazines combined an astonishing mixture of their usual ingredients with patriotic propaganda and not a little politicizing. Northern women could choose from a variety of different magazines, with an aggregate circulation of more than 250,000.

Individual circulations were harder to pin down, since in those unaudited days, such figures were not easy to certify. *Godey's Lady's Book*, with a circulation of 150,000 in 1850, appeared to be the clear leader, while *Peterson's* ranked second before the war began but possibly even exceeding *Godey's* before 1865. The others were far behind: *Arthur's Home Magazine*, with 10,000 to 30,000; *Leslie's Lady's Magazine*, not reaching 50,000 until after the war; the *Methodist-Episcopal Ladies' Repository*, 35,000 in 1864; the *Universalist Ladies' Repository*, wartime circulation unknown but probably not above 5,000 during the war; no figures at all are available for the *Sibyl* and the *Mother's Magazine and Family Circle*, but almost certainly they were insignificant.

These were Eastern seaboard magazines, most of them clustered in Philadelphia, where *Godey's*, *Peterson's*, *Lady's Friend*, and *Arthur's* all had their offices. The others were in New York, except for the *Universalist* of Boston; the Methodist-Episcopal publication, which was in Cincinnati; and *Sibyl*, issued in Middletown, New York. Divided in their views about the war, these periodicals also differed in their opinions about women's responsibilities in waging it, reflecting not only the differences in editors' political opinions but also their conceptions of women's place in society. One historian divided them roughly into "reform" and "mainstream" groups, with the latter subdivided into "moderate" and "constricted."

The reformers were on the barricades early, first calling for abolition and later offering pointed advice about the conduct of the war and about the duty of women to participate in it actively in whatever ways possible. The reformer periodicals wanted women not only to do the accepted "relief" work but to be political activists as well. "Mainstream" publications, however, wanted to draw their skirts away from the political and other controversial issues of the day. They were not critics of Lincoln or his generals or of the general conduct of the war, and while they urged women to take part in the war effort at home, they wanted them to do it within conventional limits.

There were four "reform" periodicals: *Sibyl*, which ran briefly from 1856 to 1864, *Arthur's*, and the two *Ladies' Repository* magazines. They all had small circulations, and only two of them were edited by women—*Sibyl* by Lydia Sayer Hasbrouck and the Universalist journal by Caroline Sawyer. Timothy Shay Arthur was far more noted for writing *Ten Nights in a Bar-Room* than for editing his *Lady's Home Magazine*, which he founded in 1852. A clergyman was editor of the Methodist-Episcopal journal. These editors had several things in common, however. All of them were advocates of women's rights, including education, and they were against both slavery and alcohol.

Not surprisingly, these reformers viewed the war as a great moral issue of right-thinking Northerners against a degenerate South, ruined by slavery. To abolish the peculiar institution was their single-minded goal, and consequently, their magazines were highly critical of what they considered Lincoln's lackadaisical conduct of the war, thus allying themselves with the most ardent abolitionists of all stripes before Emancipation. Nor did they have much sympathy for Lincoln's generals, or even for his wife, whom they thought extravagant in a time of national peril.

What could women do about these assorted problems? Become activists, the editors advised in their magazines, not only by volunteering to work in hospitals or on farms but also by exerting as much political pressure as they could. That would eventually require suffrage, and *Sibyl* was a strong advocate of this step. Since all the editors were activists themselves, it did not seem unreasonable to them that sensitive, thinking women should join their ranks.

Unfortunately, it is impossible to know how much of an impact these magazines had on their relatively small readerships, how much word-of-mouth activism was generated by them. But surveying the war effort as a whole, it seems clear that they inspired a minimal amount of direct political activism. Women, in general, appeared to believe that their role was in the hospital, the factory, or wherever else they were needed to replace men. The reform magazines were speaking to the already converted as far as anything else was concerned.

By contrast, the women's magazines with large circulations took a traditional approach to the proper role of women in the war, regarding any political activity as inappropriate. Indeed, the editors took pains to avoid almost all controversial subjects, although at the same time they did not try to avoid the war. *Leslie's* and *Godey's* were the most conservative of the periodicals for women in this regard, carefully advising their readers to participate in the war effort within established limits.

In between the reformers and the traditionalists were such magazines as *Peterson's*, *Lady's Friend,* and *Mother's Magazine and Family Circle,* edited by both men and women who had never been reformers nor showed any interest in activism. They largely confined the woman's wartime role to keeping up spirits on the home front and, through letters, bolstering morale at the front lines. These moderate editors sanctioned nursing and consciously economical homemaking as legitimate patriotic activities, plus the lending of a hand wherever needed to care for those sick and wounded soldiers sent home. The work of women in the Sanitary Commission and in hospitals was held up as exemplary.

These moderate women's magazines did not appear to be much interested in abolition, or in slavery itself. Reading the fiction they printed, women would not have known that the country was being torn apart. These romantic, sentimental stories were usually not concerned with the war, and when they were, dealt not with serious issues but with the complications it caused in women's love lives. The articles were studiously unpolitical. Kathleen Endres, a researcher who has explored the subject, believes that what the moderate magazines printed could be taken as a reflection of how most middle-class women viewed the war.

Stereotypes, founded in fact, were set up in these magazines: the noble, self-sacrificing mother, the valiant protector of family values until the men returned home. *Mother's Magazine* extended the stereotype to real life by supporting editorially the mothers who had organized themselves to help in the war effort. Like others in the women's magazine category, *Mother's* did play a role in helping to rally the home front around useful activities but tempered its efforts with large quantities of sentiment and blatant propaganda. The bottom line, so to speak, was the time-honored one: women's place was in the home.

Leslie's underwent a change while the war raged on. Frank Leslie was editing it personally, and there had been a good deal of factual war coverage, accompanied by excellent illustration. But after the future Mrs. Leslie, then a red-headed editor named Miriam Squier, took over a major share of the magazine's direction in 1863, the war began to take a rear seat, supplanted by more entertaining (in the old sense) stories and pictures. For a time, at the beginning of the conflict, Leslie had tried to do the impossible and not take sides, hoping to retain his Southern readers, but that approach did not last long. By 1862, he was belaboring the Confederacy and its leaders, as much as any other loyal Union editor. Oddly enough, the magazine failed to endorse Emancipation. As for women generally, *Leslie's* advocated the traditional approach—follow wherever the men led, practice frugality, and uphold moral standards.

Nowhere was conservatism more entrenched than in *Godey's,* a reflection of Louis Godey's character. Sarah Hale, its editor, simply refused to acknowledge in the magazine the Confederacy's existence. In her editorials, she exhorted readers to keep their faith in God but never advised them how they could help the Union cause she supported.

At least it could be said for women's magazines that they did not constitute a stereotypical monolith, even though they were Union supporters. They saw themselves primarily as morale builders, and they must have been correct in anticipating what their readers wanted because none of them suffered any significant circulation losses during the war, except for whatever Southern circulation they had at its beginning.

In assessing the role of the media in the Civil War, it appears that magazines were probably the least influential. Newspapers were the prime carriers of war news, notwithstanding the notable contributions of the weeklies, and they also carried news of that other daily struggle, in Washington and elsewhere, over the conduct of the war. Books were important not only for the flood of information they provided but also for entertainment. A tidal wave of paperbacks, often packed by the bale and sent to the front, performed much the same service as the Armed Services Editions in the Second World War. Paperbacks were equally popular at home as a release from wartime tensions.

Magazines, however, with the exception of the weeklies, were not a chief purveyor of news and information from the front, and what they provided in the way of home entertainment was, for the most part, awash in sentimentality. Perhaps their most valuable contribution was the effort made by the three prestigious monthlies, *Harper's,* the *Atlantic,* and the *North American Review,* to keep literature alive in a disintegrative period of national life. But it should not be forgotten that a less visible portion of the magazine industry—the business, technical, and professional periodicals—easily matched the book industry in providing quantities of useful information, as well as helping to hold together diverse constituencies suffering, like all the others, from the war's impact.

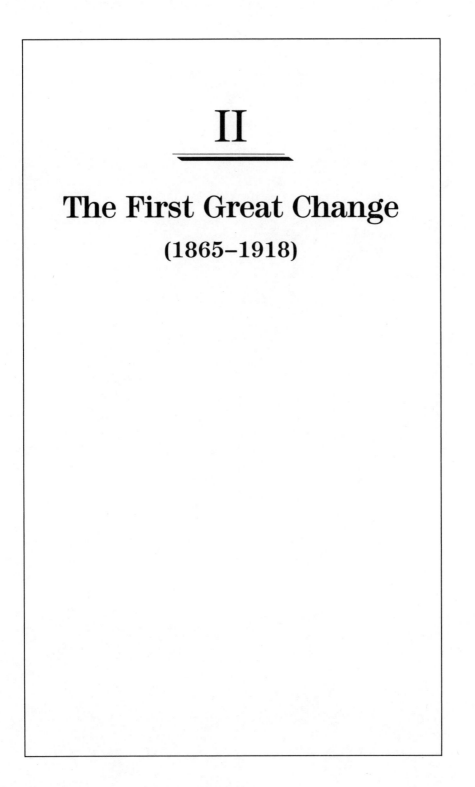

II

The First Great Change

(1865–1918)

7

New Audiences, New Editors, New Magazines

After the silence at Appomattox came the uproar of an America transforming itself from a primarily agricultural country into the industrial colossus of the West. Between 1865 and 1918, the United States we know in this century was created, beginning with industrialization and ending with the Great War. In between, we suffered cycles of boom-and-bust, fought an imperialistic war in Cuba and the Philippines, created great wealth and extreme poverty, and developed a distinctively American culture.

The magazine industry became an enormous mirror, reflecting all these changes, from the grandiose to the minute. Diversification had begun before the Civil War, creating the so-called "Golden Age," and it continued afterward until there was scarcely a single interest in American life that was not represented. As those interests continued to multiply in this century, the magazines kept pace on a scale not equaled anywhere in the world.

All the factors present in the antebellum days combined once again to make the postwar surge possible—technological developments in presses, stereotyping, and engraving; the remarkable growth of education; the rise of public libraries; and a public to whom, generally speaking, reading was a habit. That audience had grown, too, as a result of the war, for soldiers who had never had the time or opportunity to read books and magazines found distraction in them during the hurry-up-and-wait pattern of armed conflict. Paperbacks and magazines had helped to fill idle hours.

The immediate postwar boom in magazines can be seen in the statistics. There were only 700 magazines in 1865, but their number had nearly doubled, to 1,200, by 1870. There were twice that many by 1880, increasing again to about 3,300 by 1885. In short, the number of periodicals had multiplied more than four-and-a-half times in just two decades. The mortality rate was high, as always. Estimating an average life of four years for an individual magazine, Mott deduces that there must have been about 8,000 or 9,000 magazines between 1865 and 1885. This can be no more than an educated estimate, however, because the Census Bureau failed to distinguish sharply between newspapers and periodicals and kept redefining its categories from decade to decade.

One of the most significant aspects of our striking national growth was the continuing spread of magazines. William Dean Howells accurately described this phenomenon: "As soon as the country began to feel life in every limb with the coming of peace, it began to speak in the varying accents of all the sections." That did not mean a decline in the old centers of publication, however. New York was still preeminent, as it was in book publishing as well. In 1880, a fourth of all magazines were published there, including two-thirds of those with a 100,000 or more circulation.

Thus, a pattern was set up which persisted for some time, at least in several aspects. New York magazines tended to reflect the life of that city, while those elsewhere were constantly writing about it (particularly about the personalities in its political and cultural life), creating the love-hate relationship that still exists today between the now celebrated "Big Apple" and the rest of the country. Periodicals outside New York wrote endlessly about the city's corrupt government, its snarled transportation, the dangers of its streets, and the more seamy aspects of its vice and crime. All this was editorial fodder not only for the sensational journals but also for respectable reviews. The *Nation* pointed with horror to New York's estimated 6,000 streetwalkers, 12,000 "criminal women," and 773 brothels, all in a city of less than 1.5 million. *Harper's Weekly* cited the city's 9,250 saloons, most of which paid no attention to Sunday closing laws.

By contrast, Boston, the second city in magazine publishing, was losing commerce rapidly to New York and had slid back to a state of quiet respectability, proclaiming itself as still the nation's cultural capital but boasting only one magazine (the scarcely cultural *Youth's Companion*) in the 100,000 class. Philadelphia, the third largest magazine center, with *Lippincott's* as its chief literary ornament, was being challenged by Chicago but still considered itself as the home of fashion magazines, with *Peterson's, Godey's,* and the *Ladies' Home Journal.* It also was the leader in medical, legal, and religious periodicals.

In the prostrate South, not much could have been expected in the first years after the war, but amazingly, at least twenty or more literary publications were begun in the first four years, and there was considerable activity in religious and other class magazines. There were five distinct centers: New Orleans, Atlanta, Baltimore, Richmond, and Louisville. Reversing their hardline stands of the war years, Northern magazines were now full of sympathetic articles about the South—notably *Scribner's* brilliant survey series of 1873–1874, "The Great South," by Edward S. King, with numerous illustrations.

Both East and South, however, were giving the developing West more attention, as its rapid growth in industry, transportation, political power, and culture increasingly demanded. In postwar America, Eastern Americans especially were astonished to discover in their magazines that the Middle West was settled and already powerful, while the West, still a romantic region to many, was growing at a surprising rate. All this was discussed endlessly in the magazines, often with wonder, giving Americans a new sense of their country's huge dimensions and its diversity and some feeling of national unity.

Middlewestern cities were becoming magazine centers, too, although not likely to rival those in the East. There were a few women's periodicals in St. Louis, and in Chicago, largest of the new western cities, the Lakeside Publishing company was issuing nineteen different magazines in 1870 in several fields. Many of these were storypaper weeklies, supplementing Lakeside's production of cheap paperbacks. But there was also the *Current*, founded in the 1880s by an exceptional journalist, Edgar L. Wakeman. It printed contributions from such literary figures as Joaquin Miller, E. P. Roe, John Burroughs, James Whitcomb Riley, and Opie Read. Unfortunately, it lasted less than a decade, expiring in 1888. Chicago was also producing a great many home journals and class magazines. There was magazine publishing in Cincinnati and even in Toledo, where Petroleum V. Nasby was issuing his famous weekly, the *Blade*.

In the Far West, San Francisco was the only publishing center of any importance, with most of its distinction centered in the reputation of the *Overland Monthly*, which Bret Harte made notable in its early years. It did not long survive his departure in 1870, dying five years later. Individual writers like Harte were likely to be the life of these small periodicals. Ambrose Bierce, for example, wrote a "Town Crier" column for Frederick Marriott's *California Mail-Bag*, which did much for the reputation of this newspaper-magazine and for his own. He, Mark Twain, and Harte were the writers who made the reputation of the *Wasp*, founded in 1876, and Bierce's column "Prattle" did the same for the *Argonaut*. Bierce wrote in his first column that he intended "to purify journalism in this town by instructing such writers as it is worthwhile to instruct, and assassinating those that it is not."

Magazines everywhere in the country shared the same problems, in greater or lesser degree, and some suffered from the same old difficulties—problems of circulation, advertising, and editorial content. Mott points out that any study of circulation is complicated by the fact that none of the figures, in these days before auditing, can be certified as accurate. Competitive publishers did not want to disclose them, and the competition also led to outrageous lying. George P. Rowell, the pioneer advertising agent, attempted to compile a collection of newspaper—and some magazine—circulations, but when he sought the numbers at *Harper's Weekly*, the brothers Harper were so affronted that they broke off relations with his agency. Not all publishers were dishonest in this respect, but they were extremely sensitive just the same.

The most that could be said accurately about circulation figures was that they continued to grow. The 385,000 that *Youth's Companion* boasted in 1885 was unprecedented. The literary journals, like *Harper's* (both monthly and weekly), now exceeded 100,000, matched by *Scribner's*, which later, as did the *Century*, reached a circulation of 200,000 in 1885, unheard of for a literary magazine. In that same year, the *Ladies' Home Journal* was already at 270,000 and climbing, while Ebenezer Butterick's dollar fashion magazine, the *Delineator*, was at 165,000. Even cheap storypapers, and sensational publications like the *Police Gazette*, had circulation figures of 100,000 and more.

The leadership of *Youth's Companion* in the circulation race could be attributed to its system of premiums—soon widely copied. Premiums were given to new subscribers, to those renewing, and to club combinations. Most of those solicited could not bring themselves to refuse such offerings as books, pictures, clothes, tools, machines, pianos, even church bells, although they favored chromos over all the others. Premiums became so widespread and so popular that the *Literary World* half seriously advised its readers that young couples could furnish their first house entirely with premiums if they only subscribed to enough magazines.

In the end, however, premiums proved to be no more than a momentary craze. They had existed long before; the *Companion* had brought them to a peak of popularity; and by 1885 they had been so abused as a promotion device that they faded away, except in women's magazines and the mail-order journals, which went on using them effectively for years.

Clubs and combination rates presented a different problem. News dealers were hurting from this practice in the 1870s, but distributors like the American News Company and its lesser rivals became big businesses. A far-reaching effect of greatly increased newsstand sales was the quick rise of a soon common practice among publishers—dating their publications so that they would appear on the newsstands at the time of cover date or earlier. By 1870, this had become general practice.

Newsstands themselves, however, were not large splashes of color and images, as they have become in our time. Magazine covers were usually gray, buff, or brown and carried no pictures, with one exception. The highly popular storypaper magazines, small folios, splashed sensational woodcuts on their front pages—what would later be called a self-cover. These scenes presaged the pulp magazine covers of a later era—locomotives plunging off tracks, duelists, firemen rescuing people from burning buildings, and women threatened by obviously villainous men. The storypapers were a striking contrast to the standard magazines on the stands.

Some facts of magazine life had not changed in the postwar world. At a time when prices generally had risen from 50 to 200 percent above their prewar level, payments to contributors did not rise proportionately with the cost of living. The *Atlantic* increased its page rate from $6 to $10, along with *Lippincott's* and *Scribner's*, but *Putnam's* stayed at $7. "Brand names," as we call them now, got more. The *Atlantic* paid Lowell $300 for a long poem, and Mark Twain got two cents a word from *Galaxy*.

It was still impossible for most writers to live by writing, a situation not greatly changed today. To earn $2,000 a year, considered a comfortable income, a writer would have to sell an article every week. But again, by 1885 such well-known authors as Henry James and William Dean Howells were getting $5,000 for a serial. Fees on the story weeklies and magazines like *Leslie's* were better than most, the latter paying $8,000 for a serial and the New York *Weekly* giving the popular novelist Mary Jane Holmes from $4,000 to $6,000 each for the serials she turned out so prolifically. But, as had been the case before, the lofty literary journals paid little or nothing to contributors.

Nevertheless, something resembling a class of magazine writers had come into existence by 1870, and along with it, a distinctive style of magazine writing. The kind of eighteenth-century prose that had characterized periodicals before the war could still be seen in some magazines, particularly the reviews, but a brighter, more journalistic kind of writing was infiltrating many of the others. It was not an easy transition. E. L. Godkin, the *Nation's* editor, one of the great stylists of his day, wrote of his own magazine's style: "It has been so far rather heavy, and I find it difficult to lighten it. . . . In fact, it is very difficult to get men of education in America to handle any subject with a light touch. They all want to write ponderous essays if they write at all."

It was the influence of editors like Godkin and those at the helm of magazines like *Putnam's, Galaxy,* and *Atlantic* that brought a gradual change in writing style. By 1877, an observer in the *National Republic* could say, "Writing for the magazines has become a profession, employing a considerable number of trained experts." By that time, only the religious quarterlies were still adhering to the style of Samuel Johnson—without benefit of his talent, of course.

With this professionalization came the disappearance of anonymity, so long the practice, although the *Nation* continued to carry unsigned articles into the late 1870s. But the *North American Review* had begun publishing authors' names in 1868, and the practice spread rapidly. Godkin had a reason for not changing at once. He wanted his magazine to have a common style, as news magazines would later.

Authors still suffered from censorship of various kinds and from the tyranny of editors. Anything sexual was fair game for the blue pencil, and if that made us a nation of prudes, as an English critic charged, it was only the price to be paid "for being, on the whole, the decentest nation on the face of the globe," said the *Century's* editor, Richard Watson Gilder, in a xenophobic moment.

Old abuses continued. Henry Adams, editor of the *North American Review,* gave his new assistant editor, Henry Cabot Lodge, an article by a local historian and instructed him: "Strike out all superfluous words, and especially all needless adjectives." Authors were often told to cut large portions of their serials; titles were changed at will; and many editors felt free to rewrite or even add their own material to an article or story without informing the author.

Yet a new generation of great editors was rising. There was Godkin, making the *Nation* into a magazine whose influence far exceeded its small circulation. And there was Dr. Joseph G. Holland of *Scribner's,* a model for the new breed. He was close to his audience, which happened to be upper middle class. A writer himself, although not a particularly good one, he possessed those two essentials of the first-class editor: wide interests and an inquiring mind. At the *Atlantic,* there was Howells, no doubt a better editor than Holland or his own predecessor, Lowell, but unfortunately not able to solve his magazine's financial problems. He was admired, nonetheless, for excellent reviews, his charismatic personality, and his general grasp of American life.

Among the other notable editors of the day were Henry Ward Beecher of the *Christian Union,* a better contributor than editor; Henry C. Bowen of the *Independent,* an able publisher as well as an editor; Charles Eliot Norton, another editor of the *North American Review* (which had to await Allen Thorndike Rice, the prototype of the modern editor, who took the reins in 1878, to lift it to a new life with boldness and originality). These were the men replacing Godey, Peterson, Leslie, Arthur, and other prewar editors.

The magazines the new breed of editors were putting out became increasingly specialized. By 1885, there were periodicals for every occupation, activity, or interest, and that specialization has persisted until it has become the prime chracteristic of the industry. In the early postwar world, diversification grew because of the tremendous proliferation of interests among Americans. For example, whereas once there were only limited audiences for scientific, technical, and trade journals, now scientific education in schools and the interest generated by the great number of important new inventions provided a broad general readership. Much keener also was the need for information among those who had to keep up with the bustling development of new industries. There were journals devoted to pure science and to applied science, and there was a rapid proliferation of technical publications, trade magazines, and trade newspapers hardly distinguishable sometimes from the periodicals. For citizens who wished to keep generally abreast with developments in the pure and applied sciences, there was the *Scientific American* to inform them. It was one of several general science magazines founded before 1885.

The Centennial Exposition of 1876 was a powerful stimulus to this wide public interest, as it was to the new industries themselves. Americans were excited by what was happening and proud that their country was the leader. They lived in a world abounding with new inventions: the sewing machine, the typewriter, the incandescent lamp, the phonograph, the telephone, and soon the flying machine, the automobile, and the motion picture.

Trade journals covered every kind of commodity, and the professional magazines expanded remarkably in such fields as medicine, law, and finance. Agriculture, a profession of sorts, had a growing list of publications, many of them reflecting agrarian discontent over the geographical, industrial, and mechanical changes that were taking place.

Education was discussed widely not only in the standard magazines but also in a rising educational press. There was much to discuss: Froebel's theories about the kindergarten; overworked students in the public schools, a common complaint; and the great debate over the sciences versus the classics in college curricula, set off by Charles W. Eliot's article "The New Education" in the *Atlantic* in 1869, the year he became president of Harvard. He wanted the standard Greek, Latin, and mathematics curriculum replaced by a system of electives to include pure and applied science and living European languages. The resulting controversy raged through the pages of many a magazine.

(Yale's president, Noah Porter, defended classical education in the *New Englander*.) By 1882, Eliot had won the argument, and the elective system had been adopted by the best universities.

One of the most welcome advances in the magazine business was the postwar development of magazines for children, a move in which the United States was far ahead of other nations. The old wooden, moralistic journals were replaced by publications more in tune with children's real interests. The dime-novel publishers, whose output was often read surreptitiously by children, produced magazines more likely to be accepted. While the general output had been mainly religious, the new magazines were now about evenly divided between secular and Sunday School publications.

Youth's Companion remained the circulation leader, but by far the most distinguished of the new magazines was *St. Nicholas*, whose writing, art work, and general excellence raised it to a cultural level never equaled. Mary Mapes Dodge, better known for her classic *Hans Brinker and the Silver Skates*, edited *St. Nicholas* with distinction for thirty years.

On a broader level, there were publishers who knew how to exploit the new children's market. Foremost among them was Frank Munsey, whose weekly *Golden Argosy* was proclaimed as "freighted with treasures for boys and girls." His inaugural issue in December 1882 featured *Do and Dare*, an Alger serial. Munsey had to compete with the cheap weeklies for boys that had followed in the wake of the dime novels. They were as cheap in production as they were in literary quality, but they were immensely successful. The print was small and the pages garish, but boys loved these stories of other boys who did exciting things in the manner of Alger's little heroes.

Another revolution had occurred as well. While art had been distrusted before the war as foreign and probably immoral, in 1880 the *American Art Review* could offer this report: "Within the last ten years, a great change has taken place in public sentiment. The arts are no longer regarded as comparatively unimportant to our national growth and dignity, and an ever increasing enthusiasm has replaced languid interest or indifference." Again, it was the Centennial Exposition that played a major role in this shift of taste. Most of the visitors who packed its galleries were seeing famous pictures for the first time and partaking of a world they knew little about. The magazines served to spread that influence more widely by reproducing hundreds of these pictures in their pages.

On a broader level, the works of Louis Prang and of Currier and Ives had made chromolithography one of America's most popular mediums, stimulating interest in art itself. In the space of little more than two decades, the nation acquired an art public where one had virtually not existed before. A dozen or more art magazines flourished in the East.

Electrotyping revolutionized the printing of pictures. A landmark was the publication by Appleton of *Picturesque America*, a book issued in parts every month between 1872 and 1874, reflecting that house magazine's excellent illustrations of city and rural life, of scenes of natural beauty across the continent, and of people of all kinds captured by artists in their social contexts.

Scribner's and *Harper's* were also pioneers in the new depiction of America. In 1879, Josiah Holland wrote, "From the moment *Scribner's* began to avail itself of the art of photographing pictures upon the wood, a great development took place, because that presented at once to the public the work of the best artists. The men who had hitherto been shut away from us could draw and paint their pictures, which could then be photographed upon the block." It was not long before engravers were able to reproduce textures, brush strokes, and the effects of charcoal, pencil, clay, wash, or oils.

By 1884, such magazines as *Century* and *Harper's* were using about 15 percent of their space for pictures. Other magazines were noted for various kinds of pictorial representation they offered. People looked for the best work of all kinds in the *Aldine* and in Appleton's American edition of the *London Art Journal*, but they found the best etchings in the *American Art Review*, the best color work in Frank Leslie's *Popular Monthly*, and the finest lithographed cartoons in *Puck*, where the noted Joseph Keppler reigned. Examples of nearly all the best aspects of illustration were to be found in *St. Nicholas*, *Scribner's*, and the *Century*.

The other arts were not neglected by the burgeoning magazine industry. There were numerous national and local musical journals, many articles about music in the general periodicals, and music departments in such leaders as the *Atlantic*. There was a steady increase in the popular appreciation of music, and there were a hundred or more musical journals extant by 1885. The one with the longest life, lasting well into this century and known to generations of teachers and students, was *Etude*, begun as an eight-page monthly "for pianists and teachers of music" by Theodore Presser in Lynchburg, Virginia, in 1883. Presser soon moved it to Philadelphia, enlarged its size, and in time extended its musical range to violinists and singers. It became the bible of independent music teachers and a constant source of printed music for students as well as of information about what they were playing.

The theater had emerged into a state of respectability at last by 1885, not only extremely popular with the public but also discussed in all the leading magazines, reviewed by the literary journals, and written about by such noted figures as Lawrence Barrett, Henry James, and Brander Matthews. The Round Table took a disdainful view of critics—"so-called bohemians . . . long-haired and dirty-nailed men and unhooped and uncombed women"—but they could hardly fault the critical work in the magazines of such writers as William Winter, Matthews, Twain, and Howells. Howells went so far as to review in the *Atlantic* the work of a new burlesque troupe from London.

A great interest in sports was rising in America, along with its new taste for culture, and that inspired another category of magazines, following the familiar pattern of specialization, although there were general sports magazines as well. The great bicycle craze that began in 1868 and continued to the end of the century produced a whole new series of magazines. Winslow Homer was one of the first to illustrate the new Era of the Velocipedes (as the contraption was at first called); in *Harper's Weekly*, he depicted the New Year of 1869 riding in on one. As the shape of the vehicle changed, the enthusiasm waxed and waned, but with the coming of the "safety" bicycle (meaning two wheels of equal size) in 1885, bicycle magazines sprouted up everywhere.

In the general magazines, fiction began to dominate, as it did in the book business, until there was a national craze for novels; these were being denounced from the pulpit and elsewhere as immoral and a waste of time. In 1872, the *Literary World* concluded, "Fiction, to a very large extent, constitutes the reading of the masses, and in whatever education they receive from books (outside of school books, of course) it is the instrument."

But as Mott has been careful to point out, the magazines were not overwhelmed by the floodtide of fiction. It might dominate the "family" weeklies and monthlies, but it was usually properly restrained in the important general magazines to one serial novel at a time, although *Harper's* often had two and the *Atlantic* occasionally had three in progress simultaneously. Two- and three-part stories were often seen in many periodicals, and two short stories were considered the usual quota. All told, about a third of the editorial content was devoted to fiction in those magazines that carried it. It might be added that while the quality of fiction was relatively high in novels, the short story had not yet attained its full stature as an art form. Few writers were attempting it, in fact; no one had really done it successfully since Poe.

If short fiction had a general fault, it was its sentimentality, but that was being diminished as first-class writers did more in this form. It was not exactly a barren period for the magazine short story when one span of twenty years (1865 to 1885) could produce Twain's "Jumping Frog of Calaveras County" (the *Saturday Press*, 1865), Harte's "Luck of Roaring Camp" and "Outcasts of Poker Flat" (*Overland Monthly*, 1868 and 1869), Aldrich's "Marjorie Daw" (*Atlantic*, 1873), and Frank Stockton's "Lady or the Tiger" (*Century*, 1882). In the early 1880s, the *Century* introduced a new generation of Southern short-story writers—George Washington Cable, Thomas Nelson Page, and Joel Chandler Harris, among others.

Elsewhere in magazine pages, there was a smattering of poetry; it was not an important time for poets. The familiar essay flourished in those departments which foreshadowed the *New Yorker*'s "Notes and Comment," including the "Editor's Easy Chair" in *Harper's*, "Topics of the Times" in *Scribner's*, and "Casual Cogitations" in *Galaxy*.

Another development in postwar magazines was the increasing appearance of humorous writing. *Appleton's* made this observation in 1874: "America has of late years bristled with humorous writers as doth the porcupine with quills. . . . What persistent jokers!" The jokers were more admired abroad than at home, where the "coarseness" of Twain and Josh Billings was frequently deplored. Cheap-joke periodicals had been common enough before the war, but there was no class periodical like *Punch* in America, and no comic magazine had been successful.

The drought ended in 1877 with the founding of *Puck*, whose editor was H. C. Bunner, a true comic writer, and which displayed the splendid Joseph Keppler as cartoonist. This weekly was a success for decades, full of excellent satire, until finally its famous logo was all that remained, embalmed in the masthead of Hearst's Sunday comics. A *Puck* cartoonist, J. A. Wales, began a new humor magazine called *Judge* in 1881 after a quarrel with *Puck*'s owners, and in 1883, a group of young men from Harvard founded its longtime rival, *Life*. These periodicals were not comic papers in the old sense but journals of satire and criticism, addressing a new and more sophisticated audience.

All told, the magazine in America had developed into something original and distinctive by the 1890s. That change could be seen in magazines like *Judge* and *Life*, for example, which might be of the same genre as *Puck* but did not remotely resemble it. In 1885, periodicals in this country had developed well past their English counterparts in nearly every respect—a lead never to be relinquished.

Between 1885 and the end of the Great War, magazines expanded and reached a status as national institutions not even visualized before. As Mott so aptly puts it, "Of all agencies of information, none experienced a more spectacular enlargement and increase in effectiveness than the magazines." But as he also warns, it would be a mistake to exaggerate the "growth in value of magazines as popular informers and interpreters." Since Mott wrote those words, however, some media historians have challenged his view and assert, on the contrary, that the importance of magazines in national life should not be underestimated in this period.

A major factor in the changing role of magazines (one that Mott himself noted) was their reaching out to new audiences. For decades, periodicals had been primarily class magazines, leisurely and literary, in sharp contrast to the populism of newspapers. And as long as the general publications had sold for thirty-five cents a copy, they were being read by upper-middle- and upper-class people with money and predominantly conservative views. As George Price once put it in a memorable *New Yorker* cartoon, such readers were "interested spectators of the passing show."

But then a radical change occurred near the end of the century, when these well-illustrated, well-edited general magazines brought their prices down to fifteen cents, ten cents, and eventually to a nickel. This move had profound consequences. Primarily, it made mass-market products out of magazines, moving them into competition with the newspapers for the advertising dollar. It also enabled them to compete with the papers in purveying and interpreting the news. Readers who cared to go beyond what their local editors offered could choose from among several publications devoted to analyzing and amplifying events: the *Arena, Forum, Review of Reviews, Public Opinion*, and the *Literary Digest*, to name the leading ones. Of these, the *Digest*, a forerunner of *Time* and *Newsweek* (more "newsy" than "literary") became the clear leader, its longevity extending to its spectacular demise in 1936.

Newspapers were not without their own resources to counter this invasion. If it was becoming increasingly more difficult, in some cases, to draw a line between newspapers and magazines, the papers could do something their rivals could not. Led by Joseph Pulitzer's New York *World*, newspapers in the 1880s began to develop Sunday editions, replete with comics, more pictures, more analytical articles, and soon a great many more advertisements than the daily editions carried. Moreover, they were directed to a more or less captive audience on Sunday, when the whole family had the leisure time to read them, at least theoretically.

The ten-cent magazine was not a new idea, and prices had once been even lower. Several weeklies had sold for as little as three cents, and monthlies for ten cents, between 1820 and 1830, but none had been successful. Later, there had been attempts to publish dollar-a-year monthlies, but they also were failures.

It was the intense competition among magazines in the 1880s that caused revolutionary price changes. First to leap the price barrier was *Lippincott's*, which eliminated illustration, broadened its editorial appeal, and charged twenty cents, down from thirty-five. At about the same time, *Scribner's* emerged with a different challenge—a fully illustrated quality magazine, designed to compete with *Century* and *Harper's*, but selling for only a quarter, ten cents lower. It was a gamble, but it paid off in three years. Another new entry was *Cosmopolitan*, which offered a fully illustrated product at twenty cents. It was so deficient in editorial content, however, that it might have died if a great editor, John Brisben Walker, had not rescued it in 1890; the price was raised to twenty-five cents a year later. At the same time, *Munsey's Weekly* became a direct competitor by changing its name to *Munsey's Magazine* and offering plentiful illustrations and a twenty-five cent issue price.

The price revolution came at a propitious time. The nineties may have been the Gilded Age, remembered for the opulence and display of wealth that made the phrase "Gay Nineties" a lasting part of the language, but in reality, it was a decade in which America's new great fortunes existed side by side with widespread poverty. Financial panics in 1893 and 1897 added to the unrest that produced destructive strikes and social disruptions. In this climate, a ten-cent cover price had improved chances to find customers, particularly if the editorial mission appeared to be to crusade against the sorry state of society.

The price war began to resemble a shark-feeding frenzy in the early 1890s. For the older, loftier monthlies like *Harper's* and the *Atlantic*, a price had to be paid for survival. In the new mass-market atmosphere, they had been compelled all but to abandon the pure and lofty tone they had maintained earlier, but at the same time, they did not compromise on the quality of their fiction and poetry. These quality-conscious monthlies also had to contend with newspaper competition from the Sunday papers. Pulitzer's *Sunday World* had trumpeted, "Magazines Made Obsolete!" but Lorimer's *Post* replied with a definition of its own: "A good magazine is a good newspaper in a dress suit." The competition had become more direct as newspapers began to sell their literary and pictorial supplements separately, which the *Nation* had been doing for some time.

Even eminently respectable establishment members had to break new ground, shocking some readers. Under such editors as Walter Hines Page and Bliss Perry, the *Atlantic* plunged into political controversy, talked about social reforms, and joined mildly in the general exposure of corruption in government. The *Century* had the advantage of having an able journalist as editor, and Gilder steered it in a similar sociopolitical direction. *Scribner's* traveled down the same road, but *Harper's*—the only holdout—refused to shape its content to the news.

In these closing years of the century, magazines truly came into their own as interpreters of the nation's thought and feeling, of its increasingly complex life. There was more of everything, more to write about and discuss every day. At the same time, magazines reflected the enormous gulf that had opened up in the country, with wealth, accomplishment, and Manifest Destiny on one side, and poverty, human misery, and social unrest on the other.

In such a complicated world, it appeared, more and more magazines were needed to record what was happening. In 1885, about 3,500 periodicals had been published, but by 1890 there were more than 4,400; in 1895, about 5,100; and as the growth rate slowed for a time, a little more than 6,000 in 1905, or very roughly about a quarter of those being published today, depending on whose figures are accepted. The mortality rate was still high, as it has always been. About 7,500 magazines had been started between 1885 and 1905, but more than half that number had either expired or been merged. Still, nearly 11,000 different magazines had been issued in these two decades.

The key to success for all these periodicals, as always, was circulation, and in an intensely competitive business, reliable statistics were as hard to get as ever. A cynical writer in the *Western Plowman* described this problem in 1885: "The editor was dying, but when the doctor placed his ear to the patient's heart and muttered sadly, 'Poor fellow—circulation almost gone!' he raised himself up and gasped, 'Tis false! We have the largest circulation in the country!' Then he sank back on his pillow and died, consistent to the end—lying about his circulation."

In 1905, pressure from advertisers and a growing number of agencies began to make it impossible for any magazine of consequence to hide or falsify its figures. These pressures led to founding of the Audit Bureau of Circulations in 1914 and consequent honesty in reporting circulation figures. By that time, circulations in the millions were beginning to be almost commonplace. After the *Ladies' Home Journal* reached the magic million figure in 1903, others followed, and ten had circulations of half a million by 1905.

There were some departures from the categories of available magazines as the new century began. Among them was the predecessor of *Playboy* and the entire category of "skin" magazines. It was Blakely Hall's *Metropolitan Magazine*, begun in 1895 but not a major publication until after 1903. It had begun with pictures of "the nude in art" and representations of "artists' models," as well as unsubtle expositions of "the living picture craze," but the nation was not yet ready for this kind of slightly disguised sex. Hall understood this after two or three years of struggling and shifted gears, introducing such contributors as Theodore Dreiser and Alfred Henry Lewis. *Metropolitan* had become respectable, but a seed was planted.

Another aberration was *Everybody's* magazine, founded in 1896 by the

New York branch of Wanamaker's department store in Philadelphia. A ten-cent periodical, it was sold in 1903 to other entrepreneurs who turned it into a general magazine that made its fortune by muckraking. The affair demonstrated that an unrelated institution like a department store, even one as successful as Wanamaker's, was not yet qualified to operate a magazine.

Among the weeklies, old and new, *Harper's* continued to be the leader through the turn of the century. Its closest competitor was *Leslie's*, edited by the flamboyant Mrs. Leslie after her husband's death, leaving his business bankrupt. She borrowed $50,000 from another widow, Mrs. Thomas K. Smith, to revive it but sold the property in 1889. In other hands, it survived until 1922, when it was merged with *Judge*.

Aside from Cyrus Curtis's two winners, the most promising magazine as the new century began was *Collier's*, which began life as *Once A Week* in 1888, a profusely illustrated miscellany selling for the odd price of seven cents. It was founded by an enterprising Irishman, Peter Fenley Collier, whose P. F. Collier Company became one of the largest publishers of reference works. The magazine had difficulty establishing an identity at first. As a miscellany, it had reached a 250,000 circulation in 1892 when its publisher, dissatisfied, gave it a different spin to put the emphasis on news and on photojournalism of the kind begun by Leslie and then flowering. In order not to miss any possibilities, fiction and other miscellaneous items were added, so that the magazine could truly say, as it proclaimed, that it was offering a potpourri of "Fiction, Fact, Sensation, Wit, Humor, and News."

By this time, with the name changed to *Collier's Weekly* in 1895, the magazine had made a substantial impact on the reading public, particularly on journalists, many of whom wanted to write for it. The founder's son, Robert J., became editor in 1898 and presided over *Collier's Weekly's* truly brilliant coverage of the Spanish-American War. Young Collier set the magazine on a steady course, characterized by equally steady circulation and advertising increases. Its spectacular success was due not simply to its aggressive muckraking but also to the quality of the journalists themselves—such noted reporters as Will Irwin, Frederick Palmer, Richard Harding Davis, and Samuel Hopkins Adams.

With their help, *Collier's* became a national institution, getting additional assistance from Peter Finley Dunne, whose "Mr. Dooley" appeared for the first time in its pages, and from stories by Conan Doyle about the immortal detective Sherlock Holmes. All this, of course, would not have been possible without superb editors, notably Norman Hapgood, who guided it through the muckraking years, and Mark Sullivan, a reporter-historian.

Just before the Great War, *Collier's* had nearly reached the one million circulation mark, and this figure increased because of its first-class coverage of the conflict by such writers as Ring Lardner, William Slavens McNutt, and Fred Palmer, with occasional articles from Winston Churchill. The war gave *Collier's* the last push it needed. At the war's close, circulation was over 2,000,000, and the magazine itself bulked to a hundred pages. But this was the prelude to a near disaster. The first two decades of the century had been a golden period, however; under Hapgood, it had more influence on national affairs than any other magazine.

In this full flowering of magazines in America between 1885 and 1918, the chief ingredient, as in any era, was the presence of great editors. Some of the best in magazine history were working in this period, particularly in the nineties. There were veterans like the *Century's* Gilder and Alden at *Harper's*, but there were also such extraordinary new ones as Edward Bok, George Horace Lorimer, Walter Hines Page, Sam McClure (many believed he was a genius), Frank Munsey, John Walker of *Cosmopolitan*, Roger Burlingame of *Scribner's* and Albert Shaw of the *Review of Reviews*. And that by no means exhausted the list.

These men worked for comparatively low salaries, although they were not heard to complain. Bok, Gilder, and Alden were all paid $10,000 a year, but Aldrich at the *Atlantic* got an insulting $4,000. Bok thought he was being paid liberally, and in fact, it depended on what was being compared. Executives in other businesses were often paid much more, but a young businessman under thirty in New York seldom got more than $2,000 annually. Editors were often jealous because their better-known contributors were making more money, but at the same time, they realized how essential it was to work harmoniously with them, and notable friendships were struck up—Gilder and John Burroughs, McClure and Ida Tarbell, Alden and Howells, among others.

Writers were still complaining about low payments and about the dead hand of censorship, not to mention overzealous editing. Magazines remained prudish on the whole; it would take the Great War to liberate them, at least partially. Sometimes the writers consented to censorship for the sake of publication, as did both Twain and Kipling. But Julian Hawthorne, reading one of his stories in *Collier's* about a girl who had been seduced, discovered that the editor had gotten her secretly married.

When Tolstoy's *Resurrection* was serialized in *Cosmopolitan*, Walker edited out words or even scenes he considered impermissible and finally suspended it entirely, leaving readers hanging in mid-page, so to speak. He also inadvertently contributed to the popularity of Richard Le Gallienne's version of the *Rubaiyat* by inserting so many asterisks for deleted words that readers could hardly wait to buy it in book form.

By the turn of the century, professionalization had taken place in the magazine business. Journalists and other contributors now held most of the staff positions, and the top editors had usually trained on other magazines. More and more, too, writers thought of magazines as stepping stones to book publication. Writing itself had become a popular occupation. An estimate in 1900 concluded that 20,000 Americans were writing, or trying to write, for publication, most of them unsuccessfully. (The figure today would be in the millions.) Instead of pleading for contributions, editors were now deluged with unsolicited manuscripts. In 1890, the *Ladies' Home Journal* got more than 15,000, and the number increased annually. Of that number, only 497 were accepted. Writers were still badly paid, except for the best known.

As a final contribution to the making of the modern magazine, the cover became truly important for the first time in the nineties, along with the rise of an entire school of magazine illustrators who transformed the interior pages. The reform began in a trade magazine, the *Inland Printer*, founded in 1883, now the *American Printer*. An artist named Will Bradley, working for this magazine in 1894, persuaded the publisher, Henry Sheppard, to change the cover with each issue, an innovation seen before only in a limited way. The editor, A. C. McGuilkin, thought the change would be too expensive, but Bradley offered a plan to do it cheaply, and it was carried out. Considered a radical move in the business, it was soon copied by other magazines, both trade and consumer.

In subsequent editions of the *Inland Printer*, cover ideas were discussed further, and their progress in the industry recorded. Because of the magazine's own extraordinary graphics, it also became a focal point for reporting and debating the new ideas and technologies that were transforming magazine illustration—and all the other popular arts. The first great period of American illustration had already begun in 1890, and it continued until 1940, when other major changes occurred.

Technology was once more the spur, as it had been before the Civil War with the coming of the cylinder press. High-speed rotary presses were now in operation; the linotype machine had revolutionized typesetting; and automatic inking was a major help in high-speed printing.

What is often called the "Golden Age" of magazine illustration reached a peak between 1900 and 1910. As the historian James Best has pointed out, it was a unique period because there was little color reproduction before 1900 due to technical difficulties, and after 1910, photography played an increasingly important role in magazine illustration. Among the new illustrators, Howard Pyle emerged as the foremost and the most influential, teaching his techniques to many talented students.

Pyle, who had little formal art training, began his career in the late 1860s on *Scribner's*, to which he sold his first picture, and by the mid-nineties, he had firmly established himself as what Best calls one of "the first successful 'nativist' illustrators, whose illustrations were rooted in the soil of American history."

The spread of his influence began in 1898, when he began to teach a small group of students he had selected himself at the Drexel Institute in Philadelphia. These students came to be known as the "Brandywine school" of artists, although they preferred to call themselves members of the "Howard Pyle School of Art." There were 110 of them eventually, including such eminent illustrators as N. C. Wyeth, Frank Schoonover, Thornton Oakley, Maxfield Parrish, Jessie Wilcox Smith, and Elizabeth Shippen Green. By one estimate, in 1907 half the successful magazine illustrators in the country had been Pyle's students.

Pyle maneuvered his small army of students adroitly. When he had more commissions than he could complete, he farmed them out to the young artists. He took pains to select the best work of these people, and in time it was clear that anyone with talent who worked hard with Pyle would have a much better than ordinary chance of seeing his work in magazines.

James Best, who made a content analysis of three leading magazines (*Harper's, Scribner's,* and the *Century*) from 1906 through 1910, found that each contained an average of twenty-five illustrations, many of them full-page and some in color. The artists' names had become as familiar to the reading public as those of the contributors.

Most magazine illustration was in black-and-white, and not all the artists made a successful transition to color. Pyle, of course, was one of them and became just as popular in that medium. The "Brandywine school" of artists contributed about 43 percent of all *Harper's* illustrations, and three-fourths of those were in color. Much of this can be attributed to the fact that Pyle had a lifelong business association with the House of Harper, which inevitably involved many of his students. He had little time for other magazines.

Most illustrators did the bulk of their work for one magazine, and they were usually the most successful. James Montgomery Flagg did seventy illustrations for *Scribner's,* but he was not a Pyle student. Artists tended to specialize in content and subject, so that the "ideal American women" of Flagg and Charles Dana Gibson became not only their trademarks but national institutions, particularly the latter's "Gibson Girl." Stories requiring the depiction of foreign, exotic places often employed the talents of Edward Penfield and Jules Guerin. Stories of animal life were often illustrated vividly by Hugh Seton Thomas. Among the Brandywine artists, Elizabeth Shippen Green was known for her mothers and children, perfect Victorian specimens.

As Best sums it up, "Brandywine illustrators tried to recreate the world as it once was, while the non-Brandywine illustrators were more oriented toward contemporary scenes." For all the artists, the problem was that they were likely to be typecast, making it difficult for them to get any other kind of work. N. C. Wyeth, for example, had a particularly difficult time trying to break away from his Western landscape paintings.

The Brandywine school was important because it was the first such "center" for American illustrators, where they could learn the art that separated them from other artists. Pyle, moreover, was a superb teacher and was a great help in getting commissions for his best students. For those who benefited from this training, it was a lucrative alternative to the traditional art career and a means of attaining national recognition.

8

The Magazine as a Reflection of National Life

With the firm establishment of magazines as a national reading habit, reaching a rapidly increasing number of audiences, the industry as a whole began to be a multifaceted reflection of national life. In the periodicals could be found the opinions, attitudes, emotions, preoccupations, and interests of most Americans. The poor still had no real voice, except perhaps to be overheard and reflected upon from above, but since they could not afford to buy magazines, they may not have been conscious of being excluded.

Mott observes that it was a period between two major wars that was noted for the popularity of fiction, not only in magazines but in book publishing, the two interacting through serialization. The "craze," as it was popularly known, reached a climax about the turn of the century, when it was denounced in pulpit and press but, nevertheless, resulted in the first million-copy sales for books.

The fiction people read in magazines reflected the shaping of popular taste. At the mass-market level, it was pure escape, possible from the dislocations of an industrializing America or from the plain hard work of middle-class life. Much of this kind of fiction was written by women. *Belford's Magazine* complained in 1888, "The great stream that swells day by day in the form of prose fiction is simply appalling. . . . Every female born under the stars and stripes comes into the world prepared to write a novel." That may have been how it seemed to male editors, but in fact most of the novelists were men.

The complaint by *Belford's* was another voice in the general debate about fiction that was occupying so much space in magazines, sermons, speeches, and general conversation. The most controversial aspect of the new fiction was its realism, which was, of course, not a characteristic of the romances. A writer in *Forum* summed it up: "For a long time a wordy war has raged in the magazines and the newspapers between so-called realists and romanticists. In *Har-*

per's Monthly Mr. Howells has been asserting the importance of novels that keep close to the facts of life, and the critics and criticasters have daily attacked his teaching and practice as materialistic and debasing. . . . The ground is strewn with dead and dying reputations."

Howells did, indeed, champion realism in his "Editor's Study" column, and he was ably supported by Hamlin Garland, then editing *Arena*. Speaking for the anti-realists, as one would expect, *Munsey's* asserted in 1894, "These are the days when the romantic in literature—the strong, the shining, the imaginative, the ennobling—flourishes, and holds the ear of the world, while 'realism' and 'veritism' [Garland's word] are the languishing cults of the select few." Those who agreed gave Tolstoy's *Resurrection* and Frank Norris's *McTeague* a harsh reception at the turn of the century, even though the *Dial* had declared the argument dead with the publication of *War and Peace*.

As one would expect, the fiction craze produced the first highly successful all-fiction monthly, Munsey's *Golden Argosy*. Started up in 1896, it quickly ran up its circulation to 500,000. Not as well known but ultimately more durable was *Short Stories*, founded in 1890 and still being published in the 1950s. This all-fiction monthly began as an anthology, selecting stories from abroad, from classics, even from the Sunday newspapers. "Twenty-five stories for twenty-five cents," it proclaimed. Frederick Stokes, the book publisher, also issued an excellent all-fiction monthly called *Pocket Magazine*, with Irving Bacheller, the novelist, as editor. Well-known magazinists were its chief contributors.

The most noteworthy start, however, occurred in Chicago in May 1903, when a firm of Chicago retail merchants led by Louis Eckstein decided to go into the magazine business. Called *Red Book*, it was edited by Trumbull White, an extraordinary young newspaperman who was also something of an adventurer. It carried short stories but no serials and added a section it ambiguously called "photographic art" containing pictures of popular actresses. Within two years, it had a circulation of 300,000 and survives today as one of the leading women's magazines.

A taste for art that had been developing since the Exposition of 1876 could be seen in the quality magazines, particularly in *Scribner's*, which devoted more attention to the arts than the others. The *Century* published a series of Timothy Cole's engravings of old masters, which made him famous, and also a series of etchings of cathedrals by Joseph Pennell. There were a few art magazines, but they never attained circulations of more than 10,000.

The new American interest in amateur photography, generated by the development of the rapid dry plate and the arrival of the Kodak, led to competition for the illustrators in the magazines, aided by further developments in photoengraving. Photography arrived just in time (although it had been extant since 1842) to be important for both magazines and newspapers in covering the Spanish-American War. Soon there were also hundreds of magazines devoted to photography, designed both for amateurs and professionals. They had small circulations, however, and survived mostly because of advertising from supply houses. A few proved to be long-running. *Camera,* published in

Philadelphia and intended mostly for professionals, began in 1897 and was still being published in the 1950s.

But the age of illustration was far from over. In the general magazines, there was now much richness and variety, so much so that sometimes the text seemed less illustrative than the pictures. "The writers are only space-fillers," *Munsey's* complained. Readers loved illustration, and editors gave it to them, much of it unfortunately mediocre, carelessly engraved and printed. But when it was done well, magazines glowed with the work of such artists as John La Farge, Pennell, Howard Pyle, Charles Dana Gibson, Howard Chandler Christy, Harrison Fisher, James Montgomery Flagg, Frederic Remington, Arthur B. Frost, and Edward W. Kemble.

Pictures could even become a fad. After Palmer Cox introduced his Brownies in the pages of *St. Nicholas* in 1887, their joyful acceptance by children spilled over into becoming another national mania. They overflowed into other magazines, into advertising, and into books and newspapers. There was also a passion for portraits of the great, clipped from magazines and collected by middle-class readers. These were reflective of the national interest in the American past and in current events, rising from the spread of the adult-education movement. *McClure's* was first with its portraits (and biographies) of Napoleon and Lincoln, but soon other magazines began to print pictures of the famous, both men and women, with the reverse side blank so they could be removed without injuring the magazine. As a sideline, publishers then sold their readers scrapbooks to paste up their treasures.

People were interested in nude art, although only artists were likely to admit it, and *Munsey's* led the way in displaying the female figure in a manner calculated not to offend middle-class taste. Other magazines, particularly such rowdy journals as *Broadway* and *Metropolitan*, were first to display actresses in tights or women with extremely low-cut gowns, and other magazines began to move cautiously in that direction. "Living pictures" and the eternal bathing beauty were also common. The great popularity of the "living pictures" in theaters, where gentility was protected by arranging the slightly clad figures in reproductions of famous masterpieces, was carried over into pictures of these scenes in periodicals.

But in the end, it was photography that produced the real revolution in magazine illustration in the 1890s. Photographic plates could be produced at a fraction of the cost of previous methods, and a tremendous flood of pictures was soon available. The half-tone plate had first appeared in the *Century* in 1884 and followed in the other quality magazines. By the early 1890s, the half-tone was firmly established in the industry. Almost a third of the pictures in *Century* were half-tones in 1893, nearly half in *Harper's*, two-thirds in *Scribner's*, and almost 100 percent in *Cosmopolitan*.

In terms of editorial content, the great change in magazines by the end of the century was the widespread reflection of public interest in politics, economics, and social questions, which gave rise not only to the ten-cent muckrakers but to a changed emphasis in many other magazines. While the quality periodicals had once led off with a travel article, a historical piece, or an essay on literature or the arts, they now were printing more and more comment on political and social problems, once again narrowing the distinction between newspapers and periodicals. The newspapers felt this competition keenly. It was notable that the *Literary Digest* and *Public Opinion* reported on what the newspapers were saying, but the newspapers failed to return the compliment. It may be that this new grappling with reality helped save magazines from feeling serious effects in the successive panics and depressions of 1893 and 1897. Like book publishing, the industry seemed to be relatively immune to the sharp swings of the American economy.

Readers might savor editorials in newspapers on such subjects as free silver and imperialism, but in the magazines they could count on finding serious analyses of these subjects by well-known writers. Politicians were quick to see the advantage of magazine publication. Cleveland wrote frequently for the periodicals, and so did Theodore Roosevelt. Albert Beveridge's friend, George Lorimer, gave him an always available platform in the *Saturday Evening Post*. Most notable, however, were Mark Twain's vitriolic attacks on imperialism, particularly his "To the Person Sitting in Darkness," which appeared in the *North American Review* early in 1901.

The *Review*, in fact, attracted national attention in January 1890 with its debate between Gladstone and James Blaine on the perennial tariff question (continuing through successive issues with articles by other economists and statesmen).

Always sensitive to national currents of feeling, magazines at the beginning of the new century faithfully reflected the national mood of anticipation, even exaltation. Their pages were full of celebrations of rich and wonderful America. "Success" was the key word in the nation's life. Power was still not generally acknowledged as the name of the game, but it was, nevertheless, a reality as it had never been before, and the morals and manners of the marketplace had begun to prevail. The magazines that dealt with success and power (even negatively, as the ten-cent journals did) boasted circulations, however short-lived, in the hundreds of thousands.

Theodore Greene argues that such periodicals as *McClure's*, *Munsey's*, and *Cosmopolitan* were directed by virtue of their price and style to the middle and upper-middle classes. They also, he says, "struck the mood particularly of the generation coming of age in the '90s." These were youthful magazines— in tone, and in the age of their editors, contributors, and audiences. In *Munsey's*, for example, readers could learn about the life-styles of the rich and famous, as well as examine life in the Ivy League colleges and other appurtenances of the wealthy and powerful. Munsey balanced this with Alger-like stories (some by Alger himself) telling deprived young readers how they, too, might strike it rich.

Magazines like *Harper's* and the *Century*, as Greene points out, had been designed for those Americans who came of age in the Civil War; the ten-centers were meant for the next generation, young people who could not identify with that war and who were caught up in the exciting turmoil of expansionist, imperialistic America. The national demand was for success, and so it has remained.

It needs to be remembered, too, that in this period from 1891 to the end of the Great War, magazines enjoyed a time when they were not rivaled by radio, motion pictures, or television. They were the only national communications medium, and their audience was unlimited. But primarily, they were the voice of the vast middle class.

If one of these magazines could be singled out in this period, Greene observes, it could well be *Cosmopolitan,* and many media historians would agree. It was the most complex and the most interesting, borrowing features from its rivals but going beyond them in exploiting ideas. If the *Century* was noted for its travel articles, *Cosmopolitan* made its excursions more picturesque. If *McClure's* was up to the minute with reports of scientific and technological advances, *Cosmopolitan* would double the quantity. It carried more articles in the area of public affairs than any of the others. While it seemed fascinated with war and the military, it was also covering education better than anyone else.

There seemed to be no limit to this magazine's ambitions. It proposed to establish a Cosmopolitan University, sent a representative to the Spanish government offering to purchase Cuban independence for $100 million, offered a plan to save the nation from depression through a credit system, advocated a commission to establish an international language, and proposed a World Congress. For a magazine that had done some excellent muckraking, it carried more eulogies of captains of industry and their corporations than any other magazine, at the same time standing alone in calling for government ownership of railroads and trusts. Even Socialists were welcomed to its pages. In brief, it had something for everybody.

The man who made this remarkable magazine was one of the great editors of his time. John Brisben Walker had been educated at West Point and afterward had enjoyed several diplomatic and military adventures in China. When he was still in his twenties, he made a fortune in iron manufacturing and lost it. For a time, he was a newspaperman in Cincinnati, Pittsburgh, and Washington, breaking off to try ranching and real-estate speculation in Colorado. With the money from these ventures, he bought *Cosmopolitan* in 1894 when he was only forty-six.

With such a background, it was no wonder that he could outpace his rival editors. He knew more of the world and of people than they did, and he knew how to touch their interests in print. He had a grandiose dream for his new property, seeing it as an instrument for use in pushing society toward the solution of great problems. His approach was not conservative; he was a Utopian and printed H. G. Wells's "War of the Worlds" to prove it. A capitalist himself, he admired other capitalists but thought they ought to do something about distributing wealth more equitably.

Like all great magazines, *Cosmopolitan* was a reflection of Walker, and like Lorimer and the other superb editors, he personally approved everything that went into the magazine, from articles to fiction to advertising. But such a man could hardly be contained by one occupation, no matter how successful, and it is not surprising that he was distracted by the arrival of the horseless carriage and bought a factory in which he manufactured Locomobile steam cars, in time becoming the first president of the American Automobile Association. Starting another career, he sold *Cosmopolitan* to Hearst, and it began a new life, going on to later successes of its own.

The idealization of power and success was most apparent in *Munsey's*, owned by a publisher whose self-made career epitomized both. He had none of Walker's Utopian idealism. The vast majority of Americans, he said, "who reflect upon the subject are inclined to regard with extreme jealousy all suggestions looking to the extension of government control of what are now private concerns." His magazine hailed "The Reign of the Business Man" in 1896 and saw annexation of the Philippines as "a new field for American enterprise." His ideas about Cuba were simple. From the very day war was declared, he said that his theory was "to hold all the territory we might get, and to get all we could." Equally simple was his view of the 1900 presidential election: "To me the whole thing is a business problem from first to last." Munsey approved of trusts, in general.

It was a short trip from *Munsey's* to the magazine called *Success*, which appeared in 1898 with its comprehensive subtitle, "An Up-to-Date Journal of Inspiration, Encouragement, Progress, and Self-Help." It was a magazine devoted wholly to success and how to attain it, using those who had achieved it as models.

When *Munsey's* and the other ten-centers turned to muckraking, they were doing it with an eye to their readers and the national mood. These scandals, investigated so thoroughly and with such relish, at the same time provided a forum where readers could question their nation's institutions. Idealism had been momentarily displaced by reality.

But another reality intruded at the end of the muckraking era, and magazines swung in another direction. *Collier's* wrote in 1914, "Quite apart from the European war, it was already apparent that there were in America the strong beginnings of a swing toward reaction, a fatigue with tumult, a tendency to shut the ears to the din of agitation, a growing distaste for the harsher and noisier leaders of reform, a tolerance, almost a sympathy for their victims."

It was, in short, the end of Progressivism in the magazines. There had been a reaction against the idolatry of business, and when that muckraking revulsion had exhausted itself, magazines followed the new national mood. Besides, periodicals had become big business themselves, relatively speaking. By the time the war was over, they could command a national market for advertising measured by the millions, not the mere hundred thousands. More and more magazines were becoming entertainers, but at the same time, their business interests inevitably made them view the commercial world in a different light. The light was that cast by the corporate image, which the growing profession of public relations was showing the business world how to project. At the end of the war, magazines had entered twentieth-century organization society. They were now (employing that current fuzzy phrase) "mass media."

The evidence was in the circulation figures. In 1918, *McClure's* had reached its peak of 563,000, but the magazines of the new order were racing ahead. Both *Cosmopolitan* and *Collier's* had gone over a million, and the *Post* had surpassed everyone with more than two million. Advertising was providing an ever greater share of revenue for these successful magazines, coinciding with the rise of market research. Advertising itself had become big business. In 1914, total expenditures had been $256 million; in 1919, the figure had jumped to $528 million, with a per capita expenditure rising in the same period from $2.61 to $5.03. The American Association of Advertising Agencies had been organized in 1917.

Magazines in the best position to get a large share of all this advertising were the chains, now operating among general, women's, and farm magazines. Curtis had organized the most successful group, with his *Ladies' Home Journal*, the *Post*, and *Country Gentleman*. With this powerful array, he was able to get 43 percent of all the advertising dollars spent in the fields covered by his periodicals. Curtis's near rivals were the Hearst chain, which included *Cosmopolitan*, *Hearst's Magazine*, *Good Housekeeping*, and some trade journals; and the Crowell-Collier group, with *Collier's*, the *American*, the *Woman's Home Companion*, and *Farm and Fireside*, a group that would get 12 percent of total magazine advertising in 1920.

With business becoming America's business, as Calvin Coolidge would put it later, the major magazines became increasingly kinder to it. *Collier's*, for example, once among the scourgers of heartless corporations, now discovered humane and democratic values in them. During the war years, this spirit was evident in the appearance of institutional advertising, a new technique, which was in itself another source of revenue. Even in those magazines still devoted to public affairs, business leaders were more likely to be praised than blamed. All this was in keeping with the national mood in the century's first two decades.

That mood was particularly evident during the Great War itself, or during America's relatively brief participation in it. Business leaders who had lived through the muckraking era were no doubt surprised and pleased to see themselves depicted now as altruistic, efficient, and productive contributors to the war effort. It was not until much later that Bernard Baruch, director of War Mobilization, disclosed his struggle with big steel, among others, to make the kind of contributions that were being idealized in the magazines. As Greene summarizes it, "Through ads, advertising bulletins, editorials, articles, and editor's notes, rehabilitation of business leadership in magazines was complete."

In spite of the general hysteria, however, magazines did not give the same kind of unstinting support to the war effort that came from the press. Hearst opposed the war, for example, and so *Cosmopolitan* adhered to its new policy of entertainment and ignored the conflict as much as possible. *McClure's* was slow in rising to any patriotic fervor during the period of American neutrality, although it did run an admonitory forecast called "The Conquest of America in 1921," with the German army the featured villain. The nation was saved, of course, by the union of all political factions. As soon as the real war was declared, this journal announced itself as "McClure's Win-the-War Magazine" and gave itself over to supporting the national effort, a path followed by most of the other media. In *McClure's*, as in the majority of American magazines during 1917 and 1918, the tone was one of absolute conformity to the prevailing "Americanism" and total intolerance for any kind of dissent.

Earlier, the leading weeklies had interested themselves directly in what was happening in Europe. Mary Roberts Rinehart, who had already made her reputation in the *Post* (and in books) as the nation's foremost mystery story writer, begged Lorimer to let her go to Europe and cover the war as soon as it broke out. But there was so much pro-German sentiment in America at the time that the Allied High Command was refusing to accredit American correspondents. Mrs. Rinehart was compelled to enlist in the Belgian Red Cross, where she managed to do some excellent reporting along with her errands of mercy. The *Post* itself supported President Wilson right up to the signing of the Versailles treaty, while *Collier's* backed Theodore Roosevelt's denunciation of Wilsonian neutrality.

Greene argues that new and desperate fears held by the American middle class were reflected in magazine content: of threats to private property and to Anglo-Saxon supremacy, and of the menace of foreign ideologies. Thus, the periodicals reaching this most substantial part of the population were more than ever supportive of the government and the business community at some cost to the individualism that had prevailed. There was also, however, a strong minority dissent expressed in the more radical magazines. But these half-muffled dissenting voices only produced more paranoia in the major magazines. Early in 1918, *Collier's* ran an article, "The New America," full of foreboding about a postwar nation threatened from within by alien groups, international socialism, and agnosticism.

The *Post* pursued a surprisingly impartial course for some time, considering Lorimer's deep conservatism. Until the middle of 1915, his magazine carried articles representing a variety of views about the war, even though editorially he supported Wilson—again remarkable for a pillar of the Republican party. He sent one of his best writers, Irwin Cobb, to Belgium, where this able correspondent reported that he saw no evidence to support the Allied propaganda that German soldiers had been bayoneting Belgian children. But Lorimer later supported American involvement, after the *Lusitania's* sinking, while still insisting on peace and viewing Wilson as the best hope for achieving it. He went so far as to endorse the League of Nations editorially (later reversing himself).

No other editor and no other magazine so perfectly reflected the hopes, fears, and interests of America's middle class as did Lorimer and his *Post*. It *was* his magazine. The son of a noted Boston preacher, he grew up mostly in Chicago, served a business apprenticeship with P. D. Armour, returned to Boston as a newspaperman, and in 1897 was hired by Curtis to take over the *Post*, a new Curtis acquisition which was on the point of expiring. From then until he retired, stricken by cancer, in 1936, Lorimer made the *Post* the bible of middle-class America—from its famous Norman Rockwell covers, the weekly unsigned editorials he wrote (which readers did not know he had written), and the sentimental (for the most part) fiction down to its solidly conservative and staunchly Republican articles.

Lorimer was the effective champion of American business, and from the beginning, he made business attractive to readers both in fiction and reportage. His own contribution, "Letters from a Self-Made Merchant to His Son," drawn from his experiences with Armour, moved from the *Post*'s pages to the bookstores and became a classic (reissued in 1988). The *Post* popularized business and made it even more respectable—until the Crash, and Lorimer's disillusionment with Herbert Hoover, whom he and William Allen White had done more than anyone else over an entire decade to make president. But the alternative was anathema. Lorimer never understood, or forgave, the *Post*'s readers for supporting Franklin Roosevelt, and his decline, along with the magazine's, dated from 1932.

Knowing Lorimer's character, it is not difficult to understand his earlier disappointment in Wilson's failed peace efforts, or his hatred, after America's entry into the war in 1917, of those he thought stood in the way of winning the conflict. The *Post* surpassed all the other magazines in its wartime hysteria. In editorials and articles, Lorimer denounced what he believed, correctly, was the public's angry contempt for profiteers and also for those who still opposed the conflict. He fought against workers who were paid high wages for war work, and he called conscientious objectors pro-German grafters.

By the spring of 1918, Lorimer was seeing alien elements under the nation's bed, accurately foreshadowing the postwar Red witch-hunts. Aliens, said one editorial, were "The Scum of the Melting Pot." A lead article was titled "Our Imported Troubles and Trouble-Makers," and another examined "The Half-Baked Reformers." He did not go as far as *McClure's*, which was advising its readers to think of themselves as secret agents, listening for subversive remarks and observing suspicious actions. But the highly exaggerated fear of radicals and aliens expressed in the *Post* during the last months of the war reflected the national fears that would sanction the abandonment of constitutional rights after the conflict was over.

In his conduct of the magazine, Lorimer revolutionized the business in ways not apparent to readers. Before he took over the *Post*, writers might wait weeks or months before they could get a verdict on submitted manuscripts and as long or longer for payment. But when Lorimer began going up to New York from Philadelphia regularly, looking for writers to write about business and other American matters, he promised them yes or no answers within seventy-two hours after submission of manuscripts and payment on acceptance. He fulfilled both commitments, giving the *Post* immediate access to the best magazine writers in the country and a good many others besides. These writers could scarcely believe what was happening. The quick replies from Philadelphia inspired the story that manuscripts were intercepted in Trenton and put on the New York train, but Lorimer had simply organized the *Post*'s editorial department so well that it could efficiently handle the tidal wave that hit it. Soon other magazines were compelled to mend their past arrogant and careless ways, at least to some extent.

Whatever else they might think of him, writers, in general, blessed George Lorimer's name. Those who actually wrote for him did more than that. Many were his good friends. Kenneth Roberts, for instance, became Lorimer's partner on antique hunts, and others joined him on his occasional excursions to Atlantic City for relaxation. No editor in the business could claim such friendship, even devotion, from authors.

In significant ways, Lorimer symbolized what had happened to American magazines by the end of the war. An excellent businessman himself, he exemplified the triumph of the organization men in taking over the periodical business. Advertising had made the decisive difference. A large circulation was no longer the sole measure of a magazine's success or failure, as it had been for so long. It was now the essential ingredient in getting advertising, a specific audience for those who wanted to sell it something. All the leading magazines had been in existence for at least twenty years by the end of the war, and they were owned either by rich entrepreneurs or by centralized chain management (they preferred to call themselves "groups"), with the individual components managed by editors.

This was not the death of individualism, however. It still took great editors to make great magazines, no matter what their financial arrangements might be, or even how much advertising they displayed. But the editor was inextricably locked into a business system whose components had to follow the rules of business to survive. Consolidation was beginning to be the watchword, in periodicals as well as in industry and labor.

One thing had not changed. Anyone with a fresh, new idea for a magazine could, if he had the money, enter the magazine arena, and if the idea coincided with at least a substantial segment of popular taste, he could succeed. That would become evident in the outbreak of new ideas and new magazines that occurred in the twenties.

Meanwhile, the trend toward specialized publications was already well under way in 1890 when the *Journalist* observed, "The multiplication of trade and class papers during the past ten years has been something enormous. Almost every line of industrial endeavor has one or more organs which represent, or misrepresent, its interests. They are for the most part handsome publications, rich in heavy paper, fine press work and tasteful typography, well edited. . . . They are pretty generally successful." Many of these publications, it may be added, looked more like newspapers, but taken together, they were creating a trade press that would grow to outsize proportions in the next hundred years, not only providing genuine services but also generating a whole new field of specialized advertising.

There were other developments worth noting in serving the interests of many diverse magazine audiences. Scholars were already aware of what was happening. When the first number of the *Annals of the American Academy of Political and Social Science* appeared in July 1890, it was quick to point out the functional position of magazines and newspapers in shaping the thought of their mass audiences. "The increase in public interest," it noted, "is amply evidenced by the attention given to such problems by our daily and weekly papers and by our leading monthlies and reviews." Henry Mills Alden, editor of *Harper's Monthly*, put it another way: "The law of the magazine makes it the journal which records the social movements throughout the world." He might have added, "and every other movement as well."

In the magazines could be found the record of how charities were growing, along with social work; the tide of nonpolitical socialism, still unorganized; the progress of Marxist socialism, however limited; the Populist movement; and the rise of the new science of sociology. All of these had their own magazines. Former literary magazines now took on pressing social problems. Attacking what was already a major urban dilemma—housing—the *North American Review* had Thomas Byrnes, chief inspector of New York police, telling readers that "the lodging houses of New York have a powerful tendency to produce, foster, and increase crime."

Corruption was ever present in city government, and the *Century* was one of the leaders in fighting for reform, as well as for better housing. *Scribner's* joined in the war on flophouses and crowded tenements. Crime, however, was much more a newspaper topic; magazines were inclined to discuss it only as another social problem, although it was beginning to be a popular subject in fiction at the turn of the century, stimulated by the great success of Sherlock Holmes. Divorce was also a new subject for magazine discussion, and so was the problem of homeless children.

The magazines did not shrink back from discussing racial problems, particularly lynching, and they also pursued such topics as black suffrage and education. The *Atlantic* was particularly insistent in attacking lynching, including a notable article by Booker T. Washington. Paradoxically, the strongest assaults came from the Southern reviews. In the first issue of the *South Atlantic Quarterly*, in October 1903, John C. Kilgo, president of Trinity College, launched a frontal assault on lynching and followed it with another a few issues later.

In the industrial turmoil at the turn of the century and later, the muckraking magazines dominated public attention, but other journals gave considerable space to the problems of trusts and monopolies in noninvestigative ways. *Forum*, the *North American Review*, and others in this category were acutely aware that their readers generally supported the capitalist owners in the labor wars, but they were, nevertheless, not afraid to advocate reform, at least in some respects, while at the same time being against any changes in public policy. Their readers saw strikes and violence and the presence of aliens as enemies of the American Dream. The unions were saying that hard work and perseverance did not necessarily guarantee a piece of the dream as middle- and upper-class America wanted to believe.

In many magazines of the 1890s, there was a good deal of agitation against picketing and the closed shop. But *Cosmopolitan,* in its pre-Hearst days, was appalled by the Homestead strike, with its hired Pinkerton men. "American distribution of wealth has become frightful in its inequalities," the magazine said in an article. "The fact is, we have two separate worlds in this country, and the man who lives in what is known as the world of society has no conception of what the world of labor is thinking." *Railway Age,* oldest of the trade magazines, thought the Pullman strike was part of a conspiracy by socialist radicals to gain control of the government. On the other hand, hundreds of magazines for workers and union men were started, but nearly all of them died, as they had been rising and dying for fifty years previous and would continue to do in our time.

America has often been described as a nation of joiners, and the tremendous growth of secret societies in the later years of the nineteenth century produced an equally remarkable number of magazines to serve them. In 1890, the Masons alone had thirty journals of their own, while the Knights of Columbus and the Oddfellows had nearly as many.

Mott observes that an outstanding phenomenon of the contents in magazines during this period was the change in outlook on foreign countries. The old-fashioned travel article did not die, but it was largely replaced by discussions of economic and political problems in other countries. For a century, readers had been accustomed to the leisurely, stylized travel pieces that had been staples of the magazines, but now, with European excursions much more common and consequently more sophisticated audiences, the leading magazines carried more and more articles on foreign trade, international relations, and the political problems of Europe.

All the sciences could boast their own journals at the turn of the century. In the general medical periodicals, X-rays and the fight against tuberculosis were reviewed. All told, there were about a thousand journals covering medical and related fields—as always, issued for varying lengths of time. In his 1947 history of the American Medical Association, Morris Fishbein, editor of the A.M.A. *Journal,* wrote that the middle 1890s were "the days of competitive journalism in American medicine," and "the battle raged continuously."

Although women were having a difficult time establishing themselves in medicine, they had their own magazine, the *Woman's Medical Journal,* begun in Toledo by E. M. Roys-Gavit in 1893. Nurses were served by *Trained Nurse and Hospital Review,* begun in 1888 and later renamed *Nursing World;* and the *American Journal of Nursing,* begun in 1900, organ of the American Nurses Association and of all the similar state organizations.

Those who think the fitness cult of the 1980s, when jogging became a national pastime, was a contemporary invention could look for its real origins in the 1890s, when personal health, physical culture, and exercise, both for health and in sports, became a national obsession, rivaling today's. Its progress was recorded in all the general magazines, as well as the newspapers, and a variety of magazines sprang up to reflect this interest, most notably *Physical Culture,* a product of that eccentric entrepreneur Bernarr Macfadden. It was the beginning of what would later be a magazine empire. Begun in New York in 1899, *Physical Culture*, with its muscular men and numerous health exercises, had a bold, even sensational tone. It was meant for men, but Macfadden shrewdly launched a companion piece for women, *Beauty and Health*, in 1900. Its time had not yet come, however; it lasted only eight years. The remarkable growth of this craze spawned dozens of less successful publications.

The outstanding success among all the magazines in the science category was the *National Geographic*, founded in 1888 (and, in 1991, seemingly destined to go on forever). Its multiple initial sponsorship came from the Bell Telephone System, Gardiner Greene Hubbard, and Alexander Graham Bell. Hubbard was a rich patent lawyer, who had financed Bell when he was teaching the deaf and whose daughter eventually married the inventor. The genesis of the *Geographic*, it could be said, was Hubbard's intense interest in exploring Alaska, which led him to organize the National Geographic Society in 1888 at the Cosmos Club in Washington, D.C. This nonprofit, tax-exempt organization, of which Hubbard was elected first president, with an initial membership of thirty-three geographers and explorers, soon began to publish its own journal, the *National Geographic Magazine*, first issued (in October 1888) as a thin, octavo-size publication. It became a monthly in 1896 after being a quarterly for two years.

From the beginning, the *Geographic*, although it was originally intended only for its learned members, was written for both scholarly and popular audiences. Membership (which amounted to circulation as far as the magazine was concerned) grew slowly to 1,400 by the turn of the century. No serious attempt was made to add members.

After Hubbard's death in 1897, Bell became the society's president and determined to make the magazine more popular in the hope of attracting new members. Obviously, a full-time editor would be needed, and Bell chose Gilbert Grosvenor, a brillant young Amherst graduate whose father was a noted historian and whose cousin was William Howard Taft. Grosvenor had met Bell when he was courting the inventor's daughter, Elsie. Bell was impressed but hired Grosvenor at first only as assistant editor. Coming to work in April 1899 and thinking the job was probably only temporary, the young man married Elsie the following year. For Bell, as Bernard Block puts it, this almost haphazard hiring "was one of the most successful and celebrated hiring decisions in American magazine history."

Grosvenor was not actually named editor until 1903, but from the beginning, it was apparent that he was going to run the magazine—and do it well. From his reading, he had concluded that the kind of popular magazine he meant to edit would have to have accounts, preferably eyewitness, written in a style that would be both straightforward and entertaining. Fortunately, Grosvenor also had a particular interest in photography, the other ingredient he felt necessary to make his magazine a success. Missing from his formula was any attempt at political or social analysis; he believed only good things should be said about other peoples and countries. He wanted no controversies.

A significant change was Grosvenor's decision to include articles on a wide range of sciences and the social sciences—geology, botany, zoology, anthropology, ethnology, and sociology. In short, anything that would help describe the world. In applying this formula, he was lucky enough in 1904 to acquire fifty stunning photographs of the then virtually unknown Tibetan city of Lhasa, which he spread over eleven consecutive pages of his January 1905 issue, stirring tremendous interest. Three months later, he had another coup—thirty-two pages of revealing photographs of the Philippines supplied by his cousin, Taft, who was then in the War Department. By the end of the year, membership (and circulation) was up to 10,000.

That was the beginning of an astonishing and continuous climb in circulation that has not ceased to date, a climb without equal in magazine history. By 1914, the figure was 250,000; it passed the million mark in the 1920s, had reached two million by 1952, and in 1984 was reaching eleven million readers worldwide.

There were occasional reservations registered by parents about this family magazine, in whose pages many young boys saw bare-breasted females for the first time, but National Geographic was in much too respectable a position to be troubled by these parental grumblings. Later, there were also complaints from rival magazines, annoyed that a tax-exempt periodical was competing with them for advertising, but the Geographic sailed on serenely. Grosvenor remained as editor until 1954, continuing as chairman of the board until his death in 1966. Today the society is a large publishing organization involved in book and atlas publishing, television, and audiovisual projects, besides the magazine, which now has a companion, National Geographic World, intended for young readers.

Science's opposite number, religion, had always been an important part of magazine publishing, and in the period under discussion, religious periodicals of all kinds shared in the general growth of the industry, although never in large figures. They were primarily of two kinds: monthly or quarterly reviews, involved mostly with theology and scholarship; and the weeklies, which paid more attention to general news. Both editors and readers wanted solid, nononsense material concerning theology, metaphysics, philosophy, philology, and exegesis. The weeklies were inclined to be more lively, even acrimonious at times.

There were those who thought journalism and religion incompatible. In the Atlantic, Lowell called the religious press "a true sour-cider press, with bellyache privileges attached." But followers of the major religions saw these periodicals as a forum for the never-ending arguments and discussions about theology. Catholic papers appeared to do best in terms of circulation, especially the Boston Pilot, but four of the various Methodist Christian Advocates did well, too.

Of all the religious publications of the period, the Christian Century proved to be the most durable. Established in 1884 in Des Moines, Iowa, it celebrated its hundredth birthday in 1984. At the beginning, it was intended only for members of the Disciples of Christ denomination, but its editor,

Charles Clayton Morrison, observed one day in going over the subscription list that there were Congregationalists and Methodists on it. As a longtime advocate of unity, it occurred to him that by changing the focus of his magazine he might create an organ that would help the cause. Consequently, the magazine appeared in 1917 with the subtitle "An Undenominational Journal of Religion." That was the beginning of its new (and longer) life.

The *Christian Century*'s attitudes in the Great War showed that it was not that far distant from its secular brothers. At the beginning, it was no more enthusiastic about American participation than the great majority of other periodicals, and it continued to back Wilson's neutrality policy, although not in an aggressively pacifist way. But when American entry became inevitable, it joined the parade of unqualified support. At the end, like Lorimer's *Post* and other first-rank magazines, its position was that Wilson had betrayed the country at Versailles.

All the religious magazines of every denomination had to contend with the weakening of religious life, thus opening the door to other kinds of periodicals. Darwinian evolution was considered by many the entering wedge in the weakening process. This "new knowledge," as it was called, so vital in such fields as geology and anthropology, caused some existing religious periodicals to print what could be considered heresies. The *Independent*, for example, embraced evolution, as did the *Christian Union*, soon to become the *Outlook*. At the fringe of organized religion, the ethical-culture movement produced its own magazines.

A whole new field opened up with the arrival of magazines devoted entirely to house planning, interior decorating, furnishings, and landscape gardening. *House Beautiful*, founded in Chicago in 1896, began life as a badly printed ten-cent monthly, but a new owner, Herbert S. Stone, quickly transformed it. Stone was a strong advocate of simple beauty in home furnishings, a reaction against the fussiness of the Victorian era. By the turn of the century, his magazine was calling itself "The American Authority on Household Art." Stone's principal rival was *House and Garden*, begun in June 1901 by three Philadelphia architects, who declared frankly in the first issue that their point of view was that of their profession. It was a handsomely illustrated quarto of thirty-two pages at the start, and although it went through successive changes in ownership, it remained a high-quality publication. Both of these journals had extensive and varied twentieth-century careers.

The periodicals accompanying the bicycle craze of the nineties were succeeded by the automobile magazines, notably *Motor*, begun in 1903, Hearst's first publication. The popularity of railway travel continued undiminished as the new century began. It was reflected not only by articles in all the popular magazines but by anywhere from thirty-five to forty trade journals being published at any one time and fifteen or so monthly railway guides. Marine engineering and shipping magazines were still another category, and ten years before the Wright brothers' historic flight of 1903, the first aviation magazine, *Aeronautics*, a ten-cent monthly, appeared, only to make a crash landing a year later.

In this period when there was not a great deal to laugh about in everyday life for most people, an outpouring of humorous magazines occurred, not equaled before or since. Many were of high quality, inspired by *Life, Puck,* and *Judge* but also helped by Mark Twain's immense popularity. Most of the larger colleges had comic monthlies, of which the Harvard *Lampoon* was the most visible and durable. Modern readers would find little that was acceptable in these journals, which were replete with racial and ethnic insensitivities.

Although it was not exactly in the humor magazine category, *Vanity Fair* was the first American magazine to use satire successfully. It was to become the country's "most memorable magazine," as Cleveland Amory once called it. This unique periodical was the result of a merger when it appeared first in September 1913, and it bore the cumbersome title *Dress & Vanity Fair,* with a subtitle announcing that it would be devoted to "Fashions, Stage, Society, Sports, the Fine Arts."

After the fifth issue, "Dress" was quietly dropped from the title. Six months after it had begun, *Vanity Fair* acquired the editor who would make it famous, Frank Crowninshield. He was already well known as a supporter of the famous 1913 Armory Show, which introduced Americans to the new art. A sophisticated man with excellent taste, Crowninshield changed the tone of the magazine completely and dedicated it to entertainment, happiness, and pleasure.

Vanity Fair took pride in being the voice of the avant-garde. It supported feminism and tried to advance the cause of modern art. In the 1920s, it would have rivals with the advent of H. L. Mencken's and George Jean Nathan's *Smart Set* and with their *American Mercury.* After the Great War, Dorothy Parker appeared in its pages for the first time (writing first as Dorothy Rothschild), along with most of the famed literary names of the twenties—Robert Benchley, Aldous Huxley, D. H. Lawrence, e. e. cummings, Colette, and T. S. Eliot, among others. Beyond its literary quality, the magazine had an aura of freshness, of unpredictability. But even such qualities fail in time, and *Vanity Fair* died in 1936, a victim mostly of changing times. Revived in 1983, it catered to a different world.

9

Periodicals for Women: A New Phase

As all great wars do, the Civil War broke old patterns of living and thinking. At its end, the nation stood on the verge of tremendous change, merging from an essentially agrarian society into an industrial one, with profound dislocations and changes of every variety.

Women would have to wait until the next century and another war before they could play a vastly wider role in society, the programs of the suffragist movement and the isolated expressions of a more radical feminism notwithstanding. Before the war, magazines had faithfully reflected what were considered to be the elite woman's ordained life—to be a model of virtue before marriage and a homemaker who knew her subordinate place afterward.

Women had fulfilled their traditional roles during the war, for the most part, but afterward, as some of them began coming into the work force and the campaign for equal voting rights moved forward, the perennial historical question began to arise, "What is to be done?" Editors had pondered this question to the point of saturation before the war, but now there seemed to be more urgency and less patriarchal shadowboxing to the inquiry. The general magazines were full of what was being called "The Woman Question." *Appleton's Journal* summed it up neatly: "Having been idolized, sung and flattered through all the moods and tenses of poet's feeling, it seems at last woman's destiny is to *be considered*."

Feminists of the day might be irritated by such condescension, but they could hardly deny that their voices were being heard by editors who understood that names of women were predominant on their subscription lists, even on the lists of supposedly masculine journals like the *Nation* as well as conservative standbys like *Appleton's, Harper's,* and *Scribner's*. In fact, all the magazines might have endorsed the suffragist movement if it had not been clear that women themselves were divided on the question. Since that was true, male editors believed there was no need to hurry matters.

Equal suffrage was discussed in nearly all the general magazines, but the *North American Review* outdid the others in 1879–1880 with an article by the historian Francis Parkman on "The Woman Question" and subsequent replies, in which he prayed God to deliver America from "the most reckless of experiments" (meaning suffrage) propelled by "a few agitators."

How far women had yet to go was evident in the uproar that followed Parkman's article. Eminent suffragists wrote indignant replies—Julia Ward Howe, T. W. Higginson, Lucy Stone, and Wendell Phillips—and Parkman argued his case all over again in response. Three years later, the debate was still going on in the magazine, when William A. Hammond created a major explosion. He assured the *Review*'s readers that there were "grave anatomical and physiological reasons" why women should not be in politics and, even worse, that "a peculiar neurotic condition called the hysterical is grafted on the organ of woman." (One supposes he meant the brain.) That was enough to renew the debate on an even more acrimonious level.

Not all the magazines gave house room to such nonsense. The *Atlantic* was friendly to the cause of suffrage, frequently printing articles by its leaders, as did several other respected journals. But the comic papers attacked suffrage in the crudest possible way. In response to criticism from every quarter and as a means of binding the movement together, several journals published by movement leaders were established, but they were never substantial.

Analyzing these early suffrage periodicals, journalism historian Lynne Masel-Walters has concluded that they were badly financed and were run by women without sufficient journalistic experience. One woman usually carried most of the burden, and often it was a sideline to her work in the local suffrage organization. Matters did not change, Masel-Walters says, until the early twentieth century, when businesslike journalism professionals, frequently with independent means, infused new life into the suffrage press. Readers were usually female, white and middle class; and editors displayed a similar profile. Circulations were so small that only one, the *Woman's Journal*, ever surpassed 5,000 circulation. This magazine and *Revolution* were the best known of the suffrage periodicals.

Revolution was especially notable because it was begun by Susan B. Anthony and Elizabeth Cady Stanton, founders of the movement, although they had editorial help from Parker Pillsbury, a reformer, and financing from George Train. Masel-Walters calls it "the first major national publication concerned with feminine equality." It began soon after the war, in 1868, when Train, an eccentric millionaire with Copperhead leanings, offered Stanton and Anthony the money to start a female-suffrage paper. Putting aside some of their misgivings about his views, they accepted. Pillsbury resigned from the *Antislavery Standard* to help them because he believed that suffrage for women was then the greater need.

A mixed response greeted the weekly *Revolution* when it appeared. Some praised its devotion to equality and justice, its espousal of equal pay for equal work, and its call for the eight-hour day. Others were put off by Train's economics column, in which he advocated greenback currency, questionable financial schemes, and unlimited immigration. Since it was Train's money that had made the magazine possible, the founders were helpless.

They had another weapon, however. In 1869, Stanton and Anthony founded the National Woman Suffrage Association, protesting the exclusion of women from the proposed Fifteenth Amendment, and they used the pages of *Revolution* to promote their cause. Unfortunately, that alienated supporters who strongly favored the amendment. Led by Lucy Stone, these conservatives formed their own organization, the American Woman Suffrage Association, and began their own periodical, the *Woman's Journal*.

A journalistic battle raged. Anthony and Stanton were uncompromising in their fight for suffrage; their path "was straight to the ballot box." In *Revolution*, they called for an amendment that would eliminate sex as an obstacle to voting and countered the arguments of the anti-suffragists. It was not all ideology that filled its pages, however. *Revolution* also published poems, fiction, and book reviews. Train continued to contribute, much of that time from an Irish prison where he was being held, ostensibly for debt but more probably because of his Feinian views. In 1869, he resigned from the magazine, having the good sense to see that his continued presence might be discouraging subscribers. That left the founders with serious financial problems.

At this point, *Revolution* had about 3,000 subscribers, and even a subscription price elevated to three dollars was not enough to offset costs; nor could enough advertising be secured. A stock plan put together by friends of the publication, including Isabella Beecher Hooker and Harriet Beecher Stowe, fell through because the founders refused to change the magazine's name. Competition from the *Woman's Journal* was a final blow. In 1870, Laura Curtis Bullard, a rich suffragist, bought *Revolution* for a dollar and operated it for two more years until she sold it to the New York *Christian Enquirer*, after which it quickly disappeared.

The *Journal* enjoyed much better financing and its more conservative editorial policy attracted a greater number of readers. Founded in Boston in 1870 by Stone and her husband, Henry Blackwell, it merged at once with another suffrage journal, the *Agitator*, published by Mary Livermore, an experienced journalist, who now became the *Journal*'s editor. After three years, she resigned her post, leaving Stone and her husband to take over for the next two decades. The *Journal* was bought in 1917 by the National American Woman Suffrage Association with money left to it by the redoubtable Mrs. Frank Leslie. It was renamed the *Woman Citizen* and became the organ of the League of Women Voters after the Nineteenth Amendment passed. It took the Great Depression to kill the magazine, in 1932.

Perhaps the most radical of the journals advocating suffrage (among its other reform enthusiasms) was *Woodhull & Claflin's Weekly*, published by those formidable sisters Victoria Woodhull and Tennessee Claflin between 1870 and 1876. They used their journal as a forum for various causes and as a vehicle to blackmail prominent individuals, both socially and politically.

Tennessee claimed spiritual and healing powers, which she practiced for a time on Commodore Vanderbilt, but Victoria was the more charismatic and intelligent. In 1879, having meanwhile married and left a doctor named Woodhull, Victoria announced her candidacy for the presidency of the United States, and to support her campaign, *Woodhull & Claflin's Weekly* was launched. It not only promoted the candidate in its sixteen pages but propagandized for the suffrage cause as well. Through it, Victoria emerged as the foremost leader in the fight for women's rights.

Denouncing the double standard, the *Weekly* observed, "We hear of abandoned women, but not a word of abandoned men; and yet there are ten times the number of abandoned men than there are of women." Other leaders, like Stanton, could support such a stand, but more conservative elements in the movement were less enthusiastic about Victoria's eventual advocacy of free love. The *Weekly* was also an early muckraker, probably the first to expose corruption in the insurance and banking businesses. It was also the first to feature articles about women in business.

Both sisters became a *cause célèbre* when the *Weekly* exposed the scandalous affair between the Reverend Henry Ward Beecher and Mrs. Theodore Tilton. That meant a sellout issue, with copies scalped for large sums. In the same issue, however, another story resulted in obscenity charges brought against the sisters, and they were thrown in jail. Their trial and the harassment that followed were covered avidly by the press, but no more completely than by the *Weekly* itself. This episode was the high-water mark, however. After she was freed, Victoria's interests changed, along with the magazine's language, and it began to appear irregularly, finally disappearing quietly in 1878. Both sisters eventually married into the British nobility and lived out the remainder of their lives in England.

In the nineteenth century, several other suffrage journals appeared in such western centers as Chicago, Denver, Nebraska, and Portland. Later, as the suffrage battle heated up, local magazines appeared, along with the inevitable anti-suffrage journals, of which the earliest was *True Woman*, existing briefly between 1870 and 1873 in Baltimore and later in Washington, D.C.

Several periodicals were notable because they were published, edited, and often set in type and printed by women. Perhaps the first of these was the *Chicago Magazine of Fashion, Music and Home Reading*, begun in 1870 by Mrs. M. L. Rayne, a society woman of that city. It survived half a dozen years. Following it in 1872 was Mary Neal's *Central Magazine*, with an entirely feminine staff from editorial office to composing room. This was one of three all-female magazines in St. Louis, and there were others in Indianapolis and Philadelphia. All these publications were strenuously attacked by the conventional establishment.

The suffragist and feminist periodicals, however important they were politically, constituted only a small part of the great outpouring of mass-market magazines for women that occurred before the end of the century, when several of the twentieth-century giants were founded. *Godey's Lady's Book* and *Peterson's*, circulation leaders before the war, survived it, but they were on the decline and would soon expire. New titles were about to eclipse these old favorites.

In the last three decades of the century, the *Ladies' Home Journal, Woman's Home Companion, Good Housekeeping, McCall's, Delineator,* and *Pictorial Review* appeared for the first time, all destined to be major leaders in the next century. Their founders—Cyrus Curtis, James McCall, and William Anhelt—possessed the business skills to keep these journals afloat until the public responded. Bok and such others as Gertrude Battles Lane expanded the horizons of women's journals, turning them into publishers of high quality fiction and crusaders for social reform as well as trade journals for the homemaker. Focused on the problems and aspirations of middle-class women and their families, these new women's magazines soon outpaced the older titles.

Their rise was helped substantially by the increasing importance of advertising, a phenomenon that benefited the whole economy and had special significance for the magazine industry (see Chapter 12). Women's journals offered a particularly attractive medium to advertisers of the new brand-name goods. They were a sure way of reaching female consumers. Publishers and advertisers combined to create a series of mutually helpful practices that would continue for decades in the industry, and the women's magazines, in particular, pioneered several mass-market tactics: low prices to obtain large circulations, extensive promotional advertising for themselves, some of the earliest market research studies, and creation of a closer relationship to readers by paying attention to their ideas and opinions.

Cyrus H. K. (for Herman Kotzschmar) Curtis with his *Ladies' Home Journal* was the clear leader in the race to develop mass-market female audiences. Curtis was a self-made entrepreneur. Born June 18, 1850, in Portland, Maine, his first job was carrying newspapers on a route, but by the time he was thirteen, he was publishing his own newspaper, *Young America,* on a $2.50 press. (Sixty years later, he owned printing plants worth more than $8 million.)

When the great Portland fire of 1866 destroyed his family's home, Curtis had to leave high school after only a year and go to work, first as an errand boy in a dry-goods store and then as a salesman in a Boston department store. From there, it was a short step to newspaper advertising work on Boston papers and then, in 1872, to starting his own newspaper, the *People's Ledger,* which he moved to Philadelphia in order to get cheaper printing. He sold the paper there, became advertising manager of a weekly, and in 1879 started a new paper, the *Tribune and Farmer,* an agricultural journal.

A column in the *Tribune* was directed to the wives of subscribers, but it was no more than a conventional collection of household hints and short excerpts clipped from other papers. When Curtis's wife, Louisa Knapp, pointed out the column's inadequacies to her husband, he gave her the job of running it. Adopting a new title, "Women and the Home," she transformed it. As Edward Bok recounted later, "She wrote, she edited, she conceived, and practically tested countless domestic intricacies before she gave them to the public." Under her sure hand, the column became so popular that Curtis expanded it to a full page and then made it into a monthly supplement, appearing first in December 1883 and costing fifty cents a year if bought separately.

It was this supplement that became the *Ladies' Home Journal*. The first issue contained eight cheap folio pages of fiction; household tips and recipes; fashion news; pieces on gardening, needlework, and handicrafts; and a respectable amount of advertising, with a circulation of 20,000. The name was an accident. Curtis told the printer that the magazine would be a "kind of ladies' journal" (or so the story goes), and that was the title on the first issue, but the printer added the picture of a house between "Ladies'" and "Journal." Readers, taking it literally, began addressing the new magazine as the *Ladies' Home Journal*, according to this possibly apocryphal version, and the name stuck.

In any case, the *Journal* was a tribute to Curtis's business sense and his wife's editorial judgment. Unlike most of his publishing contemporaries, Curtis made no pretense of being an editor. His genius lay in knowing how to establish magazine properties and then finding editors for them, after which he had the good sense to leave them alone. They rewarded him by giving him, in the *Journal* and the *Saturday Evening Post*, two of the most successful magazines ever produced in America until recent times—and in the case of the *Post*, the most beloved.

Curtis built on the *Journal*'s original circulation through various devices. He implemented the "club" idea, in which four subscriptions were offered for a dollar, thus making one subscriber a salesman in obtaining three others. A master at this kind of list building, he doubled the magazine's circulation in six months. A newspaper advertising campaign, costing only $400, doubled circulation again in the following half-year.

Meanwhile, working together, Curtis and his wife steadily improved the magazine's content. Realizing that he needed a big name, the publisher went up to Massachusetts and knocked on the door of Marion Harland (her real name was Mary Virginia Terhune), then one of the best-known writers for women in America. He bought a short story from her for ninety dollars and spent more money to advertise her presence in the *Journal*. Her name attracted so many new readers that circulation reached 270,000 by early 1886.

Both Curtis and Louisa Knapp found it hard to get the best writers, however, because most of them considered writing for a woman's magazine little better than contributing to a household-hints journal. Nevertheless, Curtis persuaded Louisa May Alcott to add her name to his "List of Famous Contributors" by promising to contribute a hundred dollars to her favorite charity if she would do one piece for the *Journal*. Not many others joined her, however, except for Marietta Holley, whose "Samantha Allen" byline had made her well known, and Mary Jane Holmes, the popular novelist.

Meanwhile, Curtis used advertising adroitly to keep his magazine's name before the eyes of potential customers, and he actively solicited advertising for the magazine himself. Curtis was one of the first to see the possibilities of offering a magazine at a low price to readers, subsidizing it through an abundance of advertising. By the nineties, such mass-market publishers as Frank

Munsey, S. S. McClure, and John Brisben Walker were employing his methods. Announcing in 1887 that he meant to have "a million subscribers soon," Curtis moved toward that goal with new circulation schemes and cleverly written advertising. In 1889, taking a large gamble, he raised the subscription price to a dollar a year, increasing the size of the *Journal* to thirty-two pages. Fortunately, he was able to get $200,000 in advertising credit from F. Wayland Ayer, one of the founders of N. W. Ayer, the noted advertising agency. Ayer also endorsed Curtis's notes to the amount of $100,000 so that Curtis could buy the additional paper he needed. With such backing, Curtis was able to launch the broad, bold advertising campaign he envisaged, but it still took careful financing for a few years until, in 1891, he established a stock company, capitalized at $500,000, and the Curtis Publishing Company was truly launched.

In 1889, with the *Journal*'s circulation over 400,000, Louisa Curtis decided to give up her post as editor in order to spend more time with her young daughter, and Curtis then made the move that elevated the *Journal* to preeminence. No one knows for certain what kind of second sight led him to Edward William Bok, a most unlikely candidate for editor of a woman's magazine. A Dutch immigrant at the age of seven, he had been in book publishing and had edited the *Brooklyn Magazine* briefly before going into the newspaper syndicate business with a series called "Literary Leaves." Presumably, Curtis must have read some of Bok's syndicated columns directed especially to women. These had been extremely popular and reflected the writer's belief (and Curtis's as well) that a need existed for material targeted to women.

Otherwise, there was no indication that Bok was qualified to edit such a magazine as the *Journal*. At this point, he was a 26-year-old bachelor, with no discernible interest in women. In his classic autobiography, *The Americanization of Edward Bok*, he wrote of himself, in the third person, that "of women he knew little, of their needs less. Nor had he the slightest desire, even as an editor, to know them better." He *did* adore his mother. But he idealized the other sex, although his ideas about them were odd, to say the least. Moreover, he had a truly staggering egocentricity, and in the manner of similarly self-absorbed men, he often referred to himself in the third person, as he did in his autobiography.

Somehow Curtis understood that Bok, in spite of everything, was a man who knew what women wanted to read, but how he knew it and why Bok had this talent at all remains one of the mysteries of magazine publishing. Perhaps, as Mott suggests in an uncharacteristic chauvinistic aside, Bok made the magazine in the image of what women wanted to be instead of what they were. Or perhaps he simply had a sixth sense about his audience, as so many of the great editors seem to have, and knew, by that same "instinct" with which he often credited the ladies, what they wanted to read. In the bargain, he had an extraordinary amount of energy and ambition.

Other magazines viewed Bok's assumption of the editorial chair as a hilarious event, and he was the victim of a good deal of lampooning in the newspapers and journals. Some of it was cruel and occasionally verged on the libelous. Eugene Field, nominally Bok's friend, solemnly announced the editor's engagement in his Chicago *Daily News* column first to "Miss Lavinia Pinkham, the favorite granddaughter of Mrs. Lydia Pinkham, the famous philanthropist [actually the creator of a patent medicine "for female complaints"]" and then, an equally unlikely event, to Mrs. Frank Leslie, widow of the publisher and possessor of a colorful past.

Meanwhile, Bok quietly went about his own courtship of Curtis's daughter, Mary Louise, whom he married. Even then he continued to suffer abuse and ridicule, by that time inspired mostly by his remarkable success.

Bok made his readers feel that the magazine was their intimate friend. They wrote in about their problems and were answered in a department called "Side Talks with Girls," written for many years by Isabel A. Mallon under the pseudonym "Ruth Ashmore," and in "Heart to Heart Talks," edited by Mrs. Margaret Bottome. There was even a "Side Talks with Boys" department. In articles, readers were given views of the private lives (although far from today's intimate revelations) of great men, from P. T. Barnum's "How I Have Grown Old" to Benjamin Harrison's description of family life in the White House.

As time went on, Bok began to fill the pages of the *Journal* with the best writers of the day, foreign and domestic. By this time, he was vice president of the company, the magazine was highly profitable, and he was visiting England and France in search of new work. Kipling became a friend and contributor, but Bok, with his formidable ego, did not hesitate to expurgate what he thought might offend his ladies—for example, eliminating a drinking scene from Kipling's first contribution to the *Journal*. He edited Twain, too, in his sublimely arrogant way, declaring later that Twain had admitted he was right—hardly consistent with the author's character.

Bok's editorial genius was constantly pushing back the frontier of women's periodicals. He began publishing house plans in 1895 and held contests for best homes. Stanford White, the most noted architect of the day, declared that Bok had "more completely influenced American domestic architecture for the better than any other man in this generation."

Bok also wanted to improve interior decoration, Pullman cars, and the appearance of cities. He was one of the first to protest and fight against outdoor advertising.

Under Bok's administration, the *Journal* had surpassed every other magazine in circulation by the turn of the century. It was 800,000 in 1900, about 900,000 two years later, and finally passed Curtis's long-sought million mark in 1903. By 1912, the number was approaching two million and reached that in the fall of 1919. Sometimes there was more than a million dollars' worth of advertising in a single issue, which might run more than two hundred pages. Bok earned his salary of $100,000 a year.

The editor could look back on some notable contributions to culture in the pages of the *Journal*. Kipling's *Just So Stories* had appeared there, as well as *Puck of Pook's Hill*. He had introduced Jean Webster's *Daddy Long Legs* and the work of Kate Douglas Wiggin. Bok had even persuaded Theodore

Roosevelt to dictate a column to a newspaperman. (This was done in two sessions every month—while Roosevelt was shaving.) These Rooseveltian pronouncements appeared in a department titled "The President." Later, Roosevelt contributed another column, this time anonymously, called "Men," with his identity carefully guarded.

Bok's editing was bold and imaginative, often ahead of his time and sometimes ahead of his audience. It took real courage, for example, to come out on behalf of sex education and to permit the *Journal* to talk about syphilis; this decision cost him thousands of subscribers. It was the first time the word "syphilis" had appeared in an American popular magazine.

Most of all, Bok brought to the editing of women's magazines the concept of "service," which in time dominated all of them. As he described it in his autobiography, "Step by step, the editor built up this service behind the magazine until he had a staff of thirty-five editors on the monthly payroll; in each issue he proclaimed the willingness of these editors to answer immediately any questions by mail; he encouraged and cajoled his readers to form the habit of looking upon his magazine as a great clearing-house of information. Before long, the letters streamed in by the tens of thousands during a year. The editor still encouraged, and the total ran into the hundreds of thousands until during the last year, before the service was finally stopped by the Great War of 1917–18, the yearly correspondence totalled nearly a million letters."

An important part of this policy was the Curtis Advertising Code, begun in 1910. It protected readers from fraudulent or extravagant claims in advertising and specifically banned financial, tobacco, playing cards, or liquor advertising. The prohibition on liquor was absolute—not even wine glasses or steins could be shown.

At the end of the Great War, during which the *Journal* astounded its competitors and everyone else by ranking third among the magazines most demanded by soldiers, Bok resigned. He had created a great magazine, and he was rich. Now he wanted to write and pursue his philanthropic interests. His characteristic farewell did not run in the *Journal* but in its companion publication, Lorimer's *Post*. There, in 1919, he reviewed his life in the United States since he had landed as a penniless immigrant and titled his articles, as only he could have done, "Where America Fell Short with Me." He had made magazine history, not only as an editor, but as an advertising genius who joined his extraordinary talents with Curtis's to dominate the field of women's journals.

Salme Steinberg, a media historian, observes that Bok "outdid his readers in his faith in the myths and hopes of his adopted country—free enterprise—the possibilities of reform, and the need for civic responsibility." With significant insight, Steinberg also notes that Bok's "sentiments were all the more compelling because they reflected not so much what his readers did believe as what they thought they should believe."

Other women's journals followed a similar pattern, on a lesser scale, of uncertain beginnings followed by steady success. They copied many of the *Journal*'s popular features even as they experimented with departments and innovations of their own. *Woman's Home Companion* emerged as the one closest to the *Journal* in its content and editorial viewpoint. Originally called the *Home,* this magazine led a precarious existence until it was taken over by the firm of Mast, Crowell & Kirkpatrick, a publishing company that had its roots in the manufacturing of agricultural machinery and already owned a magazine, *Farm and Fireside,* whose general manager was John S. Crowell. After acquiring the *Home* in 1883, Crowell decided, much as Curtis had, that the women's page in *Farm and Fireside* was so popular it suggested a much larger audience, and he converted his new property into a women's magazine, calling it the *Ladies Home Companion.* Its beginnings were inauspicious. The contents were a miscellany of articles borrowed from other sources and a great deal of mail-order advertising.

Enough readers liked this format, however, to bring in more than 100,000 subscribers by 1890. During the decade that followed, the *Companion* expanded its content to include more articles on food and fashion, and improved its overall quality. By this time, it was competing directly with the *Journal,* although it was a one-sided match. Perhaps sensing that the similarity of their magazine's title to Curtis's was not helping, the company changed the name, switching "Ladies" to "Woman's." The editor's elaborate explanation was that "Woman" was an honest Anglo-Saxon word without a synonym, while the use of "lady" as a synonym for "woman" was vulgar.

A turning point for the *Companion* came in 1900, when Arthur Vance, an experienced journalist, took over as editor. He immediately expanded the magazine's contents to include many articles and features designed to appeal not only to the broad interests of women but to the remainder of the family as well. Vance also made an important move by hiring two highly talented editors, Gertrude Battles Lane and Hayden Carruth. Lane came to the *Companion* in 1903 as household editor, a title causing some merriment in her house, where she was noted for her lack of interest in housekeeping. But she had publishing experience, knew how to get the information she lacked, and like Bok, had an instinctive feel for what readers wanted.

Gertrude Lane became editor of the *Companion* in 1911 and held that post until she died in 1941, leading the magazine to first place in the women's magazine circulation wars of the 1930s. Her success prompted Joseph Knapp, who had taken over as publisher of the magazine in 1906, to give her an ambiguous compliment, calling her "the best man in the business."

What Lane did for the *Companion* was to reverse the expansiveness of Vance's formula and refocus the magazine firmly on women's topics, presenting articles from the woman's point of view. Hayden Carruth, who was hired as fiction editor in 1905, contributed by seeking out good writers, working on promoting the magazine, and taking over its widely read end-page called "Just for Fun." He also wrote poetry for the *Companion.* From time to time, the editorial staff included people who had made reputations in other fields— the popular writer Sophie Kerr and the experienced journalist Margaret Sangster.

But the story of the *Companion* in this period was Gertrude Lane's story.

The organization was her life. She had begun with the Crowell organization in 1903, at $20 a week, when she was only twenty. By the time she died, she was earning over $50,000 a year and was regarded as one of the best in her field.

In a *Companion* promotion piece, Lane told how she edited the magazine: "In editing the *Woman's Home Companion*, I keep constantly in mind a picture of the housewife of today as I see her. She is not the woman who wants to do *more* housework, but the woman who wants to do *less* housework so that she will have more time for other things. She is intelligent and clearheaded. I must tell her the truth. She is busy; I must not waste her time. She is forever seeking new ideas; I must keep her in touch with the best. Her horizon is ever extending, her interests broadening, the pages of the *Woman's Home Companion* must reflect the sanest and most constructive thought on the vital issues of the day."

That was the key to Lane's success. She recognized that the days of the cookie-and-pattern formula were coming to an end (although the end was a long time in arriving) and that to succeed, a woman's magazine must realize that women were thinking, sensitive human beings with many interests outside the home. She printed the popular writers of fiction for men, but she also bought the work of Willa Cather, Sherwood Anderson, Ellen Glasgow, Sinclair Lewis, John Galsworthy, and Arnold Bennett. She knew how to dress up her magazine with illustration, and she was always enterprising. Her acquisition of unpublished correspondence by Robert and Elizabeth Barrett Browning for $25,000 in 1935 was a publishing coup.

What happened to the *Companion* after Lane's death is an industry story often told. No matter how competent the editorial hands that take over, the passing of a great editor usually signals not only change but eventual and often fatal distress. William A. H. Birnie, who took over as editor in 1943, turned the magazine, with the help of his articles editor, Roger Dakin, into a crusading organ—"a fighting lady," they called it. The crusades were in such areas of special interest to women as marriage, sex, and children, and for a time the new formula seemed to work. Circulation climbed to four million by 1950, and an advertising peak was reached three years later. But then the competition caught up, and the decline was rapid. When the *Companion* folded in January 1957, during the general collapse of the Crowell-Collier group, it was losing several million dollars a year.

Its demise was a major casualty in the war among the women's magazines that had begun with the arrival of the *Journal* and its rivals in the late nineteenth century. Some proved to be more durable than others. One of the hardy survivors that eventually helped push the *Companion* to the wall was *McCall's*, begun by a tailor, James McCall, who had learned his trade in his native Scotland and had come to America to succeed in it. He built up a business in dressmaking patterns, known as James McCall & Co. To promote this enterprise, in 1873 he started the *Queen: Illustrating McCall's Bazar Glove-Fitting Patterns*. After his death in 1884, his widow headed the company for a time, but the wife of George Bladworth, another member of the organization, took over as editor when her husband was made president of what was now called the McCall Publishing Co.

For seven years under the Bladsworths, the magazine was called the *Queen of Fashion*. Although it was designed to promote McCall patterns, it became more of a general magazine for women than a fashion periodical. It was struggling, however and circulation had dropped to 12,000 by 1892. A new owner, James Ottley, a businessman, took over in 1893; he expanded the pages, cut pattern prices, and raised the annual subscription price of the magazine from thirty to fifty cents. These measures, but chiefly the demand for patterns, elevated the circulation to 75,000 in 1894, and three years later it was renamed *McCall's*.

The magazine now boasted it would contain "nothing but the latest and most tasteful of Dame Fashion's creations, besides articles on current topics, beautifully illustrated by photographs; household hints that are really useful and practical; bright and entertaining fiction, and literature of interest to all members of the family."

For a long time, even though its low price brought an increase in circulation, *McCall's* remained a rather undistinguished periodical, undergoing various changes in editors and format until it was sold in 1913 to White, Weld & Co., a banking firm, and became the McCall Corporation. Its new president, Edward Alfred Simmons, also headed the Simmons-Boardman Company, publishers of *Railway Age*, the oldest trade magazine in the country.

What *McCall's* was waiting for, however, was a first-rate editor, and in 1921 it got one with the arrival of Harry Payne Burton, who brought to its pages a generous sampling of the day's best writers of popular fiction and articles, most of whom had appeared in the rivals of *McCall's* as well. The emphasis was on making it a national magazine, which was done not only by concentrating on material with a mass appeal but also by using such promotional devices as referring to its national audience as "McCall Street," thus providing readers with a sense of common identification.

It could be said that the career of *McCall's*, which endures today, really began with Otis Wiese, who became editor in January 1928. Under his direction, it rose to new successes. At twenty-three, Wiese was the youngest editor of a national magazine in America. He was a difficult man but one of the authentically great editors and introduced several new ideas into magazine publishing. One was a concept he called "three-way makeup,"—the division of the magazine into three parts, each with its own cover page and content, giving readers the illusion of getting three magazines for the price of one. The three sections were "Fiction and News," "Home Making," and "Style and Beauty." This was simply a clever repackaging of what the others had, but it worked.

Wiese also revived the nineteenth-century practice of running a complete novel in every issue, which no one but *Redbook* had attempted to do in the new century. He also reached out to a fresh audience with a "youth conference" series. These and other policies increased the circulation to three million by 1940.

Perhaps the best—certainly the most quoted—promotion device invented by Wiese was the idea of "togetherness," by which *McCall's* hoped to tie up the whole family in a shiny package. It appeared to be a success. In 1956, the year this idea was launched, the magazine's circulation reached 4,650,000.

McCall's was one of three magazines to rise from the pattern business. Another was the *Delineator*, a product of Ebenezer Butterick's career as inventor of the tissue-paper pattern. Like McCall, Butterick was a tailor who produced his first patterns at his shop in Fitchburg, Massachusetts, in 1863. Coming on the market simultaneously with the invention of the sewing machine, which introduced a new era in home dressmaking, they soon became familiar to millions of American women.

Butterick's patterns were so successful that he moved to New York, and there he launched a fashion magazine in 1872 designed to promote a business already growing so rapidly that he could scarcely control it. The *Delineator* was subtitled "A Monthly Magazine Illustrating European and American Fashions." It sold for a dollar and in its pages, readers found the styles departmentalized, arranged for the seasons, for "misses and girls" and "for little folks," and such subdivisions as "hats and bonnets" and "stylish lingerie." The illustrations were woodcuts. Ebenezer pushed his main business by giving away patterns as subscription premiums.

Circulation increased steadily year after year until, by 1888, it reached 200,000, a remarkably high figure for those days. About that time, the number of pages was increased to eighty, and various changes were made in the format to make it more attractive and in the editorial department to increase the journal's scope. The *Delineator* now attempted to cover nearly everything that went on in a household; the result was another jump in circulation, to 500,000 in June 1892. The magazine had grown to two hundred pages by 1894. Even by modern standards, the *Delineator* at this point was a substantial success. Moreover, the Butterick company had established agencies to sell its patterns across the United States, and wherever a lady could buy a Butterick pattern, she could also get the *Delineator*. As time went on, the scope of this journal became broader, including both fiction and articles, although much of it was light reading.

In the early years of the new century, the editorial quality of the *Delineator* continued to improve. It printed the fiction of such writers as Carolyn Wells, Richard Le Gallienne, Hamlin Garland, Anthony Hope, and Zona Gale. This was due to the arrival of an editor unique in magazine history, Theodore Dreiser, who took over in 1907. Those who thought, or feared, that he might impart some of the realistic writing in his novels to the pages of the *Delineator* had nothing to worry about. Dreiser proved to be an editor too astute to fall into such a trap. He understood his audience and approached it conservatively. Charles Hanson Towne, who established his own reputation as a popular writer, wrote of Dreiser as editor: "Every department of the organization was under the control of Mr. Dreiser. Not a detail escaped his vigilant eye. He OK'd every manuscript that we accepted—read them all, in fact—and continually gave out ideas to the entire staff, and saw that they bore fruit."

It would have been unlike Dreiser, however, to adhere to the *Delineator* tradition of ignoring social problems. He was much too involved with them himself, and so readers were treated to articles about divorce, marriage problems, suffrage, education, religion, and the high cost of living. Dreiser also crusaded against the treatment of children in orphan asylums and for better treatment of the underprivileged young.

Not surprisingly, Dreiser greatly improved the fiction department. There was much less hammock reading and many more big names, in the fashion of the popular general magazines, who vied with each other to see how many brand-name authors they could display on their covers each month. Dreiser's *Delineator* had its share: Kipling, Conan Doyle, Woodrow Wilson, Jacob Riis, John Burroughs, Elbert Hubbard, Oscar Hammerstein, and F. Marion Crawford, among others.

In 1910, Dreiser went back to writing his own fiction, and once more, the departure of a great editor at the top would mean eventual formulaic change. Circulation went up during the twenties, along with those of most other magazines, going well above the million mark, but after another change of editors in 1926, the magazine moved away from its familiar pattern and took on a more sophisticated air, both in its art work and text. By merging with its sister Butterick publication, the *Designer,* in 1928 and lowering the single-copy price to a dime, the *Delineator* reached a surprising circulation peak of 2,450,000 in the year after the Crash. At that point, it stood in fifth place among the women's magazines, and one would have thought it reasonably secure. Nevertheless, it was sold in May 1937 to Hearst, who merged it with his *Pictorial Review,* the *Delineator*'s nearest rival.

Pictorial Review was still another magazine begun by a pattern maker, William Anhelt. Because it started later than the others, in 1899, it was able to benefit at once from the broader formula the others had taken years to develop. It carried book reviews, notes about plays, and society gossip, in addition to the conventional fashions and patterns. Arthur Vance, who became its editor, described the early years of this journal: "At its birth it was rather a thin, little infant, more or less homely in appearance, but still having within it the elements of a wonderful promise." By the time Vance took over as editor in 1908, some of that promise had already been fulfilled. From the start, it had been targeted at comfortably upper-middle-class women harboring aspirations to higher social standing. For them, the *Review*'s articles adopted a sophisticated tone, reporting on the activities of such socially prominent individuals as Mrs. O. H. P. Belmont and Mrs. Herman Oelrichs. The New York Horse Show was covered fully. As the magazine grew older, more and better-quality fiction appeared, but the *Review* failed to survive the perils of the thirties.

One other major woman's magazine begun in the nineteenth century not only survived but became a major player in the field during our own time. *Good Housekeeping* was begun in 1885 by Clark Bryan as "A Family Journal Conducted in the Interests of the Higher Life of the Household." This lofty subtitle more or less described a semimonthly that soon gained a devoted following. Its wide-ranging contents reflected both the interests of Bryan and the fact that much of its contents was contributed by readers—fiction, poetry, puzzles, household tips, fashion news, and home decorating advice. Bryan, like Bok, encouraged his readers to feel close to the magazine, to write in with their ideas, problems, and articles. Readers were paid for their contributions, not in large amounts but enough to provide small earnings and a sense of validation for work well done.

Bryan proved to be a successful editor, and by 1895, the magazine's circulation stood at 55,000. Unfortunately, he suffered from chronic illness and committed suicide in 1898. The magazine survived his death, however, and eventually passed into the hands of the Phelps Publishing Company, a firm already issuing four agricultural journals. The magazine continued to thrive under its management and soon gained recognition for an innovative service—the Good Housekeeping Testing Institute.

Established in 1901, the institute enabled the magazine's staff to try out new household products and report on them to readers. All the women's magazines ran columns about new products, but *Good Housekeeping's* institute sounded like a formal scientific establishment. Readers came to believe they could rely on the judgments of its staff.

A year after the institute began, *Good Housekeeping* announced that it would accept advertising only from products tested and approved by the institute, which would in time stamp each one (figuratively, of course) with the Good Housekeeping Seal of Approval. A logo bearing this seal appeared in the product's advertising. In 1912, Dr. Harvey Wiley, formerly chief chemist for the United States Department of Agriculture, became head of the institute; the Seal of Approval was his idea. This product evaluation gave *Good Housekeeping* a special value and kept readers loyal even when it came under attack in later years. In the new era that began after 1945, the magazine eventually became a part of the Hearst group and took a commanding position in its field.

Closely related to the magazines that began as pattern catalogs but much more upscale in their target audience, were the fashion journals *Harper's Bazar* (the present-day extra "a" in *Bazaar* was not added until 1929) and *Vogue*. Neither of these periodicals, unlike the others, was aimed at a mass market but concentrated instead on those who were simply fashion minded.

The *Bazar* was still another product of the fertile Harper factory and of Fletcher Harper's agile brain. Fletcher thought so much of his idea that he would have gone ahead alone if the other brothers, reluctant at first, had not agreed. He had been to Germany and had seen the Berlin *Bazar*, a different kind of women's magazine than had yet appeared in the United States, and he was convinced it could be successful here.

Harper's Bazar appeared for the first time in November 1867, subtitled "A Repository of Fashion, Pleasure and Instruction." It was ably edited by New York City's historian Mary L. Booth. Fletcher's confidence appeared to be justified; the magazine achieved an 80,000 circulation in its first decade. As Mott has observed, what Mary Booth turned out was "a ladies' *Harper's Weekly*." It contained English serials, plenty of double-page pictures, miscellaneous articles and stories, splendid art work, even Thomas Nast cartoons and other engravings from the *Weekly*'s artists. All of this for an annual subscription price of four dollars.

The fashions came from Paris and Berlin, sent by duplicate electrotypes. Some of the designs appeared as they were in the European capitals; others had to be adapted for the U.S. market. They were taken directly from the magazine's great Berlin counterpart, *Der Bazar*. The German magazine also sent proofs of its text matter, describing the new styles abroad, and to this the New York staff added its descriptions of U.S. designs. Readers were also offered patterns, fancywork, and discussions of household problems, interior decoration, and garden planning. But clothes were its chief attraction. Women anxious to clothe themselves in the latest fashions knew they could find them in the *Bazar*, which was preeminent in the fashion world until the late 1890s.

Mary Booth edited the magazine until her death in 1889, when Margaret Sangster took over; the game of musical editorial chairs was common then, even as it is now. With such strong hands at the helm, Fletcher Harper's success managed to survive the House of Harper's astonishing bankruptcy in 1899. Along with fashions, it was continuing to print the works of such writers as William Dean Howells and Mary E. Wilkins, while John Kendrick Bangs had been conducting a regular column since 1888.

After J.P. Morgan rescued the failing Harper brothers and sent in George Harvey to turn matters around, the new management appointed Elizabeth Jordan as editor in 1900, and the *Bazar* was changed to a monthly. Jordan, who had come to the magazine from the New York *World*, failed to propel the magazine upward, and it continued to pursue an erratic course, losing money steadily until Hearst bought it in 1913 for his International Magazine Company, an empire he was then building. His editors freshened up the magazine considerably, making it much more sophisticated and generally setting the tone it has today. Little survives from the old days now, however, except the general idea and a receptivity to good writing.

By that time, the *Bazar* had a powerful rival in *Vogue*, powerful because it had been started in 1892 by the socialite Arthur B. Turnure, whose original backers included a collection of Van Rensselaers, Stuyvesants, Astors, Whitneys, and Jays. Finances were not an immediate worry. The first editor was Mrs. Josephine Redding, and the art director was Harry McVickar. There were fashions for both men and women; reviews of the latest books, plays, music, and art; and much commentary on manners. Readers were, quite naturally, a group nearly as elite as the backers.

In spite of initial success, both McVickar and Mrs. Redding had left *Vogue* by the turn of the century. In 1901, Turnure hired a new editor, his sister-in-law, Marie Harrison. On the staff at the time was an employee who had been there since 1895, Edna Woolman Chase, the future great editor who would bring *Vogue* to the top of its field. Turnure died suddenly in 1906, leaving his creation rudderless but an ideal vehicle for an entrepreneur seeking to market fashion ideas and other goods to an elite audience. Condé Nast, a lawyer who had found a second career in advertising, was looking for just such an opportunity.

Nast had worked for ten years as advertising manager of *Collier's*. He had met Robert J. Collier when they were students at Georgetown University, and they had remained close friends. Collier had gone home to his father's book business, which also published the magazine, while Nast went home to St. Louis, where he pursued a law degree at Washington University. But Nast found the law unexciting, and when young Collier invited him to come work for the family *Weekly*, he accepted at once.

At *Collier's*, directing the advertising department, Nast was a prime factor in raising circulation figures by promoting the improved editorial material and illustrations his friend Robert was putting into the magazine. He also steadily increased advertising revenue, constantly keeping the periodical and its readers before advertisers, and segmented his market, identifying those who would be the best prospects for advertisers of particular goods.

Besides his work at *Collier's*, Nast had become vice president of the Home Pattern Company, a firm manufacturing and distributing the patterns promoted by the *Ladies' Home Journal*. His responsibility was to sell advertising space in the magazine, but the work so involved him that in 1907 he left *Collier's* to work full time at the Home Pattern Company, where he raised advertising revenue phenomenally. For example, income from the company's magazine, *Quarterly Style Book*, jumped from $1,500 in 1907 to $180,000 in 1908.

The fashion and pattern market intrigued Nast, and he began to look around for a property of his own. *Vogue* caught his eye, and in 1905, he began negotiating to buy it. Turnure's death momentarily slowed the sale, but in 1909, *Vogue* passed into Nast's hands. Its circulation at that moment was low, about 14,000, and its yearly advertising revenues were no more than $100,000. Significantly, however, its readers were still among the social upper class of New York City, primarily. Unlike his contemporaries Curtis and Munsey, Nast wanted what some entrepreneurs desire most today—quality circulation, not high figures.

At this point, Edna Woolman Chase became managing editor, soon rose to be "general editor," and made her first move by persuading Nast to make the magazine a semiweekly. Together, she and Nast pushed and pulled *Vogue* to the top of the elite fashion journals, collecting large advertising revenues in the process. Advertising pages increased, color covers were added, and more emphasis was placed on fashion and *Vogue* patterns. Carolyn Seebohm, Nast's biographer, has written of the transformed *Vogue* as "a richly embellished frieze of society, fashion, social conscience, and frivolity, picked out in gold by the confident and stylish hand of its new publisher."

Chase proved to be a spectacularly successful editor. Her name became synonymous with taste and fashion. Having joined the circulation department of *Vogue* when she was only eighteen, she was able to bring experience, judgment, and industriously acquired editorial abilities to her job as editor, where Nast placed her in 1914. She learned quickly about the elite, rich world she was catering to in her magazine, and edited the journal expressly for that world. Physically small, she was, nevertheless, a commanding presence, a demanding editor who expected a great deal from her staff.

One of her innovative ideas, netting exclusive material for *Vogue*, was the charity fashion show. Called at first a fashion fete, the initial show took place in 1914, featuring American (as opposed to Parisian) dress designers. To ensure the success of this event, Chase enlisted the help of Mrs. Stuyvesant Fish, regarded as the fashion arbiter of the day. She, in turn, obtained the support of such social leaders as Mrs. Astor, Mrs. Vanderbilt, and Mrs. Belmont. Proceeds from this event went to the destitute victims of the war already raging in Europe.

Vogue did its part during the war, carrying articles about relief efforts and sponsoring its own appeal for the Sewing Girls of Paris. It ran sober stories of sacrifices on the part of the rich for the war effort. The journal also wrote about ways to conserve food, energy, and clothing. All this was presented in an upbeat way, making activities to support the war the fashionable thing to do. *Vogue* reported a side effect of the war: the emergence of American couture, filling in for the French designers and never totally retreating when the war was over. Needless to say, the magazine subtly encouraged this development.

Early on, *Vogue* dominated its field and overshadowed completely other fashion journals, which became viable because women with means were buying ready-made clothes or else trying to make their own, needing guidance in either case. By 1880, N. W. Ayer's *Directory* had already listed eighteen fashion magazines, and the number continued to rise.

Clearly, in the period from the end of the Civil War to the close of the Great War, women's magazines had successfully diversified themselves into journals for an audience that was itself diversifying—socially, economically, and culturally. The intensely competitive world of the woman's magazine created during these decades set patterns that would persist to the present.

What were those patterns? They are worth examining a little more closely, as much for their dissimilarities as for their basic constituents. Then, as now, although in different terms, the mass media were valuable in helping readers to understand and cope with social change. Janet Mickish and Patricia Searles, who have examined images of female gender roles at the turn of the century, argue that the media "influenced people's conception of reality and guided them in the process of constructing attitudes and action."

The *Ladies' Home Journal* was a powerful factor in influencing women to resist change. Analyzing fiction in the magazine during 1905, Mickish and Searles found only conventional images projected. Women are seen as traditional "good-hearted, self-sacrificing people, embracing their 'natural' roles as wives and mothers. Subtly, the *Journal* discouraged any social change that would alter their idealized status. The implicit assumption in the fiction is that love will overcome all obstacles, problems, and miscommunications. Independence and careers are seen as undesirable if not impossible for women. Those who do work are sympathetic characters only if they are feminine in their ways, gentle, good, sweet, demure with men, carrying on their jobs in a nurturing way. For none of them is marriage and the family ruled out as a possibility. . . . Mass media characterizations and tacit messages may have played an important role in defusing and undercutting earlier feminist efforts, while giving meaning and direction to the rapid social change of the era."

Yet these conventional images, prevalent in most woman's magazines since the Civil War, were already beginning to change in *Leslie's Popular Monthly* during the 1890s. Examining this transitional phase, Donna Rose Casella Kern sees female characteristics changing in fiction, with women becoming more independent. In some stories, too, action has moved outside the home, often to faraway places. Kern believes that reflects a change in the audience, from almost exclusively middle-class women of comparative leisure to a broader market of those already working in offices and factories. While this may or may not be true in any significant sense, it is true that *Harper's* had already shifted from sentimental fiction to more serious work.

That change was a constant factor in the women's audience can be seen by the changing face of fiction in mass-circulation magazines between 1905 and 1955, wherein the portrayal of women can be taken as a barometer. A 1936 study by Phillip Wyman, of *McCall's*, showed that fiction, more than anything else, drew readers to buy magazines. Accepting that fact, Donald Makosky, who examined five magazines during the period from 1905 to 1955 found that fiction flourished in these periodicals until a decline induced by television and the growth of nonfiction magazines set in during the fifties.

The number of women appearing in fictional stories hit a low point in 1925, Makosky discovered, but then increased in all magazines until 1935, after which it decreased again in nearly all the journals except *Harper's* and the *Woman's Home Companion*. By this time the reality of American women's lives were being reflected, with some significant exceptions. One of these was the fact that women working on farms and in factories were seldom seen in magazine fiction, even as their number grew in real life. On the other hand, women in glamorous occupations like acting were depicted much more frequently than their number in reality warranted. Women were also more often shown flirting and experimenting with extramarital affairs, and the outcome of their behavior was much less severe than in either reality or previous fiction. Stories still ended happily; divorce and other domestic tragedies were virtually nonexistent.

All of this, however, did not prepare mass audiences in any considerable way for the great transformation the women's movement brought to women's magazines as well as to women's lives.

10

Magazines as Political Weapons in the Class Struggle

Stated in such classic Marxist terms, the intellectual ferment in magazines between 1895 and 1918 might appear to be the record of a radical movement in print aimed at overthrowing the capitalists. Although the object of the muckrakers and others was to expose the evils of capitalism, they did not have the slightest intention of overthrowing the government and installing communism even before the Revolution occurred in Russia. Their goal was reform of a system they saw as both corrupt and unbalanced economically.

Some of the muckraking magazines were themselves pretty healthy capitalist enterprises, run by such unabashed practitioners of the art as Frank Munsey and S. S. McClure. Their muckraking ran its profitable course as an adjunct of the Progressive movement, where their influence counted. To the left of publishers like Munsey and McClure, however, stood another small group whose chief organ was the *Masses*, a frankly ideological journal devoted to radical ideas but at the same time advocating some of the same reforms as their more respectable neighbors. And the *Masses*, in fact, was more of an intellectual forum than a serious call to arms.

All these reformist magazines were politically influential in one way or another. The major muckraking journals were an expression of populist demands for reform, to which politicians were compelled to respond, while the radical-left magazines were exponents of ideas over which the traditional intellectual left-right battles could be fought. The creation of this political battleground was the result of forces both in the magazine industry and in national political life. The great wave of ten-cent magazines, beginning in the early nineties, provided the launching pad for muckraking.

Low-priced magazines were far from new. They had existed from the beginning of the industry, long before it deserved any such title. But those that arrived near the century's end were different. Their popularity was so sudden and overwhelming that they overflowed the dimensions of a movement or a trend. They constituted nothing less than a revolution in magazine publishing and reading.

The coincidental arrival of the halftone was a technological factor of major importance in this phenomenon, and the hard times of the 1890s—first the Panic of 1893, then the one in 1897—provided an economic motivation. Readers suddenly found themselves with a whole array of exciting, well-illustrated journals, rich with color and costing only a dime instead of the usual twenty-five or thirty-five cents. They were also a prime cultural factor, stimulating the tremendous drive for self-improvement that characterized American society in this decade.

These were not the only factors, however. The sharp rise in national advertising was important, but above all, in the opinion of some historians, the tremendous popularity of the ten-cent magazines was due to their ability to reach out and embrace a new audience, an expanded middle class with less education and less money than those who had supported the thirty-five-cent magazines. These journals had seemed aloof; the new breed spoke directly to the people and were concerned with everyday life.

So, the formula was established: entertainment to begin with, on a broadly popular framework, and then articles about ideas and national events. It was an easy step from there to muckraking in the public interest, fueled initially by the virtual passion for self-improvement, particularly for adult education, characterizing the times. Large numbers of the population had an earnest desire for culture, and the ten-cent journals gave them something easy to swallow—not the high-level literary fare of magazines like the *Atlantic*, but the finest popular fiction available. *McClure's* led the way in this respect, with Robert Louis Stevenson, Kipling, Thomas Hardy, Israel Zangwill, Conan Doyle, Stephen Crane, O. Henry, and others of their stature contributing.

As for nonfiction, what more could readers ask for than Ida Tarbell's lives of Napoléon and Lincoln? But they also received informative pieces about new developments in science and transportation, articles and stories about railroading (another national passion), and writing dealing with exploration, animals in the wild, and—high on the list then as today—personalities. As steadily rising circulations testified, the demand for information and culture seemed inexhaustible. Members of reading circles, study clubs of various kinds, independent study groups, correspondence schools, and university extension courses welcomed magazines like *McClure's* and the others.

The old establishment journals were still well respected, but people bought the ten-centers. There were at least a dozen major periodicals catering to what we would call today the information market, and their titles identified the need they were supplying: *Review of Reviews, Our Day, Our Times, Self Culture,* the *World To-Day,* the *Progress of the World, Current Literature,* the *Eclectic Magazine,* and others. When *Cosmopolitan* attempted to capitalize on this demand by setting up a Correspondence University on a grand scale, the response was so overwhelming that it had to be given up.

At first, the publishers of the ten-centers did not realize they were appealing to a substantial audience of young men, even though they themselves were relatively young. But then an enthusiastic response came from such readers, who felt challenged by the stories of great men and the discoveries in science and industry. This was rich stuff for these people, who were living in a period of national ambition, optimism, and hope in spite of the economic and social upheavals of the nineties. While the country was still reeling from the near disaster of 1893, Edward Bok was writing in the *Ladies' Home Journal:* "Every success is possible, and a man may make of himself just what he may choose." This was the general conviction that prevailed, in spite of all the evidence to the contrary. "Opportunity"—"youth"—"success." Those were the magic words.

But if the times were so optimistic, one might ask, why did the ten-cent magazines suddenly become reform organs, devoted to exposing the corruption and greed behind the national façade? The answer begins with the true state of the nation, not understood by most of its citizens, at the beginning of the new century.

Looking back in 1913, when he wrote his autobiography, Theodore Roosevelt observed that government control of corporations was not yet significantly exercised, and in fact, he expressed some doubt as to whether government had any such power. Yet, as the Progressives were pointing out, the corporations were out of control, and there was a great deal of concentrated wealth. This seemed plainly unhealthy for the public good. As David Chalmers points out in his excellent study of the muckrake years, the regulatory capitalist system in America was about to be born, and with it, the modern American presidency as well.

In just nine years, between 1903 and 1912, with the muckraking magazines providing the public opinion behind political action, the Sherman Anti-Trust Act had become the instrument to attack monopoly, the Interstate Commerce Commission was serving as a model for other regulatory commissions, the Pure Food and Drug Act was safely on the books, and, for the first time, conservation was a national issue. From the exposure of corruption in the Senate came the popular election of senators. Women's suffrage had been given a substantial boost, and both the Federal Reserve system and the graduated income tax (unable to find support since Abraham Lincoln advocated it) were about to become realities. Moreover, there was agitation to eradicate the evils of child labor, to give working women more equality, and to shorten the hours of workers.

All these and other social reforms were, in part, the result of the investigative reporting done by the muckrakers in the ten-cent magazines. Ironically, nearly every one of these reforms, as Chalmers points out, remains more or less unrealized. We are still trying to "restrain private economic power and direct it toward the public good," but now, through their own mergers and consolidations, the magazines have become not instruments of reform, for the most part, but private economic powers in themselves.

The muckrakers had a powerful effect on the middle class, arousing its concern about social ills and shabby politics. Before they launched their attacks, voices of dissent and disenchantment had been heard, but the expert journalists who wrote for the ten-centers cut through the national confusion and gave Americans a full report on what was happening to them without their knowledge or consent. In this period, they were the national media voice, just as radio would be later, followed by the news magazines in the 1930s, and television after the Second World War.

The sheer scope of this reporting was impressive. In a space of only nine years, a relatively small group of reporters turned out more than 2,000 articles exposing the nation's corruption and confusion. The stories were detailed and factual, accompanied by supporting editorials and cartoons, and the editors were not above selling them in the most sensational way possible through striking covers and what would later be tabloid headlines. Newspapers joined in the chorus, too, but they were local voices; McClure, Munsey, and the others covered the nation.

The common theme, implicit in everything displayed, was betrayal. America's citizens had been betrayed by those who had supposedly made the country what it was, by the men who had built the great corporations and who had made the free market a national religion. In this betrayal, they had been abetted and aided by politicians at every level, but particularly those in the highest positions of power. Reading these revelations, most Americans felt both shock and outrage, and as one exposé followed another, they developed a fascination for what was being told to them and were scarcely able to wait for the next revelation.

There were some defections in the ranks of the magazines. Munsey was never a serious muckraker because he was a model of the success story himself and had an abiding admiration for those who were rich and successful too. Nor was John Brisben Walker's heart in it; his *Cosmopolitan* was only in the game for competitive reasons until Hearst bought it. At the time, Hearst considered himself a champion of the people and took up the muckrake enthusiastically.

McClure was the man who led the way into muckraking and was so successful that all the others had to follow. That claim is not unarguable, but analysis provides abundant evidence that *McClure's* was the chief wielder of the muckrake as far as the magazines were concerned. One should not overlook, however, the fact that the newspapers had been doing substantial raking since the days of the Grant administration. Pulitzer's *World*, William Rockhill Nelson's Kansas City *Star*, and Victor Lawson's Chicago *Daily News* had all been rooting out the sin and corruption in which big cities are ever wallowing.

McClure, however, knew how to create a pattern of scandalous revelations combined with concern for serious social issues. His muckraking was designed to reach a somewhat higher level of readership than the daily papers, pushing back boundaries at the upper and lower extremes of the middle class. In essence, it was not much different than James Gordon Bennett's original and highly successful prescription for his New York *Herald*—scandal and politics, in alternate doses.

Becoming a muckraker, however, was not one of McClure's many original ideas for his magazine. When he hired Ida Tarbell in 1892, the writer who would soon become the first great muckraker, it was not to wield the rake but to do her immensely successful serial biographies of Lincoln and Napoléon. It was not until just after the turn of the century that McClure sensed that President Roosevelt's trust-busting speeches had stirred up a wide public concern about monopoly. Sitting down in conference with his editors, he proposed that the magazine do a series on one of the trusts. The editors agreed that it should be about oil, and Ida Tarbell was assigned to it. She was a logical choice since she had grown up in the oil country of western Pennsylvania.

Once possessed by the idea, McClure decided to double his chances for increased circulation by sending another of his writing regulars, Ray Stannard Baker, to examine the situation in the Pennsylvania anthracite fields, where a bitter strike was under way. McClure asked Baker to find out particularly why some of the miners did not support the strike. Baker approached the task with some bias toward management. In earlier pieces, he had demonstrated his nearly unqualified acceptance of private industrial collectivism, which seemed a romantic aspect of American life to him, and he had already celebrated the individualism of nonunion workers. But Baker was a first-class reporter of formidable integrity, and he did not selectively gather facts to support his bias but tore away the top layers of the labor situation to expose the appalling conditions in the mine fields.

McClure sent another of his editors, Lincoln Steffens, to look into municipal corruption in St. Louis, and what Steffens found not only made a sensational series of articles but also inspired him to visit other metropolises, where much the same kind of thing was going on; that resulted in Steffens's eye-opening "Shame of the Cities" series.

Suddenly *McClure's* was deep in the muckraking business, as the celebrated issue of January 1903 confirmed. There, Steffens's lead article, "The Shame of Minneapolis: The Rescue and Redemption of a City That Was Sold Out," ran cheek by jowl with Ida Tarbell's exposé of Standard Oil's operations and Baker's report on the anthracite fields.

By this time, McClure was well aware of what he had embarked upon. In an editorial on the last page, he charged that these three articles were "an arraignment of American character," particularly "the American contempt of law." Everybody was involved, he said, setting the moral tone for all the muckraking that would follow, "Capitalists, working-men, politicians, citizens—all breaking the law, or letting it be broken." No one should look to lawyers, judges, or the church for help, he said; they were just as corrupt as everyone else, and the purblind colleges could not be relied on either. What would be the end result? Nothing less than the loss of people's liberty.

The "corrupt, lawless" people read these words with fascination. They smote their breasts and cried, "*Mea culpa, mea culpa*! Tell us more about how awful we are." That was exactly what McClure intended to do, aware that others were already joining the parade. Steffens, having left the cities naked to their enemies, moved on in a second series about the states—Missouri, Illinois, Wisconsin, Rhode Island, Ohio, and New Jersey—to demonstrate that their governments were no better than those in the cities.

As Chalmers points out, Steffens's technique was to personalize the people he wrote about, drawing a series of human pictures that made reading the essentially dreary facts of corruption almost as good as reading a novel. His message was always the same: America was a corrupt nation, and special privilege lay at the root of it.

Ida Tarbell's approach was different. Her research was thorough, her writing sober, and the end result was an inexorable march of facts that showed in impressive detail how Standard Oil conducted its business. But she was also fair, in a sense. She displayed Big Oil in all its legitimate greatness but argued that too high a price had been paid for it—inflated charges to the public, corporate behavior that was at best unethical and most probably illegal, the ruthless use of power to stamp out opposition, and, in the end, destruction of competition in the oil business. The case she made was substantially the one made by the government in its attempt to break up Standard Oil's monopoly, which it seemingly did by virtue of the Supreme Court's historic decision of 1910.

Tarbell's exposé, told in eighteen installments between November 1902 and the spring of 1904, was not only a story that held the nation's attention, but its mastery of detail, supplemented by the photographs and sketches accompanying it, was (and remains) a valuable contribution to industrial history. Its impact on the public was extremely impressive. Never before had a magazine been able to create such general indignation, well beyond its own readership. John D. Rockefeller became a national symbol of commercial greed, an image it would take decades of public relations and philanthropy to alter. Public reaction was so intense that it not only led to the government's suit against Standard Oil but also inspired antitrust suits in twenty states.

At first, the goal of the muckraking of *McClure's* was to get lawlessness under control by arousing citizens to put pressure on officials to enforce the laws. But then the emphasis shifted to the abuses of the public good by giant corporations, especially their impact on the poor and less fortunate. That meant putting public pressure on national and state governments to regulate, which involved first removing their fingers from corporate pockets.

When *Everybody's* joined the muckrakers in 1904, it did so with a series that was talked about everywhere in the country but whose content had considerably less merit than those in *McClure's*. These were the revelations by Thomas W. Lawson of high-flying insider trading and stock market manipulation, an oft-told story down to the present day. "Frenzied Finance" gave a phrase to the language, but these articles, although they made spectacular reading, were deeply flawed by Lawson's own boasting and by his sieve-like memory.

In a sense, Lawson was picking up where Tarbell stopped. His self-proclaimed subject was "the power of dollars" and how the banks, insurance companies, mines, manufacturers, colleges, and even churches were plundering innocent Americans. He located the core of this corruption at No. 26 Broadway, Standard Oil's headquarters, and he named Henry B. Rogers, a Rockefeller partner, and the National City Bank as the chief agents of corruption. Unlike Tarbell and the writers for *McClure's*, however, Lawson had no ideas for turning out the rascals and protecting the public. He seemed interested only in spilling out his own lurid story. But what the series did was to point an accusing finger at the great life-insurance companies, and soon Pulitzer's *World* did its own investigative report, greatly amplifying what Lawson had written, but with more cautious attention to detail. The result was an investigation by the New York State Legislature of the Equitable Life Insurance Company by a special committee whose chief counsel was Charles Evans Hughes, then at the beginning of his career.

McClure's did not intend to concede the life-insurance inquiry to either *Everybody's* or Pulitzer. Burton J. Hendrick was assigned to look into this business, and in 1908, he began a six-part series called "The Story of Life Insurance." It was not a pretty one. Hendrick went on to become an investigator and a biographer of the men who controlled the great American fortunes, particularly those involved with railroads and city traction systems. His insurance series resulted in another New York State investigation and a tightening of the laws.

By 1906, the results of three years of muckraking, principally in the magazines, were evident in the growth of the federal government's regulatory power over railroads, packing houses, and the food and drug industries. The public demand that produced this result was unquestionably inspired and maintained primarily by the magazine muckrakers, supplemented by the newspapers.

The war on big business was not entirely one-sided, however. The corporations fought back in various ways, and one of the most notable counterattacks was Senator John F. Dryden's 1906 campaign to silence or manipulate the press. As it happened, Senator Dryden was also president of the Prudential Insurance Company. Robert Reynolds has investigated this previously undocumented episode, virtually ignored by historians of the muckraking era, and finds its origin in an article titled "Our Millionaire Socialists," by Gustavus Myers, in the October 1906 issue of Hearst's *Cosmopolitan*. In it, Myers examined the current phenomenon of a few rich men who professed to be converts to socialism. It was not long on the stands before it was hastily withdrawn and a new edition substituted. Myers's piece had vanished, and a short story had been inserted in its place.

Dryden, as it turned out, was running for the Senate in New Jersey, where he faced a preferential primary for the first time. By using Prudential advertising contracts as a blunt instrument, he had persuaded *Cosmopolitan* (not a difficult task) to suppress the Myers article and, with it, a part of David Graham Phillips's "Treason of the Senate," then causing a stir. Myers had helped Phillips research the piece and in doing so had uncovered some unsavory facts about the senator. Dryden's suppression list also included a profile of the other New Jersey senator. According to Reynolds, *Cosmopolitan's* business manager, George Von Utassey, had been quite frank about it. He told Myers that the material had to be suppressed because Prudential had just bought another $5,000 worth of advertising space in the same issue.

This was not the worst of it. Through September and October of 1906, Dryden had arranged to publish in thirteen leading magazines self-congratulatory advertisements for Prudential disguised to look like regular editorial material. The magazines included such leaders as *Harper's* (both weekly and monthly), *Lippincott's, Munsey's, World's Work, North American Review, Literary Digest,* and *Leslie's Weekly.* Dryden, in effect, had literally flooded the quality magazine world with self-serving propaganda, at the same time suppressing articles he believed would embarrass him. *Collier's* was the only magazine to refuse the pseudo-editorials.

There were even more damaging facts. The advertisements had been written by several eminent writers and editors, including Isaac Marcosson, later the *Post's* famed interviewer of noted people, then on the staff of *World's Work;* Orison Swett Marden, founder and editor of *Success* magazine; and Alfred Henry Lewis, editor of *Human Life.* Others with a hand in this bogus operation included Dr. Nathan C. Shaeffer, president of the National Education Association, and Broughton Brandenburg, president of the American Institute of Immigration.

"With the exception of Lewis," Reynolds says, "Dryden presumably hired those six writers for reasons other than their ties to exposé journalism . . . yet at some point in their careers most of them participated in some aspect of the muckraking movements . . . Money may have attracted muckrakers to act as apologists for Dryden." So, one might conclude, a few muckrakers were not above creating some muck themselves, for the same reason as the corporations and their entrepreneurs—greed.

Sometimes abuses of the public were so flagrant that muckraking overflowed into magazines not ordinarily given to it. One of these abusers was the patent medicine business, a $59-million industry in 1900 when it came under fire. The companies involved in it actually spent more money on advertising than what it cost to make their products. One of the clauses in these companies' advertising contracts specified that they could cancel contracts if newspapers carrying their ads printed adverse reports about their products. For that reason, some papers refused to sign, lucrative as the contracts were.

Edward Bok, who felt strongly about the subject, had banned patent medicine advertising from the *Ladies' Home Journal* since 1892, and in 1905, the *Journal,* a most unlikely muckraker, joined with *Collier's* in a broadside attack on patent medicines. Bok's zeal carried him a little too far. In an editorial listing harmful ingredients in patent medicines, he misstated the facts about Dr. Pierce's Favorite Prescription, for which the *Journal* had to issue a retraction and pay damages. To protect against any further mistakes, Bok took on a young lawyer-journalist, Mark Sullivan, who was assigned to conduct the *Journal's* crusade within the law of libel. Sullivan produced a report so long that Bok decided it was too much for the *Journal*, but having paid for it, he realized a small profit by selling it to Norman Hapgood at *Collier's*.

That put Hapgood's magazine into the muckraking business, and the editor enhanced his position by hiring another young journalist, Samuel Hopkins Adams, to investigate the patent-medicine business more thoroughly. The result was a year-long series called "The Great American Fraud," reporting that Americans were being fraudulently relieved of $75 million every year for nostrums containing all kinds of dangerous ingredients, including alcohol, opiates, narcotics, depressants, and stimulants. It was this exposé that so roused the public that the Pure Food and Drug Act was passed by Congress.

By this time, all the muckraking magazines were hard at work. Samuel Merwin attacked the meat packers in *Success*, and Charles Edward Russell added further ammunition with an eight-part series, "The Greatest Trust in the World," in *Everybody's*. *World's Work* published a smaller exposé of the meat packers. The packers polished off in print, Russell took on the railroads for *Everybody's*. He plunged into the assignment with a literal vengeance, his father having lost his newspaper after it had dared to attack them. In ten years, Russell wrote more than 150 articles, many of them collected in books, in which he exposed a good many evils, including the railroads' nourishment of the Beef Trust.

Everybody's even sent Russell abroad to see how monopolies were operating in other countries, and he found enough new material to make a three-year series of articles confirming everything he had written before. Russell had become an ardent Socialist by this time and advocated government ownership of transportation. No one magazine could contain his advocacy. He spilled over into *Cosmopolitan, Hearst's, Success*, and *Hampton's*, but he failed to convince President Roosevelt, who wanted to regulate the railroads, not own them.

Russell did have a convert, however, in fellow muckraker Ray Stannard Baker, who also came to believe that regulation was not enough. In an article in *McClure's* "Railroads on Trial," Baker embarked on a personal crusade against special privilege. He almost became a Socialist but never joined the party, as Russell did.

Inevitably, the muckrakers were led to Congress as the source of capitalist evils. Russell had looked down on the Senate from the press gallery and saw a "chamber of butlers for industrialists and financiers. . . . a row of well-fed and portly gentlemen, every one of whom, we knew perfectly well, was there to represent some private (and predatory) interest." In fact, Joseph Keppler had portrayed them as just that in his *Puck* and *Judge* cartoons, as did Opper in Hearst's New York *American*.

Hearst himself was in Congress at the time as a representative, infrequently seen on the floor. He was going through his socialist phase, in which there was much more cynicism than real conviction. He had his eye on the presidency and planned to use his newspapers and the recently bought *Cosmopolitan* to stir up further public discontent over the trusts and big business. Consequently, he listened to Russell's ideas with interest and would have hired him, but the writer was just then leaving on his European tour. Hearst found a more than capable substitute in David Graham Phillips, a writer who was already known for flaying the rich in a novel, *The Plum Tree* (running serially in *Success*), some of whose characters were senators.

With Hearst's blessing and money, Phillips began an investigation that resulted in his sensational "The Treason of the Senate," whose first installment was in the February 1906 issue of *Cosmopolitan*. It was of no visible help to Hearst's political ambitions, but it increased the circulation of his magazine by 50 percent, and he obtained more mileage by reprinting it in his newspapers. It was the most bitter of all the muckraking attacks, and it provoked a response proportionate to the offense from those assailed and their supporters. No other muckraking series was so savagely counterattacked, and the nation found the president himself leading the rally. It was Roosevelt's anger on this occasion, in fact, that led him to go beyond "hysterical," the description of zealous journalism he had used before, and coin the word "muck-rake" at his speech for the annual Gridiron Club dinner. He repeated it later, referring to "the man with the muck rake" in an address at the laying of the new House Office Building's cornerstone.

His was one of those felicitous phrases that politicians, or their ghostwriters, sometimes manage to inject into common speech, and it may have been the beginning of a public reaction against what the muckrakers were doing. Chalmers and other historians of these years have argued that much of this reaction was justified, that the evils exposed by the muckraking magazines were not as black as they were portrayed, and that even Hearst thought Phillips's articles should have been better documented. But if Phillips was intemperate, as later historians generally agree, he did raise questions of morality in government that are still with us as urgently as they were then.

After three years, in 1906, muckraking had become something of a profession, just as investigative reporting did after Watergate. Those who practiced it, many of whom had helped Roosevelt into the White House, were hurt and angered by his contemptuous responses to what they were doing. The gauntlet had been thrown down, yet there was a tendency among some writers to be a little intimidated about picking it up. At *McClure's*, the talented stable that included Tarbell, Baker, Steffens, White, and Dunne, along with some of the editors, joined together, resigned *en masse*, and bought *Leslie's Monthly Magazine*, which they renamed the *American Magazine*. They promised it would be "a journal of uplift, looking for the hopeful as well as the underside of society." Some concessions were obviously being made.

Although he was already looking toward other speculative interests, Sam McClure was not yet ready to give up muckraking, despite this mass defection. He assigned Burton J. Hendrick, a Yale man who had been an *Evening Post* reporter, to dig into the life-insurance business again, as well as to investigate the makers of great fortunes, the railroad monopolists, and related subjects in the public interest. Besides Hendrick, McClure took on George Kibbe Turner, another ex-reporter, to look into the continuing evils of urban life, particularly political machines, crime, public service corporations, and the liquor interests.

Turner had his own solution for these problems: to instill a widespread sense of ethics in business. His fellow muckrakers did not join him in this pious hope but simply went on attacking big-city bosses and the robber barons in *Human Life, Cosmopolitan,* and *Collier's,* although in a more conservative way.

The last great wielding of the muckrake came in 1909–1910 over protection of the environment. Much of the land in America, with all its natural resources, was then in the hands of large corporate interests. American policy had always been to open up the country for private development as rapidly as possible, and the developers, ranchers, mining and lumber interests, mostly in the West, saw no reason to alter things. Roosevelt, however, was a conservationist. He called it "preservation." T. R. wanted national regulation of land management, with much of it to be held permanently by the government. But William Howard Taft was president now, and Secretary of the Interior Richard Ballinger wanted to open up public lands for development, bringing him into a historic confrontation with Chief Forester Gifford Pinchot.

The muckrakers jumped eagerly into this battle, particularly over Ballinger's intention to sell much of the public land in Alaska. *Collier's* and *Hampton's* led the assault, and it was a campaign that made the former one of the leading magazines in national influence, boosting its circulation to nearly a million. B. B. Hampton, an advertising man who had bought *Broadway,* junked its "snappy stories" and renamed it for himself, hired Theodore Dreiser as his first editor, but was predictably soon running it himself, with a formula combining quality fiction with social criticism.

There was not much left for the muckrakers to rake, however; they had exposed nearly all the problems of an expanding industrial, urban society at the upper levels, and they now turned to describing the condition of the underclass who were the victims of these broad social trends. The problems were the same as now: children, poverty, slums, courts and prisons, racism, working mothers, and juvenile crime as the result of poverty. The titles of these articles typified them: "The Kid Wot Works at Night," "Making Steel and Killing Men," "Beating Men to Make Them Good," "A Burglar in the Making," and "Daughters of the Poor." The vocabulary was only slightly different. "Underprivileged" was not yet employed; "the less fortunate" was the cliché of the times.

Oddly enough, although the country was being wracked by the greatest industrial violence in its history during these early years of the century, most of the muckrakers took a conservative attitude toward these developments. Unions and union leaders were not highly regarded. It was easier to depict the middle class as caught between big business and big labor and much safer to concentrate on prostitution and the social effects of saloons. In *Collier's,* Will Irwin, in his series "The American Saloon," called the saloon a "poor man's club, caught in a system of crime, vice, and corruption." The brewers, he said, were the villains.

Ray Baker produced one of the landmarks in magazine journalism between 1906 and 1908 when he took on, for *McClure's,* a subject still capable of arousing passions that were more than half a century old by that time. He toured the South to examine the state of race relations and found that four decades of "freedom" had not ended repression or brought about Utopia. There was warfare between poor blacks and poor whites and outbreaks of black rioting against the old injustices. Baker was the first to make such a comprehensive survey of the subject, and it was so successful that he pursued it for two years. When Gunnar Myrdal wrote his "An American Dilemma" in 1944, he praised Baker's work and acknowledged that he had drawn from it. Later historians have not been so kind, accusing him of being an accommodationist, but they do grant that he exposed institutional racism.

There was a fundamental flaw in the work of the muckrakers, however; they had no grand vision, no set of solutions for the ills they uncovered. The weight of their articles was on the sickness, not the possible cures. Looking at the trees, they failed to see the forest—the new and highly complex society evolving in America. They also failed to understand the role of the press—that is, other than their own magazines—in a free society. While the newspapers were certainly not without fault, the attempts of the magazine muckrakers to rake over their rivals were mainly unconvincing and often grossly exaggerated. And they themselves did not get off scot-free. When Upton Sinclair wrote his celebrated attack on the press, *The Brass Check*, in 1919, he included several of the muckraking journals in his indictment. Sinclair put his finger accurately on their common fault: the muckrakers were not reluctant to expose anything or anybody—but always within careful limits, sometimes related to advertising.

One of the last muckrake investigations was Will Irwin's 1911 attack on the press in *Collier's*, "The American Newspaper," in which he made much of the pressures on local papers to either ignore or soft-pedal hometown scandals and of their reluctance to attack big companies who might be big advertisers. But by then, ironically, magazines were beginning to do the same thing, not excluding the muckraking periodicals themselves.

In the critical year of 1912, when Taft conservatism lay dying, Wilsonian idealism was rejuvenating a public saturated with bad news about its institutions, and Roosevelt's Bull Moose Progressives were bellowing in vain, muckraking was plainly going out of style. The radicals were certain that "the interests" had killed it, but later historians agree that the muckrakers had simply worn out their welcome. For nearly a decade they had delivered successive hammer blows on the public skull, creating a colossal national headache, and now people were beginning to sense how good they would feel when it stopped.

The publishers were both disillusioned and nervous. They sensed the shift in public mood, and they were also acutely aware that they themselves had become big business, whose continued prosperity hinged on profit margins and bank credit. It was simply not sensible to keep on affronting the advertisers responsible for those margins and the bankers who supplied capital. They had already been hurt by advertiser withdrawals. The exposé of the beef trust in *Everybody's* had cost it soap, fertilizer, and railroad advertising. *Hampton's* and *Success* found themselves unable to get any more operating capital from their bankers. But there were other pressures: competition in an over-crowded market, for instance, and continuing rises in operating costs, forcing some of these magazines to the wall.

Most of all, however, it was the middle-class feeling that though great evils had been exposed they were now being corrected by government regulation and the feeling of most Americans that Woodrow Wilson's idealistic vision of the nation was a great deal more comfortable than constant gloom. Respond-ing to this changed climate, it was astonishing how rapidly *Cosmopolitan, McClure's, Collier's,* and the *American* abandoned contemplating the evils of greed and corruption and switched back to the old innocuous (and also suc-cessful) formula of offering good fiction, romance, pictures of pretty girls, short stories, and articles about successful men, some of whom had only recently been excoriated in the same pages.

As historians of the era have pointed out, the muckrakers may have appeared like radical angels attacking the devils of capitalism, but in fact, they were basically full of old-fashioned American optimism, believing that if peo-ple were only informed about what was wrong, everything would be corrected and the nation would resume its triumphal march. If their rulers in and out of government abused them, the muckraking magazines asserted implicitly, the people would simply throw them out. This was what Lincoln Steffens called "practical Christianity," which was neither particularly practical nor Christian. The muckrakers believed in a political morality that would keep industrialization and all its problems under control, and this peculiar kind of rationalization, along with the others, led even William James to declare that leadership in American thought had passed from the universities to Sam McClure and his fellows—a staggering thought to other, more worldly observers.

Not that the muckraking magazines were without merit. They threw a strong spotlight on the most serious of America's social ills and alarmed the government enough to cause it to make an attempt at setting up a regulatory check on laissez-faire capitalism. Further, they invented the kind of journal-ism we now call investigative reporting, thus fulfilling what the architects of the First Amendment expected the press to be: a force able to make govern-ment accountable to the governed and to provide constant fuel to the demo-cratic machine by means of free inquiry.

When the muckrakers and their magazines abandoned the field of social reform, radicalism (or social consciousness) was left in the hands of three jour-nals whose voices were much more authentic: the *Nation,* the *New Republic,* and the *Masses.* They were not burdened by advertising or any other financial obstacle, since they were as poor as the society they championed. But they took up the banners the muckrakers had dropped and added them to the pen-nons the *Nation* had been carrying since 1865, and the others, since 1911.

Something of the spirit of the *Nation* and the *New Republic* (the *Masses* was a different matter) was implicit in a preface written by one of the *Nation's* great editors of our time, Carey McWilliams, in 1965. "No one has ever 'owned' the *Nation*," he wrote. "It is impossible to own or possess it or bequeath it or sell it or mortgage it. If it ever ceased to be what it has always been, it would simply not exist—regardless of who 'owned' it or how much money stood on deposit in its name. It is an idea, a spirit, a name without an address; it is fragile, without physical assets, but it is free and it lives."

The *Nation*'s roots were deep in nineteenth-century liberalism. It was conceived by Frederick Law Olmsted and E. L. Godkin, a young Anglo-Irish writer on the editorial staff of the *New York Times*, but the original idea was Olmsted's. As early as 1863, he had written to ask Godkin if he would be interested in helping establish an American version of the London opinion journals the *Spectator* and the *Saturday Review*. Godkin, a born reformer, was enthusiastic, but Olmsted was distracted by other enterprises (he went off to manage mines in California), and Godkin had to carry on alone.

Before he left, Olmsted had introduced Godkin to his friend Charles Eliot Norton, the *Atlantic*'s founder but not yet president of Harvard, who, in turn, put Godkin in touch with a few rich Boston abolitionists, who provided both financial support and a few writers. More help came from a not disinterested source—James Miller McKim, a Philadelphian who had helped William Lloyd Garrison found the American Anti-Slavery Society and believed its magazine, the *Liberator*, needed a strong postwar successor. Besides, he wanted to help the career of Garrison's son, Wendell Phillips Garrison, who was about to become McKim's son-in-law. A Harvard graduate (1861), Wendell became Godkin's valued assistant editor.

In drawing up a prospectus for Norton's approval, Godkin laid out plans for a weekly designed to deal with current events but with "greater accuracy and moderation than are now to be found in the daily press." It would monitor the condition of freedmen, promote popular education, help investors find openings in the prostrate South, and offer criticism of literature and the arts. Most important, it would not be "the organ of any party, sect, or body." There would be such well-known contributors as Longfellow, Lowell, Whittier, and Garrison, along with such lesser-known figures as Henry James, Jr., William James, Henry Adams, and Howells. The articles would be anonymous; this old practice had not yet died.

Godkin set up shop on Nassau Street, where the New York press was then centered. Just down the street was Bryant's *Evening Post*, which predicted the new weekly would be "a boon to intelligent and thoughtful men in all parts of the country."

The first issue's first three pages consisted of short paragraphs summarizing "The Week," in much the same way the London *Spectator* does to this day (and more recently, *Punch*). This section was a forerunner of the *New Yorker*'s "Notes and Comment." The following articles covered Reconstruction, problems in Europe, a volunteer clergyman's court-martial, a poem, three pages of book advertising, six pages of literary notes and reviews, two-and-a-half pages of comment on the fine arts, and a final four-and-a-half pages of advertising for banks, insurance companies, clothing stores, and household goods. There were no illustrations.

Capitalized at only $100,000, the *Nation's* first year was precarious, but the intellectual content of the new magazine was so high that it began to have an influence far beyond its modest circulation. Olmsted returned as assistant editor, and in 1866 he, McKim, and Godkin became co-partners as E. L. Godkin & Co. There was a brief experiment as a twice-a-week periodical, but the *Nation* quickly reverted to its weekly status. Soon it became the *Evening Post's* weekend edition, in 1881.

To trace the *Nation's* early career would be to follow the twistings and turnings of Godkin's mind as he wrestled with the problems of the nineteenth century while trying to preserve his liberalism intact. Under his editorship, the magazine continued to be one of those low-circulation publications with a trickle-down influence on large numbers of people. It was quoted frequently in pulpit and press, and its ideas were sometimes the subject of debate in Congress. In 1899, Godkin resigned because of increasing ill health and went home to England to die, calling himself a "disillusioned radical." Wendell Phillips Garrison, who had been carrying the load for some time, now became editor in fact. Even without Godkin's powerful political observations, the *Nation* remained respected. One of the reasons was the addition of Oswald Garrison Villard, son of Henry Villard, the *Post's* owner, who began at once to write for the magazine and in the early 1900s turned it more and more to liberal causes, retreating from Godkin's late conservatism. Villard was particularly interested in black causes; he helped found the National Association for the Advancement of Colored People in 1909.

Garrison resigned in 1906, after forty-one years as editor. Without undue modesty, he was willing then to let the magazine die because there was no one he could think of worthy to succeed him, but Villard had his own candidate. He was Hammond Lamont, Harvard 1886, who had been a newspaperman in Boston and Seattle and had taught at Harvard and Brown before becoming managing editor of the *Post*. There is no knowing what this essentially conservative, kindly gentleman might have done because he died less than three years later after a minor operation.

Another Harvard man, Paul Elmer More, succeeded him in 1909. More was an ascetic who had studied Sanskrit, Greek, and Latin at Harvard, where he took his M.A. in 1893, after which he taught there and later at Bryn Mawr. Retiring à la Thoreau to a hut in the New Hampshire woods, he began to write the essays that occupied his literary life from 1904 to 1925, published in fourteen volumes. Meanwhile, he spent five years editing the *Nation*, writing essays for it but finding his job "a crushing grind," perhaps because he was no liberal while Villard was more and more on the opposite side. More wanted to keep the *Nation* drab in appearance and conservative in a humanistic way. The end was inevitable; More departed in 1914.

The new editor placed in charge by the *Post's* owners was still another Harvard scholar, Harold deWolf Fuller, whose Ph.D. dissertation had been on the sources and authorship of *Titus Andronicus*. Like More, he was no liberal, and he was up against a Villard determined to make a new magazine out of the *Nation*. Villard was responsible for enticing a fresh generation of writers to contribute to its pages, people like Horace Kallen, Harold Laski, and Allan Nevins. As the war in Europe gathered force, he was in strong opposition to "war madness" and believed Wilson was responsible for pushing America toward inevitable entry. The *Nation* remained pacifist in 1917, but not publicly; it would have been ruinous to do anything else but support the war effort.

Villard and the *Post* published early in 1918 a revelation of "secret treaties" which the Bolsheviks had unearthed in czarist archives and had previously published in Moscow. The *Nation* reprinted these, with comment by well-known university professors. That was too much for Fuller, who resigned, leaving the editorial chair entirely to Villard. In July 1918, the *Post* finally divested itself of its long-time weekly edition, selling the *Nation* to Thomas Lamont, Hammond's brother. Now the magazine was independent again, as it had not been since its merger, and a new era in its history began.

Meanwhile, Villard had acquired a rival in 1914 with the advent of the *New Republic*. As Arthur Schlesinger, Jr., observed in 1964, at the fiftieth anniversary observance, this magazine appeared to be the child of the Enlightenment and of Progressivism, but if so, it was an ambiguous child, like the Progressive movement itself. "Some favored acceptance and control of industrial concentration," Schlesinger wrote, "others wanted an economy of freely competing small units. Some were internationalists, others isolationists; some avant-garde, others conventional or even reactionary." Taken all in all, Progressivism was "a bundle of contradictions held together by a common middle-class ethos and by a common moral energy."

What neither the Progressives nor the *New Republic* lacked was confidence. But it was the confidence of the Eastern, metropolitan intellectuals, not the grass-roots evangelism of Robert La Follette or Hiram Johnson. That confidence was projected by the magazine's founder Herbert Croly. After graduating at Cambridge, Croly had been an editor of the *Architectural Record*, but in 1909, he published his first book, *The Promise of American Life*, one of those remarkable volumes whose sales are far exceeded by its influence. It was truly an event in American intellectual history, hailed by Walter Lippmann, who called it "a political classic which announced the end of the Age of Innocence, with its romantic faith in American destiny, and inaugurated a process of self-examination."

In it, Croly advocated the restoration of power to what he termed a disenfranchised majority—not the chimerical classless society but (almost as mythically) one in which the classes would meet on equal terms. Croly believed in a strong federal government that would complement a broad democracy.

It was this book that led indirectly to the founding of the *New Republic*. Two of its readers were Mr. and Mrs. Willard Straight, who were then living in China. The Straights were two of those occasional aberrations among the rich—people with a highly developed social conscience. They much admired Croly's book and on a trip to the United States arranged to meet him, asking

him to be chairman of a committee that would investigate educational methods. Croly accepted, and the resulting conferences led to the idea of founding
a new magazine, for which the Straights would provide the backing.

Before the launching, Croly called a meeting of those chiefly concerned and
read from a memorandum he had prepared, outlining what he intended the
magazine to be. Primarily, he said, it would be critical, concerned more with
ideas than facts. It would be different from any other magazine in its category,
past or present, because it would try "to convert criticism into a positive
agency of progressive democratic ascendancy." Croly reminded his listeners
that it was criticism of ideas and institutions that had resulted in the founding
of America, but in the hurly-burly of the nineteenth century, this spirit had
been somehow lost. The building of America had been so successful that any
criticism of it had come to seem no more than carping and without value. The
New Republic, in a fresh version of Progressivism, would argue that this condition could not always prevail, that in the new century, after a transitional
period, a different America would emerge.

Much misinformation surrounded the actual launching of the *New Republic*. Bruce Bliven set the record straight in 1935: "For many years there has
been a tradition, in New York publishing circles, that the editors of the *New
Republic* had planned to bring out the first number early in August 1914, and
that, because of the advent of the World War, the magazine was postponed
several months while everyone waited to see whether, as numerous leading
economists believed, the war would prove a business calamity in the United
States. Like so many other traditions, this one is without foundation in fact.
Work began at the *New Republic* on August 1, according to schedule, and the
first number, dated November 7, appeared on time. It is true, however, that
the coming of the war profoundly altered the editors' plans. They had
expected to deal mainly with domestic events and from a point of view
approximately that of Theodore Roosevelt's Progressives. They found instead
that they must interpret the events of a worldwide cataclysm."

There were dedicated conservatives who thought the new magazine was a
radical sheet, although not as bad as the *Nation*. But Croly's kind of radicalism
was more in the nature of intellectual questioning of Establishment ideas and
attitudes, and it was this conception he expressed in the pages of his magazine
until he died on May 17, 1930. Certainly, those around him were not radicals
in any conventional sense. Straight was a Cornell graduate who had worked
for J. P. Morgan and the State Department. His wife was a Whitney. Walter
Lippmann and Philip Littell, who became editors in time, were both Harvard
men. Walter Weyl, another editor, was a graduate of the Wharton School, and
Francis Hackett was a literary Irishman. Together, as Schlesinger observes, all
these people "proposed to patrol the fields of politics and culture, seeking
through reasoned analysis and muted wit, to raise small insurrections, as Croly
liked to put it, in men's minds."

During the early years of the war, Croly and his editors still believed that
events could be influenced by reason. They supported Roosevelt's views and
Wilson's neutrality. Finally, like all the others, they demanded participation

in the war, and when it was over, they shared the common disillusionment. They thought Wilson's journey to Paris was a voyage of betrayal. A new writer and editor, Charles Merz, on his way to a long career with the *New York Times*, took the train to Marion, Ohio, and there, on Warren Harding's front porch, inexplicably saw the future. The *New Republic* still had a long career of intellectual vicissitudes before it.

Both the *New Republic* and the *Nation* represented uptown radicalism, what conservatives would later refer to contemptuously as the work of "limousine liberals" bathing in "radical chic." They were high-minded and intellectual, but they did not represent the growing force of socialism, rising in America chiefly out of labor unrest. Even as these magazines were being established, however, something new was taking place in magazine journalism.

When the International Workers of the World (the IWW, as it was popularly known) led 20,000 immigrant workers in a strike against the Lawrence, Massachusetts, textile mills in 1912, the result was not only a successful labor action but also a stimulating boost for the cause of Eugene V. Debs, the Socialist party candidate for president, who polled an unprecedented 900,000 votes in the November election. Real radicalsim appeared to be gaining ground, although most of this progress proved to be illusory, and hundreds of magazines suddenly appeared to preach the gospel of collective ownership and control by the workers. (The IWW itself advocated open class warfare.)

One of these magazines stood out from all the others. The *Masses*, published in Manhattan, was an unlikely combination of art and radical politics. Oddly enough, it was read by some of the same people who read the *Nation* and the *New Republic:* intellectuals who championed the lower classes but did not mingle with them and who decried conditions in factories without ever having visited one. There were never many more than 20,000 on the subscription list of the *Masses*, but they were loyal readers, and the magazine itself attracted some of the best young writing and artistic talent of the day.

In fact, the *Masses* had been in existence for nearly a year before the Lawrence strike. It was founded in 1911 by a Dutch immigrant named Piet Vlag with the intention of "bettering" the working class "whether they want it or not," as he said bluntly. Its early issues mixed discussions of socialist theory with educational pieces about art and literature. Vlag did not realize he was being condescending and no doubt also failed to realize he was publishing a dull magazine. He had to fold it within less than a year, but its young contributors, who had not yet been given a chance to display their real talents, banded together and revived it. Modeling the new magazine after the satiric European journals, they gave the *Masses* a new format and a new life inside its covers. Its political viewpoint remained the same, but it was no longer "educational."

As befitted a true radical publication, the staff and contributors worked without pay, and all editorial decisions were made collectively. The first editor was a young Socialist named Max Eastman, who had written his doctoral dissertation at Columbia on Plato but later had been fired from his post as pro-

fessor of philosophy. He had been offered the editorship in a letter signed
(collectively) by John Sloan, Louis Untermeyer, and Art Young. The invitation
had been sketched out with a paintbrush on paper that looked as though it
had been retrieved from a wastebasket. Succinctly, it said: "You are elected
editor of the *Masses*. No pay."

In a capitalist society, writers for the new magazine had to earn a little
money, so they performed other journalistic chores in order to work free for
the *Masses*. Some of them wrote for the *Metropolitan*, a popular monthly that
had abruptly announced in 1912 that it was going to "give socialism a hearing" and opened its pages to these writers. Thus, John Reed went to Mexico
and reported the Revolution. He later covered the war in Europe, teamed
with the artist Boardman Robinson. Art Young supplied the *Metropolitan*
with cartoons from his Washington studio. The writers and artists were not
too particular about who paid them as long as they could contribute to the
Masses.

Eastman's magazine also had serious rivals in its own backyard. One was
the *Seven Arts*, "an expression of artists for the community," which printed
two powerful and highly controversial antiwar articles as late as 1917—Randolph Bourne's "The War and the Intellectuals" (June 1917) and "The Unpopular War," by John Reed (in August 1917). Another socialist magazine of consequence was the *Comrade*, begun as early as 1901, which operated with
much the same formula as the *Masses* adopted later. It welcomed anyone who
wanted to criticize the social order, and in those days, that would include such
people as Walt Whitman, Thomas Nast, Edward Markham, Jack London,
Maxim Gorky, Clarence Darrow, Upton Sinclair, Eugene Debs, and Mother
Jones. Even with such talent, the *Comrade* expired after only four years (it
never really got off the ground), merging with the *International Socialist
Review* in 1905.

The *Masses* had the advantage of rich patrons, the first of whom, unbeliev-
ably, had been Rufus Weeks, a vice president of the New York Life Insurance
Company. Muckraking at this level seemed to attract rich liberals. But taking
their tainted money made the editors uncomfortable. As Floyd Dell, who was
first managing editor and then co-editor, wrote later, "Our getting money
from the rich was a sort of skeleton in our proletarian revolutionary closet."

In spite of his invitation, Eastman—and Dell, too—was paid a small salary
and he actually got out the magazine. Everybody owned shares in the Masses
Publishing Company. Eastman, a strikingly handsome young man and a persuasive talker, proved to be good at raising money in crises. He told prospective contributors, and everyone else, that he wanted to "make the *Masses* a
popular socialist magazine—a magazine of pictures and lively writing." This
credo was spelled out in more detail in the January 1913 issue. The *Masses*,
it was promised, would be "a revolutionary and not a reform magazine, a
magazine with no dividends to pay; a free magazine, frank, arrogant, imper-
tinent, searching for true causes; a magazine directed against rigidity and
dogma wherever it is found; printing what is too naked or true for a money-
making press; a magazine whose final policy is to do as it pleases, and concil-

iate nobody, not even its readers—there is room for this publication in America."

For five years as editor, Eastman put out the magazine he had promised. He and the others looked upon it as "not a socialistic magazine but a magazine of free expression," in Dell's words. But while it was never a party organ, it did support anarchists, the IWW, birth control, and free love. It endorsed socialism in its broadest possible application.

Naturally, the Establishment did not welcome it. The Associated Press filed a criminal libel suit against it in 1916, and a good many libraries would not give it shelf space; nor did it appear on New York subway stands. Magazine distributors boycotted it in Boston and Philadelphia. The Canadian government refused to admit it to the country's mails.

There was, admittedly, something in the *Masses* to offend almost everyone. It advocated not only free love but also divorce, birth control, toleration of homosexuals, and sexual satisfaction for women. It attacked organized religion, particularly churchmen who upheld plutocracy, and it liked to refer to "Christ the agitator" and "Comrade Jesus." But it also covered labor news in a depth completely unknown, even to the best newspapers, and some of its stories of strikes remain classics of reporting. It came out unequivocally for radical equality and took up every cause of the Progressive era.

One of the most remarkable contributions of the *Masses* was its art work, particularly that by Young and Sloan, derived from the rough crayon drawings of Daumier, who until then was not well known in America. The magazine's editors considered him a "people's artist," whose work celebrated the virtues of the lower classes. This borrowed technique gave the *Masses* a distinctive appearance and introduced the style of Robinson and George Bellows, those realists of what came to be known as the Ashcan School. What they drew was "life in the raw," intended to look artless.

The appearance of these artists drew others to the magazine, some out of political motives, others simply interested in rebelling against society. In common, they were devoted to a magazine they considered aesthetically and morally far above the commercial publications for which they worked to make enough money to rebel.

In 1916, however, there was internal as well as external rebellion. In the spirit of socialism, the monthly editorial meetings were democratic sessions in which everyone voted on what would be in the magazine. Eastman put up the drawings on the wall and read aloud from submitted stories and poems. Naturally, there were differences of opinion, since these sessions included not only editors, contributors, and artists, but even visitors. The discussions had become angrier and compromises less acceptable.

Eastman and Dell ran the everyday operations of the magazine on a much more practical basis, understanding that the democratic process had its limitations. They set policy, did the overall planning, and meanwhile went out to raise money. At the same time, they recognized that they were completely dependent on all these disputatious contributors who were, after all, working for nothing. But it was an uneasy situation, creating a tension that exploded in 1916 when Sloan, Stuart Davis, and others led an "artists' revolt," claiming the magazine had become a "one-man" show. They charged that Eastman and Dell had established a policy of artistic propaganda, and they objected particularly to having captions added to their drawings, although other magazines had employed that technique since the late nineteenth century. Sloan and the others thought their work was being contaminated by the captions, which they considered sloganeering. The pictures were enough, they said.

The inevitable showdown arrived, and at a March 1916 meeting, Sloan moved to abolish the position of editor-in-chief, whereupon Eastman offered to resign. By this time there was no room for compromise. The editorial board reelected Eastman, and seven artists, including Sloan, walked out. Something of the sour spirit that remained was evident in Eastman's reply to Sloan's letter of resignation: "Dear Sloan: I shall regret the loss of your wit and artistic genius as much as I shall enjoy the absence of your cooperation."

Sloan left the magazine, in company with Stuart Davis, Glenn Coleman, Maurice Becker, and one writer, Robert Carlton Brown. To replace them, the board elected Robinson, Robert Minor, G. S. Sparks, and John Barber. Pursuing its course, the *Masses* displayed more political cartooning, less social satire, and fewer depictions of the virtuous proletariat. Its chief thrust editorially was to prevent, as it said, American workers from joining their European counterparts in a greedy "capitalist war." The magazine was against the draft, the armed-forces buildup, general war hysteria, and government suppression of dissent.

In the national climate, this kind of advocacy was bound to cause serious trouble. Americans already gripped by hatred of the kaiser did not welcome M. A. Kempf's cartoon depicting three nude women, representing England, Germany, and France, in the embrace of a skeletal Death immersed in a dark sea identified by the caption "Come on in, America, the blood's fine!" The public had no more tolerance for Robinson's cartoon showing Christ being shot as a deserter.

When Congress passed the Espionage Act of 1917, virtually nullifying the First Amendment, the end was at hand for the *Masses* and any other dissenters. Now it had become a crime to oppose the government, and the Post Office was given arbitrary power to prevent the mailing of anything promoting "treason, insurrection, or forcible resistance to any law of the United States." Exercising that power, the postmaster general declared the *Masses* "unmailable" and a month later revoked the magazine's second-class privileges permanently on the grounds that it was no longer a monthly publication because it had missed an issue. In December, the *Masses* had to give up.

The editors might possibly have been able to save it, but for some reason, no doubt their blinkered convictions, they did not take passage of the Espionage Act seriously. In fact, the business manager had taken an issue of the magazine to the chief censor, George Creel, who adjudged it innocent of violating the law. But the postmaster general did not agree, and even though the magazine got a temporary stay in the courts, it was too late, and the government vindictively invoked the conspiracy clause in the Act. Even then, the editors could not believe what was happening to them.

But the full seriousness of the situation struck home when Eastman, Reed, and several others were arrested and charged with conspiring to interfere with the draft. The trial did not take place until April 15, and by that time, the editors were producing a new magazine called the *Liberator,* which backed Wilson's prewar views, advocated a negotiated peace, and seemed to be mailable.

There were two farcical trials, both resulting in hung juries. Judge Learned Hand presided at the first one, and the defendants were Eastman, Merrill Rogers, Dell, Young, and a young poet, Josephine Bell, who, as it turned out, had never met her alleged co-conspirators. They were defended by Morris Hillquit, a Socialist who had run for mayor, and one of the city's most distinguished lawyers, Dudley Field Malone. Outside the Federal Building, in City Hall Park, patriots gathered to sing patriotic songs.

It was one of the more unusual trials in legal history. Eastman, on the stand, discussed the nature of patriotism for nearly three days, while Dell favored the jurors with his views on war, militarism, conscientious objectors, and related topics. Socialists who heard Eastman were outraged, charging that he had betrayed them by his testimony that his editorials were no different from Wilson's prewar views and that he had changed his mind about the war after American entry; he had gone on to advise the prosecutors that they would be better employed chasing down spies, profiteers, and supporters of Prussianism.

Art Young had trouble staying awake during the trial, and after being prodded to consciousness, he drew a sketch of himself asleep, with the caption, "Art Young on Trial for His Life." On the stand, he was unshakably good-natured. When the prosecution introduced his cartoon "Having Their Fling," in which press, pulpit, politics, and business were shown in a hysterical war dance, with the Devil leading an orchestra playing on war instruments, he said he had drawn the Devil as leader because General Sherman had said, "War is hell."

In the end, hell saved the defendants, in a sense. Only one juror held out for acquittal, assuring his fellows that he would not change "till hell froze over." This holdout permanently excused him from jury duty in New York City, although he had served regularly before.

By the time the second trial took place, the war was over and the national climate had improved. John Reed even came back from Moscow for the occasion to stand trial with the others, an act dismaying Leon Trotsky, who considered such behavior un-Marxian. Leslie Fishbein, who has studied these "Rebels in Bohemia," summarizes the second trial: "Max Eastman defended the antiwar St. Louis declaration of the Socialist party, Floyd Dell praised conscientious objection, John Reed justified the class war and recounted his experiences under Bolshevism, while Art Young expressed his disapproval of all wars."

Once more the courtroom drama had its farcical moments, one of which

may have nullified Prosecutor Earl Barnes's final eloquent, partriotic perora-
tion to the jury. Barnes was concluding, "Somewhere in France a man lies
dead. He is but one of a thousand whose voices are not silent. He died for you
and he died for me. He died for Max Eastman. He died for John Reed. He
died for Merrill Rogers. He demands that these men be punished." At that
moment, Art Young, who had appeared to be taking one of his usual courtroom naps, roused himself and cried, "What! Didn't he die for me, too?" With
that, the jury retired and disagreed, this time eight to four. Judge Martin J.
Manton had to dismiss this case also.

Judge Hand had issued a temporary restraining order against the Post
Office on behalf of the *Masses*, but the postmaster general still refused to per-
mit access to the mails. However, after he had denied second-class privileges,
Circuit Judge Hough supported him and stayed the injunction (although the
Circuit Court of Appeals eventually, though too late, reversed Hough's
decision).

That was the death blow for the *Masses*. It was not the death of the revolutionary impulse that inspired it, however, and the *Masses* rose from the
ashes to appear again in the 1920s. As David Oshinsky has written, it "existed
in a special era for American radicals, an era free of Bolshevik orthodoxy, a
time of innocence, high hopes, and unlimited possibilities. It could explore the
parameters of radical culture and politics with a spontaneity and humor
denied to its successors."

11

Black Periodicals
Between Two Wars

None of the black publications begun before the Civil War survived it, and it seemed for a time as though they would have great difficulty starting again. The hopes of Reconstruction faded quickly as blacks saw that the end of slavery did not necessarily mean freedom and that legal and social inequalities were going to be maintained. But such adversity could only provide the same primary reason that had led to the rise of black magazines before the war—protest against injustice.

One by one, they began to appear once more, in a reprise of the antebellum scene, with some of the same people and institutions involved, for the same reasons, and operating in much the same way. Writers who had contributed to the earlier journals surfaced again, with Douglass and Frances Harper emerging as the most prolific. Like those before the war, many of the new magazines had institutional support. Others, issued by editors and publishers alone, were largely concentrated in the Northeast, although these independents also sprang up in the South and West. But the ones with organizational support survived longest, some of them well into this century, including the A.M.E. Church Review, founded in 1884; the A.M.E. Zion Quarterly Review (1890); National Notes (1897), published by the National Association of Colored Women; and the Journal of the National Medical Association, begun in 1909.

Those surviving without institutional support included the Colored American, Howard's Negro-American, McGirt's, Alexander's, the Voice of the Negro, and Horizon. Conditions had improved somewhat for those who published and contributed to these journals. Some editors were paid salaries, and a few even began paying contributors if they were people the magazine particularly wanted. For example, such writers as Charles W. Chesnutt and Mary Church Terrell commanded modest sums from the Voice of the Negro.

While the distribution process was improving rapidly for white publications in the postwar world, the black journals still had to struggle. In cities, they had to rely on subscription agents who were not always reliable, and the old problem of delinquent subscribers was not much improved, even though subscription prices remained generally at the old dollar level for most. It was still nearly impossible to get enough paid-up subscriptions to make publishing profitable.

This situation compelled black publishers to follow the lead of their white counterparts at the turn of the century and to pursue the advertising dollar, a pursuit that had seemed futile in earlier days. It was a do-or-die effort, and those journals that did not join the hunt aggressively soon expired. The advertising came from black businesses and institutions primarily, but now the publishers reached out to white corporations as well, getting advertising from such companies as Coca-Cola, Underwood, various railroads, A. & P., Doubleday Page, and the Century Manufacturing Company. Following along in time were Fels-Naptha soap, Runkel Brothers breakfast cocoa, and a company making detachable bicycles.

As before, however, it was the church that played a major role in black magazine development. Ministers served as editors, defining what they considered important issues. But inevitably, as debate and controversy increased, secular voices were raised too, and arguments over what to do about the plight of blacks filled the pages of all the magazines, regardless of sponsorship. The debate over the best course for blacks to pursue generated its own schisms in and out of the churches.

The strongest church voice was the *A.M.E. Church Review*, a magazine of such vitality that it survives today. It intended, as it declared, "to produce a periodical that would give to the world the best thoughts of the race, irrespective of religious persuasion or political opinion." As time went on, these "best thoughts" encompassed biography, poetry, history, religion, fiction, art, economics, and social issues. The emphasis changed from editor to editor, but the magazine commanded respect as a broad-based general periodical, publishing some of the best black writing and thinking.

A notable event of 1890 was the appearance of the first magazine published in the South by blacks, it was called *Southland*. Two schoolteachers, Joseph Charles Price and Simon Green Atkins, were publisher and editor respectively. Atkins had been a founder of the North Carolina State Teachers' Association, a co-editor of that organization's journal, *Progressive Educator*, and active in the education department of the A.M.E. Zion church. Price had trained to be a minister, but after serving some time in London as a delegate for the church, he returned to the South to become president of Livingstone College in Salisbury, North Carolina. In 1890, he was elected leader of two national organizations for blacks, but at the apex of his career in 1893, he died of Bright's disease.

For *Southland*, Price and Atkins devised the motto "Not the Old South or the New South, but the Southland as it is and ought to be." That included, as its pages disclosed, articles on education, current events, reprints of articles about race relations, a women's department, and politics. Its approach to Southern racial problems was generally conciliatory, relying on the tenets of Christianity to achieve a society where blacks and whites could live and work together peacefully.

That issue, naturally, was increasingly important to most blacks, but there was disagreement. Should the policy be accommodation to the status quo and working within the system for betterment? Or should there be continuing protest against inequities and oppression? Booker T. Washington had emerged as the chief accommodationist, and his views had attracted the support and financial backing of powerful white leaders. To advance his beliefs, Washington sought to get positive coverage of them in the media and even tried to exert direct influence on some newspapers and magazines, sometimes to the extent of acquiring partial ownership.

Washington was not above applying pressure on editors and publishers. Abby and Ronald Johnson, historians who have explored this aspect of his life, assert that by 1900 the black leader "essentially controlled *Alexander's Magazine* and *Colored Citizen* of Boston, the *Colored American* of Washington, and, among others, the *New York Age*, all of which propagandized his own approach to race relations." Eventually he gained complete control of the *Colored American* and attempted to influence, both directly and indirectly, the positions of the *Voice of the Negro*.

Inevitably, a small but growing number of blacks rose up to oppose Washington—including among the magazine writers and editors W. E. B. Du Bois, Pauline Hopkins, J. Max Barber, Charles Chesnutt, and Paul Laurence Dunbar. They pointed to the continuing lynchings, suppression of political rights, and segregation, declaring that accommodation would achieve nothing except to establish blacks as second-class citizens. Blacks must speak out and protest, they said, and defend their rights. At the turn of the century and on into the twentieth, black magazines were preoccupied with the struggle between these two radically different points of view.

Two of the journals with the largest circulations were caught up in this fierce debate before the Great War introduced other issues. The *Colored American*, with a peak circulation of 17,840, and the *Voice of the Negro*, with 15,000, found themselves at the center of the controversy. The former, appearing in 1900, was published in Boston by the newly formed Colored Cooperative Publishing Company. One of its founders, Walter Wallace, had started as the magazine's editor, but Pauline Hopkins, one of his most talented staff writers, became editor about 1902 or 1903. (Like many other journals, the *Colored American* did not display the name of its editor on the masthead in any consistent way.) Hopkins had been hired to run the women's department, but she was so evidently talented and prolific that her rise to the top was rapid.

Founded with the intention of providing black writers and scholars with a forum for their work, it was natural that this magazine should develop a strong feeling of racial pride and solidarity among its readers. As a general magazine, it printed the usual mixture of politics, history, biography, and travel, as well as fiction and poetry. Hopkins herself contributed three serialized novels and ten short stories.

Another Bostonian, William Stanley Braithwaite, served as the journal's book editor and later went on to become a nationally known critic. Other outstanding contributors were Thomas Fortune, a rising journalist; the dean of Howard University, Kelly Miller; and the novelist Sutton E. Griggs.

But in spite of the efforts of all these talented people, the *Colored American* lost money. Subscription agents paid slowly, the Co-operative's stock failed to sell rapidly, and it was soon necessary for three sympathetic Bostonians—William Dupress, John F. Ransom, and William G. West—to rescue the sinking journal. They bought both the publishing company and the magazine. The new management retained Hopkins as editor and insisted the magazine remain in Boston, but their efforts failed too, and the New York publisher John C. Freund along with Fred C. Moore, who was secretary and organizer of the National Negro Business League (founded by Washington), bought the property. This brought the journal under Washington's influence.

That acquisition was a forecast of modern times. Washington, looking for outlets to publish his views, had bought a few shares in the Co-operative in 1901 but sold them in an effort to divert attention from his rumored attempts to influence the media. When Moore bought the company, however, $3,000 of the $8,000 purchase price came from Washington's pocket; Moore provided the remainder. Washington's participation was kept secret, although readers no doubt wondered why the magazine printed so many laudatory articles about him and his institute as time went on.

From 1904, when the purchase was made, until 1909, when the *Colored American* expired at last, Moore depended on Washington for regular contributions to the magazine's coffers. In 1909, however, Washington decided that this journal would never attain the prominence and influence he required to advance his views and withdrew his support, occasioning its prompt death.

Moore had published the magazine in New York, where he already owned the *Age*. Hopkins lost her job in the move there because her activist views were contrary to accommodationist theory. Readers were told that she had left because of ill health, but those who followed her work found her again in the pages of the *Voice of the Negro*.

Her departure hurt the *Colored American* badly. Fiction continued to appear, but it was of lesser quality, and the articles on literary criticism and theory virtually disappeared. The tone of the journal was now unmistakably accommodationist, reflecting the hidden influence of Washington, and consequently did not have the bright edge Hopkins had given to it. Nor was Moore able to attract new talent of any consequence. Black writers might not be aware of Washington, the hidden angel, but they could see for themselves what the magazine had become.

Voice of the Negro, the deceased *American's* chief rival, was begun, oddly enough, by a white man, A. N. Jenkins, a manager in a subscription–book publishing firm. He later brought in a partner, John A. Hertel. It was Jenkins's job that led him to this unlikely venture. As manager of the Atlanta branch of the J. L. Nichols Company, he handled the products of the company's division devoted to books designed for a black audience. One of its best-sellers had been Washington's first autobiography, *The Story of My Life and Work*. It was this connection that inspired Jenkins's venture.

He hired as editor John Wesley Edward Bowen, a well-known and much-respected professor of historical theology at Atlanta University. As managing editor, Jenkins took on J. Max Barber, recently graduated from Virginia Union University in Richmond, where he had edited the school newspaper. Since Bowen did not give up his university position and did his editing on the side, Barber became the de facto editor. His first issue appeared in January 1904.

The historian Louis Harlan has written that of all the black magazines appearing in the first decade of the twentieth century, this was "the most promising. . . . As its name suggested, it attempted to be the voice, or at least the forum, for all Negroes. Its signed articles scanned the whole range of black ideology, from the back-to-Africa nationalism at one extreme to the immediate integrationists at the other, with all shadings of conservative and militant attitudes in between."

There was some dissent on the staff. Bowen tended to admire Washington and his accommodationist views as well as his advocacy of industrial education for blacks, but Barber took a more radical position. In that situation, the magazine tried to steer a middle course at first, presenting both sides of all issues, but the first publications also seemed to welcome diverse ideas, as long as they were presented in a moderate way.

This approach did not please Washington, of course. His secretary, Emmett Scott, was infiltrated as a part-time associate editor of the *Voice* in the hope that this connection, as well as Washington's good relations with Bowen, would be enough to swing the magazine in his direction. Barber, however, resisted all of Scott's attempts to carry out his mission and refused to budge. When Washington complained to the publishers, they replied by telling him that they had evidence that he was part owner of the *Colored American*, the *Voice*'s chief competitor, and there was an obvious conflict of interest. Washington, of course, denied everything and did not give up. When Barber wrote editorials indirectly criticizing Washington and his policies, the powerful leader used his influence to persuade one of the magazine's chief advertisers, the Afro-American Realty Company, to cancel its contract.

Barber effected a reconciliation of sorts, whether deliberately or otherwise is not clear, by visiting Tuskegee and writing glowing articles about Washington's institute. This could only have pleased Washington. But meanwhile, he was also involved with organizing the Niagara movement, a radical intellectual organization which advocated militant methods and the use of civil rights to achieve black equality. A full account of the movement's first meeting appeared in the *Voice*; Barber had been there at the creation in Niagara Falls. Subsequent issues gave readers further information about the movement and W. E. B. Du Bois, its leader. Presumably, this seriously diminished any Washington goodwill generated by the Tuskegee articles.

The *Voice* had other critics as well. Some blacks complained because it was owned by white publishers, who by 1906 had concluded that the magazine was costing too much money and showing too little, if any profit. They withdrew their support. A desperate attempt to salvage the property was made by organizing the Voice Publishing Company and offering its shares to the public, which showed no particular interest.

In the end, however, it was a chance event that had nothing to do with its finances that ended the *Voice*'s career. There had been race riots in Atlanta in September 1906, and Barber responded to them by writing an outraged anonymous letter to the New York *World*, accusing Atlanta newspapers and politicians of employing yellow journalism and sensational methods to increase tensions and help bring on the riot by whipping up anti-black feeling. Atlanta officials were somehow able to find out that Barber wrote the letter, and they gave him a choice—serve on a chain gang or leave Atlanta.

Barber departed for Chicago, taking the *Voice of the Negro* with him. There he changed the magazine's name simply to *Voice* and sought new backers as well as advertising. Members of the Niagara movement offered him some support, but it was not enough, and eventually Barber had to sell his magazine to Thomas Fortune. But he, too, could find neither capital nor advertisers, and that proved to be the end. Barber decided to get out of journalism, blaming Washington for all his problems, and became a dentist as well as an early member and director of the N.A.A.C.P.

At this time, Du Bois was a professor in Atlanta University, where he founded several journals designed to demonstrate the necessity of a national magazine for blacks. He was convinced that such a journal, along with other magazines and newspapers devoted to black interests, would stimulate racial pride and create a sense of self.

One of Du Bois's ventures was the *Moon Illustrated Weekly: A Record of the Darker Races*, published in Memphis at a printing plant he himself had founded in partnership with two graduates of Atlanta. This periodical was heavy with reprints from other journals, but it also had original pieces, some of them taking up the old theme—opposition to Booker T. Washington. *Moon* lasted only half a year, but Du Bois learned about magazine publishing from the experience and, with two associates, founded a new magazine, *Horizon: A Journal of Color*, in January 1907. It was published in Washington, D.C., where the two new editors, Lafayette McKeene Hershaw and Freeman Henry Morris Murphy, lived and worked. Du Bois wrote a good part of the contents of *Horizon*, with added contributions from the editors. The three contributed sections titled "The Over-Look," "The Out-Look," and "The In-Look."

Du Bois refused to accept any organizational help for his venture, as a matter of principle, with the result that the editors often were able to keep the magazine alive only by dipping into their own pockets. They did so because they were deeply committed to what the magazine stood for—consistent and unremitting protest against racial injustice. In 1910, Du Bois wrote, "We are called agitators in the sense of irresponsible persons who get their chief amusement and their daily bread by making a noise; yet we must remember that some of the greatest movements in the world's history have been led by men who were also called agitators, and who were agitators in the sense that they tried to arouse the conscience of a nation or of a group to certain persistent wrongs."

Sadly, Du Bois wrote those words in the same year that *Horizon* ceased publication, not because of financial problems alone but also because Du Bois had become director of publications and research for the N.A.A.C.P., a job that left him no time for magazine editing. However, in his new post he was in a position to persuade the Executive Committee of the organization that there was a great need for a national magazine. Agreeing somewhat reluctantly, the committee pledged to provide fifty dollars a month for such a venture. With this commitment, Du Bois founded a new journal, *Crisis: A Record of the Darker Races*, the longest-lived of all the earlier black publications. He remained as its editor through 1934. Organizational backing ensured its survival.

In the first issue of *Crisis*, Du Bois offered this preview: "The object of this publication is to set forth those facts and arguments which show the danger of race prejudice, particularly as manifested to-day toward colored people." That was clear enough, but what remained unclear was whether *Crisis* would be the N.A.A.C.P.'s house organ or a personal vehicle for Du Bois. On this point, the editor found himself in opposition to the views of Oswald Garrison Villard, who had become chairman of the Executive Committee. As editor of the *Nation*, Villard appeared to think that this prestigious position should give him authority over the *Crisis*. There was a brief contest of wills that ended when Villard resigned from his chairmanship. Du Bois remained in firm control during the 1920s, when the magazine began to lose circulation and develop an alarming deficit.

The reasons for the decline of *Crisis* were not difficult to find. When it was fresh and new, the magazine had attracted a steadily growing readership, greater than any black journal had yet known, rising from 9,000 in 1911 to 35,000 by 1915. This increase was considerably helped by the fact that new N.A.A.C.P. members subscribed routinely. By the twenties, however, the black newspaper press had expanded and improved, offering serious competition, to the point where the organization had to subsidize the *Crisis*, thus weakening Du Bois's control of it. He resigned in 1934, but the magazine has continued into the present.

During Du Bois's editorship, the *Crisis* fought particularly hard against lynching, but it also hammered away on increasing educational opportunities for blacks as well as suffrage for all women. Culturally, the magazine paved the way for the literary renaissance of the twenties by publishing the works of such new writers as Jessie Fauset, Mary Effie Lee, Otto Bonahan, Fenton Johnson, Georgia Johnson, and Roscoe Jamison. Walter C. Daniel, a scholar who has examined the career of the *Crisis*, believes that Du Bois was able to build the magazine into "the single most identifiable vehicle for American black expression of the national conscience."

There were a few other black magazines of importance during this period, although they were mostly short-lived and of less consequence. In Pittsburgh during 1889 and 1890, James H. W. Howard started *Howard's Negro-American* magazine. This man of many interests was involved in both government and business, as well as journalism. In 1895, he started a second periodical called *Howard's American* magazine, moving it to New York City in 1901. Both these journals published fiction, nonfiction, and poetry and were illustrated. The *Negro-American* talked of politics and education and had a woman's page.

Its sister publication had a long, explanatory subtitle: "Devoted to the Educational, Religious, Industrial, Social and Political Progress of the Colored Race." Under this rubric, it printed articles by such well-known figures as Washington and John Wanamaker and by lesser-known writers, such as Ida Well Barnett and Mary Church Terrell. Howard made a mistake in moving this periodical to New York, however; it expired there in 1901. Later, in 1912, Howard became editor of *New Era* magazine, sponsored by the National Colored Democratic League.

A Philadelphia publication called *McGirt's* magazine began in 1903 and remained alive for six years. It was established primarily as a showcase for the short stories and poetry of its publisher, James McGirt, but it was also the official magazine of the Constitutional Brotherhood of America, an organization devoted to helping blacks organize as voting blocs to gain power. *McGirt's* promoted that purpose in its columns and, besides the owner's work, carried pieces by such writers as Du Bois, Kelly Miller, Terrell, Frances E. E. Harper, and Dunbar.

Another journal subsidized by Washington was *Alexander's* magazine, which was little more than a platform for Washingtonian ideas. Its editor, Charles Alexander, had previously published the short-lived *Monthly Review* in 1894 and after it failed had taught printing for ten years at Tuskegee and Wilberforce University. During that time, he came under the sponsorship of Washington, who sent him first to Boston, where he failed to make a success of a newspaper called the *Colored Citizen*, and then switched him to the magazine business.

Alexander described the magazine he edited as conservative, and so it was, promoting not only Washington's views but also teaching a nonactivist "doctrine of optimism." Its particular virtue was its coverage of black educational institutions. Briefly, it advocated emigration to Liberia until one of its editors, Walter F. Walker, visited the country and found it far from being a promised land. The remainder of its content was the usual mix of general magazine topics by more or less well-known writers, but there were more pieces by Alexander than by anyone else, including the most prominent writers. The editor wrote prolifically for other magazines as well and produced books in the bargain. But *Alexander's*, even with Washington's support, could not survive the usual financial problems and folded after four years.

In his magazine, Alexander had written frankly about the kind of problems, financial and otherwise, that a black journalist was compelled to face: "He must be prepared to labor under great difficulties and embarrassments for fully 18 hours out of every 24, make great sacrifices, meet with all sorts of discouragements, and sometimes suffer humiliation and deprivation in order to accomplish his purpose."

In contrast to *Alexander's*, the *Messenger*, founded in 1917, began as a socialist journal, edited by A. Phillip Randolph and Chandler Owen, who were also its founders. In their opening number, these editors wrote, "Our aim is to appeal to reason, to lift our pens above the cringing demagogy of the times and above the cheap, peanut politics of the old, reactionary Negro leaders." The latter barb was aimed not only at Washington but also at Du Bois, whom they attacked directly. The magazine's content, both articles and fiction, was radical in viewpoint.

Ironically, by the twenties both Randolph and Owen were involved in the politics they had disdained and had to delegate their editorial responsibilities to others, particularly George Schuyler and Theophilis Lewis. These editors brought more variety into the magazine, publishing works by Countee Cullen and Langston Hughes, among others. Lewis wrote a regular drama-criticism column titled "Theatre Souls of Black Folks," frequently advocating the establishment of an independent black theater and urging black playwrights to draw on their own culture for material.

From 1925 to 1928, the *Messenger* was the official organ of Randolph's Brotherhood of Sleeping Car Porters, but it continued to print a wide variety of nonunion pieces, including work by Zora Neale Hurston, Arna Bontemps, and Claude McKay. Some articles urged blacks to form labor unions. In the end, however, the *Messenger* became the victim of what had caused so many organization-sponsored publications to fold. Randolph decided that it was taking too much money from the union's treasury and suspended it in 1928.

There were a few small magazines devoted to special interests in this period, most of them reflections of a segregated society. They were directed to both personal and professional interests, and, not surprisingly, most of them dealt either with education or religion. Seven women's magazines emerged, and there were specialized journals covering music, theater, medicine, business, and agriculture. Many were organization-sponsored, others were started by entrepreneurial journalists, but all enjoyed relatively brief lives. While that could be said for white magazines too, black journals had their problems in an aggravated form, and it is a tribute to conviction, courage, and perseverance that so many appeared and survived for a time.

12

Advertising and Circulation: The Establishing of a Magazine Business

A transformation of the magazine business occurred in the last decades of the nineteenth century as a mutually beneficial relationship between the periodical and advertising industries took root and flourished. Nationally distributed journals could now offer manufacturers an effective way to advertise their goods nationwide, and the resulting abundant advertising dollars permitted publishers of periodicals to lower their subscription rates as well as their cover prices. Their pages were crowded with a wealth of stories, articles, and illustrations by the best-known writers and artists.

Magazine historian Theodore Peterson summarizes this transformation: "Advertising in large measure made possible the low-priced magazine of large circulation which emerged during the last years of the nineteenth century." Publishers could sell their products for ten cents, or even five—less than the cost of producing them—and still make huge profits from selling space for advertising. With subscription lists in hand, these entrepreneurs provided the means for manufacturers to reach large numbers of customers. Week after week, month after month, they placed their brands and trademarks before the eyes of millions of readers.

By the turn of the century, circulation and advertising revenues had reached unprecedented heights. Kodak cameras, Campbell's canned goods, Ivory soap, Uneeda biscuits, and a variety of other products were capturing the imagination and pocketbooks of the American public. Between 1890 and 1918, magazine advertising increased in both volume and dollars. The trade magazine *Printer's Ink* prophesied in 1915, "When the historian of the Twentieth Century shall have finished his narrative, and comes searching for the subtitle which shall best express the spirit of the period, we think it is not at all unlikely that he may select 'The Age of Advertising' for the purpose."

A few statistics tell the story of advertising's growth and its effect on magazines. Daniel Pope, an advertising historian, has estimated that between 1890 and 1914, the amount spent on advertising in the United States rose from $190 million to $682 million. Much of this went into magazines. The *Saturday Evening Post* became the leader in advertising revenues, which jumped from $6,933 in 1897 to $1,266,931 in 1907—the first decade of George Horace Lorimer's editorship. Revenues had topped $16 million by 1917.

The Curtis Company, embracing both the *Post* and the *Ladies' Home Journal,* saw its advertising revenue climb from about $500,000 in 1892 to almost $23 million in 1917 (including the *Country Gentleman* by this time). This kind of growth had become the order of the day as the amount of advertising in magazines ballooned without restraint. By 1907, advertising comprised more than 50 percent of an average issue of a leading monthly magazine. Ten years later, the Publishers Information Bureau estimated that gross national advertising revenues in general and farm magazines had reached almost $60 million. These burgeoning revenues led to lower prices, which in turn yielded ever higher circulation figures.

Circulations of nationally distributed consumer magazines continued to soar. Curtis's *Journal* passed the million mark in 1903, and other periodicals were not far behind. By 1918, the *Post* had two million readers and was in top position, where it remained for some time. Several factors coincided to make such progress possible: low prices, efficient distribution, relevant editorial material, and attractive illustrations.

Greatly increased advertising revenues were saving many magazines from the financial disasters that would surely have overtaken them in earlier days. The money made it possible not only to survive but to look forward to a much longer life, since it was now possible to compete for higher quality stories and illustrations. Publishing, in short, was becoming a lucrative business after a long history of boom and bust. Fortunes were being built from magazines by Cyrus Curtis, John Crowell, Joseph Knapp, and others. Inevitably, as these publishing empires grew, they no longer remained personal enterprises but became big and more impersonal businesses.

By the time America entered the Great War, a fundamental shift had taken place in the financial structure of periodical publishing. With advertising revenues underwriting a healthy percentage of costs, more attention had to be paid to attracting and pleasing advertisers and their representatives, the agents. Publishers consequently found themselves directing their marketing efforts to two different customers: readers *and* advertisers.

While advertising had appeared in magazines from the beginning, most antebellum journals had carried little of it. What did appear were little more than announcements with black print providing basic information about the location and the price of items for sale. These notices usually described only local products and services available. Circulations, too, were limited, with readerships numbering at best in the tens of thousands. When *Godey's Lady's Book* reached a peak of 150,000 during the 1860s, it was considered an astonishing figure.

After the Civil War, circulations began to rise, although the increase was slow by comparison with the later growth of the mass-market magazines in the last decade of the century. A few journals consistently maintained large circulation figures. The *Youth's Companion,* with 385,000 readers by 1885, held substantial figures for much of the late nineteenth century, while *Harper's* led the older quality journals with a paid readership of nearly 200,000 by 1891. Mail-order journals, essentially advertising catalogs, achieved the highest circulations in this period. The *People's Literary Companion* attracted huge readerships, and *Comfort* went past a million before the end of the century. They were part of a general boom in magazines, with the total number increasing from about 700 in 1865 to 3,300 in 1885.

With such large figures, many periodicals could claim to be reaching a national audience, an expansion consistent with the general industrializing of America. With the tremendous expansion in manufacturing, producers sought ways to inform customers about the availability and desirability of their products, and obviously, magazines were the medium designed to carry these messages to a national audience. They could do that because better rail transportation was reaching into every part of the continent, postal rates were lower, great improvements in print technology were being made, and waiting for the messages was a literate and interested audience. It was a marriage made, if not in heaven, at least in the counting houses of the magazine publishing business.

The new relationship between the journals and advertising in the decades after the Civil War involved three players: magazine publishers, manufacturers wishing to advertise, and advertising agents. This three-way relationship has been the basic structure of the business ever since, with recurring shifts in the balance of power.

Many producers had been reluctant to advertise their wares at first. Patent-medicine manufacturers, dominating advertising for years, were seen as charlatans by respectable businessmen, who were reluctant to have their companies appear on magazine pages. It was the advertising agents (and a few publishers) who took on the role of encouraging manufacturers to advertise in periodicals, telling them that such an action would boost their profits. This sometimes not so gentle persuasion brought soap, sewing machines, organs and pianos, typewriters, seeds, and farm tools into magazines. The targeting of special audiences made the medium all the more attractive, since this was something newspapers did not learn how to do for some time. Because of that advantage, agricultural and religious journals carried much of the first advertising, soon followed by household and women's periodicals as well as the mail-order specialists.

By the turn of the century, many manufacturers of new, brand-name goods were advertising extensively in magazines, promoting their own distinctive labels. By appealing directly to consumers (what would later be called a "pull" technique), manufacturers hoped to force retailers to stock their goods. These early advertisers had simple goals: to get their brands known, to provide information about new products, and to create demand.

An early advertiser, the Royal Baking Powder Company, well understood the value of its trademark. In 1915, it set the value of that symbol, well known to the public because of extensive advertising, at $1.5 million a letter. Those companies that had not been so astute and had stopped spending on advertising to keep their names before the public soon realized their mistake. For some, understanding came too late. Both Bozodent and Rubifoam tooth powders lost sales and went out of business after they stopped advertising. The same fate overtook Egg-O-See and Force breakfast foods when they dropped their campaigns. Even Pear's soap, a large advertiser in the first decades of the twentieth century, whose product was well known to consumers, saw its sales decline sharply when it inexplicably ceased advertising. But nearly all businesses in time came to understand the power of national magazine exposure that familiarized the public with brand names and reminded consumers constantly that the products existed.

There were many magazine publishers at first who also felt reluctant about carrying advertising, fearing it would lower the quality of their publications. George Rowell, perhaps the first great advertising agent, recalled how Fletcher Harper turned down the $18,000 offered by the Howe Sewing Machine Company to advertise on the last page of *Harper's Monthly* for a year. He preferred to follow the firm's policy in the 1870s and 1880s of using blank pages in its books to advertise its own products. Similarly, the Butterick Company, publishers of several fashion journals, carried advertising primarily for Butterick products—patterns, thimbles, scissors, and other accessories.

Most of the general advertising that did appear in magazines was clumped together in the back pages. The messages were printed in small black type, relieved only occasionally by an illustration. Even publishers who had opened up their pages to advertising in the years before the century ended did not often go beyond this pattern.

In retrospect, it is surprising that more publishers did not see immediately how logical it was to view magazines as ideal vehicles for promoting all kinds of products. One of the earliest to understand was Cyrus Curtis, who joined such mail-order kings as E. C. Allen in using advertising during the decades after the Civil War. Frank Presbrey, an early chronicler of advertising history, calls Curtis "the chief figure in modern advertising." He and Allen were joined in the nineties by such smart publishers as Frank Munsey, S. S. McClure, and John Brisben Walker.

Others took longer to see the light. The high-quality monthlies—*Atlantic, Scribner's,* and *Harper's*—had accepted a little advertising for years, but not until they were pushed by new competition did they actively solicit it. An exception was the *Century*. Business manager Roswell Smith's aggressive efforts to acquire advertising for its pages were not only successful but helped

eventually in giving messages in this journal, and elsewhere, an air of respectability. By the end of the century, most general magazines, including the older
journals, actively sought advertising.

By the time the Great War began, the selling of products had become a
necessity for most journals. This transformation was not lost on readers, and
not all of them were happy about it. As early as 1909 Finley Peter Dunne's
Mr. Dooley could put a reverse spin on what was becoming a common complaint: "Th' first thing ye know there won't be as many pages iv advertisin' as
there are iv lithrachoor. A man don't want to dodge around through almost
impenetrable pomes an' reform articles to find a pair iv suspenders or shavin'
soap."

But in spite of Mr. Dooley's fears, advertising was here to stay. Those journals able to attract it to their pages were demonstrably the best able to survive,
and consequently, it was coming to play an increasingly significant role in the
financial health of publications. Less weight was given now to the revenue
from subscriptions. The shifting emphasis of these elements might vary from
magazine to magazine, and even generally (in time) from decade to decade,
but the basic structure had been planted for good.

In the world of national magazine advertising, agents were performing a
new and crucial function. With booming growth both in the media and in the
developing national markets, an intermediary was clearly needed. The time
had long since passed when placing an advertisement was an intramural
affair, with tradesmen bringing their messages to the local newspaper, the editor knowing how much credit to allow him, and the advertiser easily able to
check on placement. Audiences and markets had expanded far beyond local
communities, making such cozy arrangements impossible.

The lack of any standard measure of advertising space compounded the
confusion of the new advertisers. Prices for space varied wildly, while intricate
rules offered reductions for repeat placements and quantity discounts. Publishers usually had rate cards, but they were often open to negotiation. Rowell
noted in his memoirs that " . . . advertising space had at that time no recognized measure or standard value. Practically, within certain limits, it
amounted to getting as much as possible and taking what one could get; and
my memory does not remind me that those who paid a low price for a large
space were, as a rule, any better satisfied than those were who paid a higher
price for a smaller space." Rowell recalled a publisher who "devoted four full
pages to impressing upon us the necessity of paying according to his schedule,
and said the price would be the sum he named, 'and not one d———d cent
less' and added in a postscript, 'Now what will you give?'"

Problems also existed for publishers. Increasingly, they recognized the need
for advertising revenues, but that meant taking time to solicit manufacturers
who might be located in distant parts of the country and whose financial
standing was unknown. However, unstandardized rates and disorganization in
the relationship between advertisers and publishers could not exist for long.
Agents stepped in as intermediaries, serving as brokers who could reduce the
uncertainty in a disorderly business.

One of the first agencies to focus on periodical advertising was Carlton & Smith, specializing in religious periodicals. It was this agency that J. Walter Thompson took over in 1878. By promoting advertising in all magazines, he became the preeminent figure of the nineteenth century.

By the 1890s, a measure of order had been established between publishers and advertisers. As national magazines gained strength, they were no longer willing to negotiate so freely with agents on the price for space. Instead, they chose to maintain steady rates. At the same time, since directories were now providing information about the kinds of magazines available and their circulations, advertisers armed with these figures felt less need for agents. They could deal directly with the publisher through their managers in charge of advertising sales, a position now becoming common. It was an ironic twist, since the agencies had provided the directories themselves.

Facing a fight for survival as some advertisers began to question the need for agencies, these middlemen began to expand their activities. Beginning sometime in the 1880s, they started to write their own copy, a commodity previously supplied by the advertiser. They also started to create ideas for campaigns, to assign art work and layouts, and to develop market-research capabilities. By the 1920s, agencies were orchestrating the entire marketing strategy for clients and their products. This development proved to be critical in the evolution of the relationship between publishers and advertisers.

Newsstand sales of magazines gained in importance as the nineteenth century came to a close. Before that time, publishers had been encouraged by the lowered postal rates of 1879 and 1885 and the creation of the rural free-delivery system in the 1890s. These developments had favored subscription sales by mail, but newsstand sales began to get more attention, in part because subscriptions had become routine. Mechanized subscription and renewal systems were in place at major publishers by the early twentieth century. Consequently, competition for customers shifted to the newsstand.

A sign of the publishers' growing concern for advertisers was the appearance of promotional pamphlets, written by the journals' advertising departments and fueled by data collected through market research. Another major change, one Bok claimed to have pioneered in the nineties, was to end the grouping of advertisements at the back of the magazine; now they were mixed with the editoral matter. It was not long before advertisements were being placed next to related editorial material, particularly in the women's and farm journals. Such innovations in layout were standard practice by 1920, with advertisers increasingly demanding particular placements as they became more aware of their power.

In the nineties, the appearance of advertising began to get more attention. Agencies and publishers combined to help advertisers create more attractive and eye-catching copy, designed to stand out against other messages competing for attention. Among the more memorable results were the advertisements for Ivory soap, with charming illustrations and a slogan that became a part of the language, "99 and $^{44}/_{100}$ths per cent pure," and, almost as familiar, the accompanying assertion, "It floats." Products still sold today were launched

by this new advertising, including (besides Ivory and Campbell's) Heinz cat-
sup, Quaker Oats, Postum's Grape Nuts, Welch's grape juice, Cream of
Wheat, and Knox's gelatin.

Another effect of advertising growth was to standardize, in a sense, the size
of magazines; summer issues were skimpy, and the Christmas issues overflowed with pages of advertising, much as today (although publishers and
advertisers have learned how to lessen the effect of what once were off
seasons).

Eventually advertising came to exert a strong influence on the appearance
of magazines. When they were deciding on a page size, publishers had to consider what dimensions would most effectively showcase the advertisements.
Usually, they selected one of the common formats, so that advertisers could
use the same plates for copy that they used for other journals. But as the twentieth century advanced, the design and color of advertisements often overshadowed the editorial material, forcing art directors to improve the overall
appearance of the magazines. Those with the most advertising dollars could
afford to hire the most talented directors.

Editors, as well as agents and product manufacturers, claimed that magazine advertising promoted the welfare of consumers, displaying to them the
tremendous variety of available goods. *Woman's Home Companion,* among
others, ran editorials touting the virtues of advertising, explaining that by
encouraging greater consumption on everyone's part, lower prices for individuals would be the result. Others stressed the value of buying brand names.

Some manufacturers benefited more than others from the magazine
medium. The automobile industry, for instance, advertised extensively from
the beginning. Many of the new home appliances becoming available in the
early decades of the century—electric refrigerators and washing machines—
were known to the public through magazine advertising. But for the American consumer, as for the journals themselves, advertising remained both a
blessing and a problem for some time.

III

Developing New Audiences
(1919–1945)

13

The Fall and Rise
of Empires

In the aftermath of the Great War, change was the watchword of the day. The witch-hunt for Bolsheviks (and Reds in general) subsided as frenetic patriotism dwindled into disillusionment with Wilsonian idealism. The war had been a belated addendum to the nineteenth century, and the cultural revolution of the twenties was beginning—that decade known as the Jazz Age or the Era of Wonderful Nonsense. Old restraints were loosened, the boundaries of social behavior redrawn, and a spirit of optimism swept the nation as the stock market boomed.

All this was reflected in the magazine business, which itself was busy leaving the past behind and striving to interpret the present profitably. Old institutions were declining, disappearing, or changing. Relatively new ones were thriving, and several startling new ventures got under way that would soon become institutions themselves.

One of the forces propelling changes in the magazine industry was the growth of advertising, greatly accelerated by the new prosperity and by the great industrial surge that occurred after a brief period of depression. Advertising had been growing as the decisive factor in periodical publishing since the turn of the century, but now it was coming into its own as the fuel that made most magazines solvent, no matter how high their circulations might be.

It was to be 1930, however, before the advertising business would possess a magazine devoted to its own. G.D. Crain's *Advertising Age* was called a newspaper *(The International Newspaper of Marketing)*, but it was another example of that hybrid form that has always caused classification problems—a periodical that is part newspaper, part magazine. By stretching a little, it could be defined either way. *Advertising Age* was the cornerstone of a new, Chicago-based publishing empire that in the late 1980s (known under the name Crain Communications) was to comprise some twenty newspapers and magazines, with a circulation of more than 80,000 and more than 3,000 pages of advertising. More remarkable, the group would still be family owned.

Although it now has competitors, *Advertising Age* stood virtually by itself for many years as a trade publication in the field. In its pages can be found not only the history of the advertising business from 1930 to the present but a series of important glimpses into how the American commercial system operates. As early as 1935, *Advertising Age* had a ready-made audience of executives, media buyers, and others involved in a business that had expanded massively in the twenties, and even more massively after the Second World War. A virtual trade-news explosion kept its columns filled.

The success of *Advertising Age* testified once more to the necessity of having a strong individual at the helm in launching any new magazine venture. Crain was an optimist who believed prosperity was really around the corner, just as Herbert Hoover had assured him in 1930, but he intended to help it along by attending to every detail of his venture. Like Lorimer, he read and edited galley proofs of his newspaper-magazine himself, and he did it for forty-one years. He thought of his publication as a newspaper, and he considered it to be far above the usual trade publications of its kind, which were often little more than public-relations tracts. He wanted information, hard news, and he was not reluctant to undertake investigative reporting of the industry he served.

As a pioneer in modern trade journalism, Crain had founded *Hospital Management* in 1916 with only a few thousand dollars and his perennial optimism. He moved into the advertising field because he believed that *Printer's Ink*, the only trade journal of consequence then serving it, was not covering the news. A month after he had launched his first magazine, he started another called *Class* (later, *Class and Industrial Marketing*), aimed at taking some of the audience away from *Printer's Ink*. It soon had a circulation of 2,500, enough to make it viable in that field. Its success led him to launch *Advertising Age*.

In the euphoria of 1929, there were others who believed, as Crain did, that the business world was not being adequately covered by existing magazines. The burgeoning empire of Henry Luce started two new publications almost at the same time—*Tide*, intended to be a news magazine for the advertising world, and *Fortune*, directed to an audience of business executives. McGraw-Hill, already the leader in industrial publications, also started a new magazine, *Business Week*, in September 1929.

As Paul Sullivan, a researcher in this field, has pointed out, Luce, McGraw, and Crain injected something more than fresh enterprise into the magazine business with their ventures. Trade journals were not highly regarded in those days, and the phrase itself was sometimes more of an epithet than a description. These three enterprisers were ahead of their time in insisting that it was possible to have editorial excellence in such a publication—and proving it.

Luce and McGraw were better situated financially, but Crain had the courage of his convictions. He started *Advertising Age* with an initial press run of 7,500 copies, which were mailed free to a list compiled from the Standard

Advertising Register. At a dollar per year, it took a long time to build a respectable subscription list. There were only twelve pages in the first issue, a somewhat bigger tabloid than we know today. In this brave survey of advertising, there were only twelve paid advertisements, with a total of 300 inches. But readers must have felt the vigor of its editorial page at once and liked the news and features that were offered. Hard news was what Crain thought they wanted, and he was right.

What is quite surprising is that these business publications survived the Depression. Perhaps this proved Coolidge everlastingly right when he said that the business of America was business, but it also proved that quality in the editorial product will attract and retain readers through both thick and thin; these magazines not only maintained themselves during the Depression but spread their influence further.

In the twenties, the magazine business had grown large enough and powerful enough to become the object of governmental attention. The new freedom, expressed in their pages, was subject to censorship moves, which had been only haphazard and infrequent in an earlier time. The censorship drives of Anthony Comstock and John S. Sumner had been directed primarily at books. Authorities—federal, state, and local—now looked at some magazines as greater threats to morality than the flapper and Prohibition gin. Not only were there indiviudal censorship cases; there was also interference with distribution. The witch-hunts of the early twenties generated a sense that magazines could be propagandists, as they often had been in the past, but this time on behalf of what were considered more sinister forces.

In spite of such opposition, those entrepreneurs who came to the twenties with empires already built, or those about to build one, must have felt a sense of power. They could look with equanimity at the failure of smaller, individual magazines that had been popular but were now declining. *Puck*, whose incisive articles and pointed cartoons by first-class artists had delighted a generation since it was founded in 1877, died at the end of the Great War. It was the victim of two even better rivals, *Life* and *Judge*, and of an already veteran empire builder, Hearst, who bought it in 1917, changed it from a bright, distinguished fortnightly with a circulation of nearly 90,000 into a monthly, and dropped it entirely in September 1918. Some of *Puck*'s bright young stable of writers, however, would make new careers in the twenties—George Jean Nathan, John Held, Jr., Franklin P. Adams, and Arthur Guiterman.

As *Puck*'s competitor, *Life*'s roll call of literary and artistic talent was impressive, and it was different from its rival because of its crusades, which were not so much muckraking as assaults on such annoyances as Anthony Comstock, crooked judges, and rigid Sabbatarians. *Life*'s founders, John Mitchell and Andrew Miller, had conducted the magazine through the war (as ardent Francophiles, they supported it wholeheartedly), but Mitchell died in 1918 and Miller the year after. In 1920, Charles Dana Gibson, the artist who had sold his first work to the magazine in 1887 for four dollars, bought a controlling interest for what was said to be something more than a million dollars, but early in 1925, Miller's widow asked that a receiver be appointed for the company and that Gibson be put on a reasonable salary. This suit was no sooner filed than it was withdrawn.

By this time, Robert Sherwood had become editor and had gathered around him a stellar collection of contributors that included Adams, Robert Benchley, Percy Crosby, Will James, Rollin Kirby, Dorothy Parker, Frank Sullivan, and Gluyas Williams. In 1929, however, Sherwood left the magazine and Norman Anthony, who had been editing *Judge*, succeeded him. But then, as so often happens when magazines begin to decline, editorships began to change rapidly. Anthony departed to start *Ballyhoo*, aimed at a larger audience, but before leaving, he had changed the style of *Life*. Under subsequent editors, the magazine's circulation dropped from 227,000 in 1922 to less than half that by 1930. Gibson retired from the presidency in 1928, and three years later, *Life* became a monthly. It was now in the hands of new owners, and the end was near. In 1936, its name was sold to Time Inc. for $92,000, and its subscription list was absorbed by *Judge*.

That magazine had hit its circulation peak of 25,000 in 1922, the same year Anthony became its editor and that *Leslie's Weekly*, another nineteenth-century veteran, was merged with it. In spite of illustrious contributors—Heywood Broun, Ring Lardner, Nathan, and William Allen White, among others—*Judge* went bankrupt in the dark year of 1932 with liabilities of $500,000. Members of its staff raised $17,000 and bought it, changing it to a monthly.

By this time, however, it was challenged not only by the failing *Life* but by Harold Ross's brash *New Yorker* and several other humor magazines. In 1936, *Judge* was bought by Monte Bourjaily, a fomer executive of United Features, who at that moment was trying to save another publication he owned, *Mid-Week Pictorial*. He bought *Judge*'s title from the printers, who now controlled it as the result of unpaid bills, but a year later, the printers got it back. Bourjaily had not been able to revive the dying veteran, nor could others who attempted it in subsequent years. It became nothing more than a caricature of itself and disappeared at last almost without being noticed.

Even sadder was the fate of Sam McClure's ebullient muckraking periodical, once one of the most popular in America. After he sold it, it was suspended for six months in 1921, for eight months in 1924–1925, and for another six months in 1926. In a quixotic mood, McClure bought it back in 1924 but was not able to restore it to health. He left it to die, retreating to his home in the Murray Hill Hotel, where he worked on a history of freedom and other books until a heart attack carried him off at nintey-two in 1949.

The peripatetic Hearst bought *McClure's* in 1926, adding a subtitle, "The Magazine of Romance." About a year later, he thought better of the subtitle and hired a new editor, Arthur S. Hoffman, who had made a tremendous success of *Adventure* magazine, by then one of the leading periodicals for men. Hoffman left after only a year, and Hearst sold *McClure's* in 1928 to James R. Quirk, publisher of the pioneer movie fan magazine *Photoplay* and of a less-distinguished periodical called *Opportunity*. In Quirk's hands, *McClure's* became a "snappy story" magazine briefly until it was merged with *Smart Set* and lost its identity at last in 1933—a sad end for the periodical of which Ida Tarbell once wrote, "It was a magazine which from the first put quality above everything else."

Another old empire was beginning to change under the impact of shifting configurations in the magazine world as the new century progressed. An aging William Randolph Hearst, who had been a maker and shaker for so long, was now having to deal with internal struggles for power, which began in the late twenties, in his vast organization. The man who emerged in 1931 as the new power behind the tottering throne was Richard Berlin, who had risen from advertising salesman to be executive vice president and general manager of the Hearst magazines. That brought him into conflict with Ray Long, who was editor of *Cosmopolitan*, editor-in-chief of all Hearst magazines, and president of the International Magazine Company, as the periodical group was called.

Long, a small and rather elegant man, was a brilliant editor and had what seemed a secure position in the organization. As one of the highest-paid magazine editors of all time, at $185,000 a year, he was believed to be one of Hearst's favorites. He had taken *Cosmopolitan* in hand in 1918 and saved it from extinction, filling its pages with fiction and articles by authors whose famous names graced the cover each month. He appeared to be in an impregnable position, but in 1931 Berlin succeeded in getting him fired.

The official release said merely that he had resigned to become chairman of the board of Richard R. Smith, a small publishing house he had helped to launch, but no one believed this story, particularly those on the inside. He had been known to argue fearlessly with W. R. himself without ill effects, but apparently he could not argue with Berlin, who considered Long his only obstacle to complete control. Coups were common in the corporate world but not so common in the magazine business of those days, and this coup sent tremors of panic through the entire Hearst organization, accustomed though it was to jungle politics. Other executives had watched with fear for their own jobs as Berlin deftly moved in on Long's perogatives, reducing his authority, even helping spread a rumor that the flamboyant editor might be losing his mind. It was this kind of pressure that finally led Long to understand that either he or Berlin would have to leave, and since Hearst, who had no sense of loyalty whatever, did not support him, he had no option but to resign.

After that, his career went steadily downhill—a long year in the South Seas, trying to pull himself together; the bankruptcy of his publishing house; and a little work in Hollywood, editing *Photoplay* for a time. It was no use. Out of work and broke, he shot himself in New York in July 1953, only four years after Berlin's coup.

Berlin replaced Long with Harry Burton, who had been editor of *McCall's*, but Berlin now took control of the magazines himself. Lindsay Chaney and Michael Cieply, who have examined Hearst and his empire in the later years, report that at this stage in his career, Berlin was regarded as a hard-working, serious man, "ruddy-faced and slightly nervous . . . he nevertheless never lost his salesman's charm, prompting the frequent observation that 'he always makes a terrific impression when you meet him.'" He was also, as those who had to deal with him knew, an utterly ruthless, ambitious man. He lived for his work and his career.

Whatever he may have been personally, however, did not detract from his ability. He ran the American Hearst magazines, operating separately from those in Britain, as the master of a tight ship. The magazines turned in satisfactory profits year after year as a whole, even though one or the other might lose money in a particular period.

Nevertheless, after a decade of depression and some recovery, set back again by the recession of 1937, the Hearst empire was in a delicate condition by 1939. Berlin himself was momentarily shaken by the loss that year of *Pictorial Review*, on which he had lavished more than the usual concern. Its collapse was a major shock both to Berlin and to Hearst, since it was the largest magazine, with a circulation of 3,000,000, ever to expire in this country. Of course, Berlin could console himself with the thought that it had already been in precarious shape when he bought it from George S. Fowler in 1934, but he had intended to use $2 million of the money the other Hearst magazines were making to promote *Pictorial Review*. The financial problems of other units in the empire were too serious, however, to make that move possible, and without such help, Berlin could not save it.

With such a dismaying collapse, the Hearst magazine organization now found itself in much the same kind of situation the Crowell-Collier group would face later on. When such a major failure occurs, gossip and rumor can undermine the whole structure by shaking the confidence of advertisers and banks, making revenue hard to get and endangering credit. Joseph P. Kennedy, a shrewd investor, sensed what was happening and offered to buy the entire Hearst magazine organization for $14 million. An internal battle was touched off, with Berlin fighting to save the magazines and others willing to accept the offer, since the banks were demanding money and Hearst needed $7 million to pay off notes.

In this crisis, Berlin appealed to John Cuneo, the Chicago printer who had the contract for Hearst magazines and was one of the creditors. Berlin asked for a little breathing space, another deferment, which Cuneo granted. He also suggested that A. P. Giannini, president of the Bank of Italy, might be helpful and offered to introduce them that very day at the La Salle Street railroad station, from where Giannini was about to leave for San Francisco. After a friendly conversation in Giannini's private railroad car, the banker invited Berlin to talk with him in New York later. "A good California boy should not be in the hands of those thieves," Giannini said, meaning the reluctant New York banks. The result was an $8 million loan to Hearst, which saved the magazines and considerably furthered Berlin's standing in the organization. Without much fanfare, he was made president of the Hearst Corporation.

The crisis year was 1940. Hearst still owed ten million dollars to a group of Canadian paper producers, and they were pressing for the money; by this time, Hearst's publications were the largest users of newsprint in the world. But Berlin had a personal friend in the Canadian banking world, Morris Wilson, president of the Royal Bank of Canada, which happened to hold the notes of several papermaking companies. With Wilson's backing, Berlin made a deal—a one-year moratorium and an amortization of the debt over five years. The producers were angry and reluctant, but they had to agree; their own financial futures were now involved.

Still, the corporation was not out of the woods. Large-scale bank refinancing was required, and the bankers who had been overly generous with their loans to Hearst in the past were not about to extend any substantial further credit. Moreover, the Hearst stockholders were both anxious and irritated; lawsuits had been filed. It was the time, too, when the Federal Trade Commission was charging *Good Housekeeping* with misleading advertising over its Seal of Approval. The corporation was trying to sell some of San Simeon's land, and it was closing a few newspapers in the Hearst chain.

That Hearst survived was due not only to Berlin's particular abilities and those of a few other loyal executives who also had to deal with internal maneuvers by some who wanted to wrest control from the chief himself, but also to the quiet workings of that same aging chief. Hearst had once been the organization's primary liability because of his frequent and massive raids on available cash, but he had been curbed, as Chaney and Cieply tell us in their survey of the later years. W. R., who was well over seventy when his empire began to come apart, "remained in his retreat at Wyntoon through the years of the crisis, dictating orders to papers that only occasionally failed to respond and, unknown to the bankers, directing a quiet guerrilla war against the alien force that had invaded his institution."

It was something of a miracle that in 1945, with the crisis finally past, Hearst still controlled sixteen newspapers (mostly large city dailies with combined circulations of more than five million); King Features, the largest such service in the business, with fifty-two million readers; and *American Weekly*, the Sunday supplement that had become the most widely read magazine in the world, with a circulation of 9,927,000; radio stations in Pittsburgh, Milwaukee, and Baltimore; the International News Service, still lusty though headed for oblivion; Metrotone Newsreels, also doomed, however; and eight magazines, including the highly profitable *Good Housekeeping, Harper's Bazaar,* and *Cosmopolitan.* At this juncture, the magazines alone were producing annual revenues of more than $50 million. The Hearst empire would shrink further in time, but the magazines have continued to be leaders.

In contrast to the disarray and losses in the Hearst organization, a much smaller empire, the one constructed in the nineteenth century by Cyrus H. K. Curtis, continued to do well after the Great War. In the twenties, Lorimer's *Saturday Evening Post* was established as a dominant force in middle-class culture, but the *Ladies' Home Journal,* after Bok's departure, had slipped badly, and a succession of editors had not been able to slow its decline. In fact, it was not until 1935 that Lorimer brought in a husband-and-wife team, Bruce and Beatrice Blackmar Gould (the first such combination since Frank Leslie and his wife) to run the *Journal.* The Goulds rescued it from the dull service-and-fiction formula it had fallen into and restored its luster, making it the largest-selling women's magazine in the world until they retired in 1962. Both Goulds died in 1989, he at ninety-one.

The tremendous success of the *Post,* however, which was at its all-time peak in the twenties, led Curtis Publishing to launch new publications. Curtis himself, now an old man, spent less and less time in the office, sailing the Atlantic on his yacht and enjoying his Maine summer home. A year before he died in 1933, Lorimer became president of the company; he had already been running it, in effect, for some time.

But Lorimer, too, was heading toward the exit. He had fought against Franklin Roosevelt's election in 1933 with all of his editorial skills and had been defeated—betrayed, he thought, by his own readers. He continued to fight the New Deal bitterly, and for the first time, the *Post* found itself out of touch with its middle-class audience. Lorimer's retirement at the end of 1936 and his death in the following year put a period to the *Post's* remarkable career as well as his own.

Under Wesley Stout and a succession of other editors, the *Post* struggled on to near extinction, but before that, the Curtis organization was far from being in serious trouble. The *Journal* was doing so well in 1938 that the company was encouraged to launch *Jack and Jill,* for children under ten. Somewhat to everyone's surprise (the launching had been tentative), this magazine grew rapidly from its initial run of 40,000 to 726,000 by 1955. Another venture was *Holiday,* a big, slick travel magazine that came out in March 1946. For the first few months, it appeared to be a disaster, but then the company brought in a new editor, Ted Patrick, the kind of editorial genius Curtis had been so adept at finding in the beginning. Under his direction, it became one of the best of its kind, a highly successful property, until it, too, succumbed to competition and merger.

Crowell-Collier was the third major empire begun before the Great War, and it remained basically strong in the twenties, although it was not without troubles. As the decade began, the company owned its big three—*American, Collier's,* and *Woman's Home Companion*—and another magazine called *Farm and Fireside,* (renamed *Country Home* in 1929 but suspended ten years later). In 1920, the company introduced a new idea with *Mentor.* Each issue was devoted to one subject discussed by a variety of specialists. There were big names in its pages: Luther Burbank, Roger Babson, Dan Beard, Fritz Kreisler, and others covering a broad range of knowledge. But it was clear that it would never attract a large circulation, which was the company's business, and although it did relatively well, Crowell-Collier suspended it in 1930.

This organization may have been the first magazine publishing company to diversify. Much of its revenue came from books, particularly *Collier's Encyclopedia,* and from its Sunday newspaper supplement, *This Week.* In time, the company also acquired some broadcasting properties. These diversifications were holding Crowell-Collier together at the time of its spectacular collapse in the 1950s.

The shifting ground in the magazine business between 1920 and 1945 was, of course, not solely the result of a long boom-and-bust cycle. A major reason why old empires were having to change to survive—and why even established independent magazines were struggling, with some failures—was the arrival of new people and new ideas on the magazine scene. That meant competition both for circulation and for advertising dollars. Success or failure rested on the ability of entrepreneurs and editors to identify new audiences and to understand why and how the old ones were changing. Then, too, there was the first chilling threat from another medium. Radio seized the public imagination in the twenties as nothing had before and, by bringing country and city together through network broadcasting, created huge national audiences.

For magazines, this was serious competition indeed. But they still retained the prime advantages of portability and the possibility of re-reading as often as desired. Absorbing a magazine's contents could be stretched over days, even weeks, and there were important things radio could not do that periodicals reserved for themselves. But radio programs were now competitors for advertising, and although that would not be a serious competition until television added sight to sound, magazines felt it in the late twenties, the thirties, and the forties.

The threat of radio aside, in the twenties, and early thirties, several of the old, established favorites that had dominated the business for a long time were on their way out. Death came to the *Century, St. Nicholas, Judge,* and the *Delineator.* Older readers were shocked to see the swift decline of such staples as the *North American Review, Living Age, Forum,* the *Independent,* and the *Outlook.* The latter had been one of the most respected magazines in America, with many distinguished contributors in its columns, but in 1924 its circulation was only 84,000, and in 1928 it merged with the *Independent.* The union came too late. These two fine old periodicals died together in 1932, victims of the Depression.

But if the twenties and thirties were times of destruction for old patterns and formulas of magazine making, they were also years when innovators were appearing who would transform the business—men like DeWitt Wallace, Henry Luce, and Harold Ross. They were introducing magazines of a kind never seen before and winning audiences attracted by fresh approaches. The newcomers understood changing tastes; many of the old-line entrepreneurs did not. With the innovators and the builders of new empires, magazines lived through a transition period that ended after 1945 with an unprecedented explosion of periodicals that is still going on today.

14

Henry Luce and the Growth of an Information Press

Modern magazine publishing begins with two men, Henry Robinson Luce and Briton Hadden, who together founded the giant publishing conglomerate known as Time Inc. (since 1989, Time Warner, following its spectacular merger with Warner Communications). They were the first publishing entrepreneurs after the Great War to break away from the past and offer something radically new in the field of news and information. In doing so, they founded an empire that could stand as a prototype of the contemporary corporate organization in the magazine world.

Their beginnings were modest indeed, according to Time Inc.'s authorized biographer, Robert T. Elson, whose company history, while not without its flaws and biases, is one of the few relatively honest accounts. Luce himself was unabashedly biased about a great many things, but the candid company history was his idea. He assigned it to Elson, a 25-year Time Inc. veteran, and instructed his writer to be truthful, warning him not to suppress anything relevant or essential. Luce died just before the book was published, but it is unlikely he would have found much to quarrel about within its pages.

The whole Time Inc. colossus, according to Elson, began at Camp Jackson, North Carolina, during World War I (this shorthand itself is one of *Time's* contributions to the language) when Luce and Hadden, who had left Yale as sophomores to take officers' training, decided to form a partnership someday, work together, and start a magazine. Later, Luce recalled that he and his friend began their decisive conversations one night as they slogged through the mud to barracks, and that exchange went on endlessly. They were not immediately specific about what they would do; they simply recognized each other as kindred visionary spirits.

Luce and Hadden were temperamentally unlike and had few other interests in common, although they were children of the McKinley era and had grown up with Theodore Roosevelt in the White House; he was a hero to both. They were products of the era of excessive national optimism, members of the elite ruling class, yet not the sons of rich men. The phrase "upper-class WASPs" would have described them.

Luce, born April 3, 1898, in Tengchow, China, the son of a missionary family, had religion added to his Rooseveltian patriotism. Both parents told him about America, and his mother read to him from the Bible and a child's history of the United States. He entered this country for the first time at seven and a half when his father came home on furlough. Back in China again, Luce saw his first prose in print when he wrote a letter to *St. Nicholas* describing his Chinese home. Educated first in a British-run boarding school with a strict religious tone, he found himself among a minority of American boys, including Thornton Wilder. Naturally, he edited the school paper.

More ingredients were poured into the Luce mold: Hotchkiss on a scholarship, wrestling with a stammer, sailing from China to Southampton at fourteen, and entering the cathedral grammar school at St. Albans; afterward, Paris and Switzerland and on his own through Italy. In later years, Luce admitted that this background, particularly his early years in China, gave him a romantic view of the United States that might have been too idealistic, but the reality never appeared to alter his original view.

Hadden had his roots deeply in Brooklyn Heights, then as now one of the best parts of the borough. He was born there on February 18, 1898, to a young stockbroker (whose father was president of the Brooklyn Savings Bank) and his wife of twenty-nine. Hadden's father died when he was six, and his mother was remarried to a successful doctor. They summered at Quogue and Westhampton, and Hadden attended Jonathan Buckley's select kindergarten, Brooklyn Heights Seminary, and Brooklyn Polytechnic Preparatory School— a proper track to success.

At Hotchkiss, Luce and Hadden were rival editors of student publications, in which the outlines of their later project already appeared. Hadden's *Weekly Record* foreshadowed *Time*, with world news on the front page, school news, and a current-events column designed, he said, to provide a condensed record of important events for those Hotchkiss students who had no time for the daily papers. Luce's *Literary Monthly* forecast *Life* with a new fourteen-page illustrated section, supplemented with articles by prominent Americans; Luce got the first one from Roosevelt. Moving on to Yale, both men were on the *Daily News,* and by that time, their mutual plans for the future were taking shape.

When the war was over, the partners finished Yale and split up for a time. Luce returned to Europe for a spell at Oxford, coming back to the United States to be a reporter for the Chicago *Daily News;* Hadden went directly to the New York *World.* By chance, they were both offered jobs on the Baltimore *News.* Reunited there, they resumed their planning for a magazine they intended to call *Facts.* After three months of intense work, they left their jobs and moved up to New York—an ambitious pair, Hadden not yet twenty-four and Luce six weeks younger. They had little capital and no experience. Their single asset was a typewritten dummy, which they showed to Samuel Everitt, treasurer of Doubleday Page, then in the business of magazines as well as books.

Everitt thought well of their project, but not well enough to buy it. He encouraged them to show it to Henry Seidel Canby, the current editor of the New York *Post*'s literary supplement. He had been their English professor at Yale. It was Canby, a born editor (he later made the *Saturday Review of Literature* famous), who provided the partners with the missing ingredient. They had to develop a style, he told them, "and when you do that, you will have added an appreciable something to American prose."

It was a remarkable forecast of what Luce and Hadden would do with *Time*. Parts of the "Timestyle" they developed would filter permanently into the language. Canby further advised the partners to take a year for organization and estimated it would cost a quarter of a million dollars to start the magazine. That meant scrounging for capital. At the moment, they were living mostly on hope. Hadden was still staying with his family in Brooklyn, while Luce did the same thing in a house on West 122nd Street, where his family was living while his father was on furlough from his missionary post in China.

Hadden produced enough money to rent office space in a remodeled house at 141 East 17th Street for $55 a month. They equipped it with $48.70 worth of furniture, sharing a common ashtray—an old iron soap kettle. There the partners wrote their prospectus, which proved to be a historic document. They argued, correctly, that Americans were poorly informed (which they are, ironically, in spite of the Luce and other publishing empires). The partners did not blame the newspapers, nor the weekly magazines already engaged in the information business, then including *Century, World's Work,* the *Literary Digest,* and *Outlook,* among others. The reason people were uninformed, said the prospectus, was "because no publication has adapted itself to the time which busy men are able to spend on simply keeping informed." Luce and Hadden intended to make this process easy for everyone.

There was some argument about the name. Neither man liked *Facts.* Luce recalled later that the name *Time* came to him in a flash of inspiration while he was coming home on the subway one night, reading advertisements that contained such slogans as "Time to Retire," and "Time for a Change." He could not remember which one turned on the light bulb over his head, but the next day, Hadden agreed that this should be the magazine's name.

In keeping people informed, the prospectus declared, the emphasis would not be on how much was between *Time*'s covers but on how much of its contents could be conveyed to readers' minds. They would amass facts on subjects of general interest from every source they could find and from this mass of material produce a hundred short pieces, none of which would be over 400 words long. They would be departmentalized for easy reference. In short, the partners intended to organize the world's news and give it to readers in short, easily digestible doses. Brevity was the key word. That would make it different from the other weeklies, as would the fact that the presentation would be news, not comment.

In the prospectus, the partners drew a comparison between *Time* and its most obvious competitor, the *Literary Digest*. The *Digest* selected its subjects arbitrarily and treated them at length, quoting liberally from newspapers and other sources. *Time,* by contrast, would cover all the news, briefly and in its own organized way. There was another important difference. The *Digest* took pains to quote opinion on both sides of every public question. *Time* would do that too, but it would also "clearly indicate" which side it thought was right, or at least more correct. Both Luce and Hadden had definite opinions on the state of the nation and the world, and they did not intend to create an "on-the-other-hand" kind of publication, in the manner of *Forum* or the *Digest*.

Without using an editorial page then, the new magazine would convey its proprietors' beliefs. And what were these beliefs? Fundamentally, a distrust of "interference" by government, a Republican viewpoint deriving from their backgrounds. This was coupled with an equal distaste for the increasing cost of government, another sound Republican principle. They also expressed an interest in new ideas but at the same time maintained an abiding respect for the old. They did not intend to argue these matters, just "to keep men well-informed." Only the news, they said; but when it came to talking about what the news meant, they could hardly avoid controversy. Already, Luce's American Century notions were in gestation.

The partners conceived of their audience as a million or more people with college educations and perhaps tens of thousands of others who might have missed college but were intelligent enough to want the news anyway.

When they returned to Samuel Everitt with their prospectus to ask his opinion, they were advised that the next step was to learn something about direct-mail techniques. He sent them to the head of the company's direct-mail operation, W. H. "Doc" Eaton, who assured them that their project had no chance whatever but explained circulation to them anyway. Having absorbed this lesson, the partners tried out what they had learned on one of Doubleday's own magazines, *World's Work*, with a circular sent to 7,000 people on its subscription list. They got an extraordinary 6.5 percent return. Later, *Time*'s success contributed to the demise of *World's Work*.

As the magazine moved toward launching, Luce and Hadden employed the old-boy network more extensively than it had ever been used before. Harvard had supplied the *Atlantic* and other literary magazines in the past; now it was Yale's turn. Luce had already persuaded one of his former classmates, Culbreth Sudler, who had been working in his father's Chicago outdoor advertising firm, to be *Time*'s first advertising manager. Now, in an effort to get capital, he appealed to another young advertising man, Samuel W. Meek, formerly managing editor of the Yale *Daily News*, who, in turn, introduced them to his brother-in-law, John Wesley Hanes (Yale 1915), a partner in the Wall Street firm of C. D. Barney & Co. Hanes was not optimistic about their prospects, but he did show them how to work out a financial plan that would entail selling preferred shares at twenty-five dollars each and, instead of a quarter of a million capitalization, begin with a more modest hundred thousand.

Luce and Hadden thought they might be able to persuade ten rich Yale classmates to contribute $10,000 each, but as Elson observes in relating these events, the friends were all too busy trying to prove to their fathers that they were good businessmen. Nevertheless, some were willing to take a chance. One, the first, was the partners' classmate, Henry Pomeroy Davison, Jr., who had gone from Yale to J. P. Morgan & Co., where his father was senior partner. He subscribed $4,000 worth of stock and also passed on their plan to another Morgan partner, Dwight Morrow, who put in $1,000 himself.

The partners were quite frank about their offering. They made no attempt to disguise the obvious fact that their venture was a gamble with high odds against success and declared they would take no money from anyone who could not afford to lose it. On that basis, they got $5,000 from another Yale classmate, Seymour H. Knox, of Buffalo; another contribution of $2,500 came from William F. Griffin (Yale 1912), who sent the prospectus to his friend Robert A. Chambers (Yale 1917), who put up another $2,500.

Samuel Meek, who had started this fund-raising, advised Luce and Hadden to sell charter subscriptions to important people who would also be willing to endorse the magazine, and the energetic partners soon had an impressive list of such people. Hadden went to see Newton D. Baker in Cleveland, who said he would buy a block of stock if *Time* would support the League of Nations. Hadden said no, but he persuaded Baker to become a charter subscriber and endorse the magazine.

It was slow going. By midsummer, the partners had raised only $35,000, a third of the money they needed. But then came a breakthrough, accomplished by another Yale recruit, Wells Root (1922), who was working for them as a $12-a-week helper. Root appealed to his classmate William Hale Harkness, who not only bought $5,000 worth of stock but sent the partners to his mother, Mrs. William L. Harkness, who read the prospectus in her Fifth Avenue apartment and subscribed $20,000. She also brought in other members of her family. Mrs. Harkness sent the Luce-Hadden offering to her son-in-law, David S. Ingalls, a member of Skull and Bones at Yale and then studying law at Harvard. Ingalls told them he thought the *Digest* was too firmly entrenched to be displaced, but his wife, Louise, bought $5,000 worth of stock, and her cousin Edward S. Harkness, a philanthropist, bought another $5,000. By the end of October, $85,975 had been pledged.

Impatient, the partners decided to go with what they had. They sought out Judge Robert L. Luce, Henry's distant cousin and also a Yale graduate, who drew up their incorporation papers without fee. *Time* began with 4,000 shares of $25-par-vlaue preferred stock, paying a cumulative 6 percent; 2,000 shares of Class A common, with no par value; and 8,000 shares of Class B common, also with no par value. Seventy original stockholders held shares ranging from one to Mrs. Harkness's 800. Forty-six of these shareholders were Yale men, and fourteen had been classmates. The incorporation papers were dated

November 28, 1922—or forty-two weeks after the partners had left their jobs in Baltimore. They were still a little less than $15,000 short of their financial goal, but they were ready to begin at last.

The stockholders must have viewed the first issue of *Time* with dismay. The mailing system fouled up, giving some subscribers no copies and others two or three. This was explained by the fact that some of Hadden's debutante friends had done the wrapping. Nor was the editorial product immediately engaging, as the newsstand returns quickly confirmed: half of the first 5,000 put out went unsold. Moreover, people were cancelling their trial subscriptions, and those who had not cancelled seemed reluctant to pay. Advertising, too, was not coming in as expected. In its first year, *Time* lost $39,454. It was "the hardest year of my life," Luce said later.

During this rough period, the office was moved to a large room partitioned off into cubicles in the Printing Crafts Building on Eighth Avenue. According to legend, a coin was flipped to see who would be editor, but there was no doubt that it would be Hadden. That was his bent, as it was Luce's for business. Another key figure was hired, Manfred Gottfried, at that time a Yale senior taking honors in English, but then the old-boy network broke down, and a Harvard man, Roy Edward Larsen (1921), joined as circulation manager. At first, Larson had no desk and had to work out of the Harvard Club's library. Sudler had departed as advertising manager, but instead of hiring a new one, the partners took on an established publishers' sales representative, E. E. Crowe (Yale, naturally, 1903). Thomas Martyn became first foreign-news editor, soon after finishing his Oxford degree work, and the first national affairs editor was Alan Rinehart, son of Mary Roberts Rinehart. The full-time writing staff was completed with one of Gottfried's Yale classmates, John A. Thomas. By that time, it was necessary to move again, and the change was made to 9 E. 40th St., into a space sublet from Crowe.

After a pair of trial runs, the first issue of *Time, The Weekly News-Magazine*, appeared on March 3, 1923. It was thirty-two pages including the covers, looked undernourished and not at all impressive, and sold for fifteen cents. But the partners had done what they promised; that is, to reduce the news into twenty-two departments; the whole thing could be read in an hour. And it was like no other magazine. The first cover subject, in a series that would eventually be famous, was Joseph G. "Uncle Joe" Cannon. He was retiring at eighty-six, after twenty-three terms in Congress.

The bulk of the news was in the eighteen columns devoted to national affairs, with foreign news taking up fifteen columns in its coverage of fourteen countries and a separate Latin American column. The remainder of the magazine's content was substantially what it would always be—books, theater, music, finance, education, law, the press, religion, medicine, science, crime, aeronautics, art, cinema, sport, miscellany, and milestones. Everything was anonymous, except for a literary column signed J. F., for John Farrar, editor of the *Bookman*, who wrote it gratis. Hadden and Luce were listed on the masthead as editors.

While the staff writers were able enough, part-time contributors like Farrar considerably illuminated the magazine. Stephen Vincent Benét, for example, reviewed books; and Archibald MacLeish, who at that time was in Boston practicing law during the day and teaching it at night, covered education for a modest ten dollars a week. Other former Yalies, friends of the founders, also sent in pieces. Luce wrote the religion department himself, which, appropriately enough, he did on Sunday.

The other media were not impressed with *Time*'s debut. It got four paragraphs in the *New York Times* and less than that in the *World*. However, with the aid of a little judiciously placed advertising, the magazine's circulation showed signs of picking up after the first month or so. By 1924, the circulation guarantee was set at 30,000, raised to 35,000 in October. By the end of that year, circulation was 70,000, and more important, it was beginning to attract advertising.

Although they had just barely established themselves, the audacious young publishers undertook to launch another new magazine in 1924. Thomas W. Lamont, who then owned the New York *Post*, sold it to Cyrus Curtis, along with Canby's weekend edition, the *Saturday Literary Review*, which had a circulation of no more than 10,000 but boasted an influence that was substantial and national. Upset by the sale, Canby resigned, along with his staff, including Amy Loveman, William Rose Benét, and Christopher Morley. Lamont, who had admired the *Review*, was dismayed by the prospect of its disappearance and offered to help Canby and his staff in establishing an independent literary magazine. He approached Roy Larsen at *Time* (he had been a Harvard classmate of Lamont's son) to see if there was any interest. Larsen went to the circulation manager of the *Post*, who told him that if *Time* would publish the new magazine, he could start with the entire subscription list of the *Post* supplement.

A deal was cut. Time Inc. would publish the magazine, for which Lamont would put up $50,000, and Canby, together with his associates, would operate it. Moving out of the *Post*'s quarters on Vesey Street, the dispossessed supplement journeyed uptown to the third floor of an East Side factory where *Time* was then located. There Canby and his talented staff began to turn out the *Saturday Review of Literature* on August 2, 1924. It was a magazine with a long and exciting career ahead, but not under Luce's wing.

In 1927, Time Inc. launched still another magazine, a monthly digest of advertising news which the company's agency, Fuller & Smith, had been distributing free. Luce turned it into a regular trade journal with its own advertising, beginning publication in January 1928, a year in which Luce and Hadden momentarily traded jobs. *Tide* was almost a carbon copy of *Time*, with the same kind of style and organization but with a blue instead of a red cover. It was irreverent and controversial, with one section actually labeled "Controversy."

Toward the end of 1928, Hadden was stricken with a streptococcus infection that affected his heart, and on the same day that he had put the first issue of *Time* to bed, February 27, he died six years later, in 1929, at thirty-one. The shock was great, particularly for Luce, but the recovery was so swift that in the same year, Time Inc. began to create another major magazine. Luce had been thinking about it for some time. He wanted a periodical that would cover business in a journalistic fashion, telling people how it operated, with its audience the very people who were managing it.

Luce went about this project in a way that displayed his talents as a creator of magazines. In September 1928, he installed *Time*'s business writer, Parker Lloyd-Smith (a Princeton man) and a researcher, Florence Horn, in a sound-proof room, just large enough for their desks, on W. 45th Street. They were called the Experimental Department, and Luce bombarded them with memos and telephone calls. He gave them long lists of possible articles, enough to supply the magazine for some time if they had all been carried out. These ideas covered a broad spectrum: "The Rothschilds," "Household Budgets for Incomes of $5,000 to $100,000," "Who Is Now Making Money Out of Munitions?," "The Meat Business—Range to Eater" (shades of the muckrakers!). He suggested the names of a few possible writers, and they would have been hard to surpass—Ernest Hemingway, Philip Guedalla, Joseph Hergesheimer, and Calvin Coolidge, among others.

The Experimental Department did an experimental article about International Telephone & Telegraph, then in the hands of Sosthenes and Hernand Behn, brothers who did not care to have their business written about by anyone. They refused to provide even biographical data about themselves. Nevertheless, the Department produced a story so detailed and revealing that Lloyd-Smith could write to Luce with satisfaction, "There is no competition. Established business magazines will not or cannot handle stories as we think they should be handled." The I.T.&T. article was the model for all the articles that followed.

By February 1929, after five months of experimentation, everything was ready. *Fortune* had been chosen as the name of the new magazine because Luce's wife had liked it best. He told the board of directors that it would cost $400,000 to start, which seemed to dismay no one but the dying Hadden. However, Luce had been honest about it, as usual. He described it as a fifty-fifty gamble.

As events turned out, the odds were even longer. Recalling the moment in 1943, Eric Hodgins, a Time Inc. executive, described events in a *Time*-like way: "Almost on the eve of *Fortune*'s publication, the whole economy of the United States clapped a hand over its heart, uttered a piercing scream, and slipped on the largest banana peel since Adam Smith wrote *The Wealth of Nations*." *Fortune*'s first issue, he noted, had appeared in February 1930, just four months after the Crash, when America still was not sure what had hit it.

Nevertheless, it was an editorial product to be admired. Advertisers recognized that fact by placing 779 pages of ads in its first year. The design was by Thomas M. Cleland, one of the best designers in the country, who chose an eighteenth-century Baskerville body type, produced by the English Monotype Company. The paper was designed to eliminate glare, and the unusually thick cover was almost the consistency of cardboard. It was even hand-sewn.

After its auspicious debut, *Fortune* picked up speed rapidly, even in the teeth of the Depression. Luce saw same photographs that 24-year-old Margaret Bourke-White had taken for an Ohio steel company and hired her on the spot. But he gave up his earlier idea of soliciting contributions from well-known writers, and the magazine began to be completely staff-written. "We made the discovery that it is easier to turn poets into business journalists than to turn bookkeepers into writers," he said, a puzzling comparison, but to prove it, he took on Archibald MacLeish and the lesser-known poet Russell Davenport, who had been a fellow member of Skull and Bones at Yale. To this unlikely assemblage of business writers was added the even more unlikely Dwight MacDonald.

In spite of hard times, the new magazine sold for a dollar a copy, and most readers had to admit it was a handsome, interesting periodical that was worth the price. Assessing it later, Walter Davenport wrote, "The most notable fact about the early *Fortune* was its daring; and this was a product, chiefly, of two lines of force—Luce's journalistic experience and Parker Lloyd-Smith's journalistic inexperience. *Fortune* could never have been founded by professionals alone. Luce was a professional all right, but Lloyd-Smith was the precise contrary."

Luce had two major obstacles to overcome before *Fortune* could be accounted a success, however. One was the Depression itself, which grew immediately worse, and the other was the attitude of the big businesses and big businessmen his magazine was explaining to the public and to the business community itself. At first, proofs were shown to companies for factual correction before stories were printed, but some large corporations were still not cooperative. Anonymity remained prevalent in the business world, as it had been in those long-ago days of 1906 when Isaac Marcosson had given readers their first full-length view of a major business figure when he profiled the elusive John Archbold of Standard Oil.

Difficulties were overcome simply because Luce was putting out a superior product, which he kept making better. The impetus came not only from his own genius but also from the highly talented people he kept adding to his staff. A major addition was Ralph McAllister Ingersoll, managing editor of the *New Yorker*, another new face of the twenties. Ingersoll had followed Luce and Hadden through Hotchkiss and Yale, several classes behind, and did not know them. A brilliant, volatile, energetic man, he had been a proper foil for Harold Ross in launching the *New Yorker*, but after it was established and the initial excitement had worn off, he was susceptible to new challenges. Luce gave him one, making him associate editor of *Fortune*.

Another factor in Luce's success was the changing climate in America. Entertainment had been the key word in magazine publishing during the Jazz Age, but sober Depression times had generated a demand for facts, particularly about what was happening to the country; that especially benefited both *Fortune* and *Time*.

In late 1933, at the dismal bottom of the Depression, Luce was fertile with new ideas. He talked of an English edition of *Time*, of a new magazine for women, of one for children, of a picture magazine and of one dealing with sports—even a daily paper. His Experimental Department was now headed by John S. Martin, who had been managing editor of *Time;* that position was assumed by John Shaw Billings, who would become another chief figure in the organization.

By this time, Luce had acquired the first magazine he had not created himself, *Architectural Forum,* which, as Elson tells us, "had a salutary influence on American architecture and an influence on American life out of all proportion to its limited circulation." Luce, who had always been interested in architecture, got it for the proverbial song. He had been thinking of creating a magazine in this field, and some steps had been taken. Along with $10,000 for exploratory work, Luce turned this project over to a new man, Charles Jackson, another key figure who would be with him for thirty years. A dummy was put together for a magazine to be called *Skyline: The Weekly Newspaper for Architects.* It looked possible, but Luce hung back; he thought the time was not right, particularly because he saw that other architectural journals were suffering.

Then in 1931, opportunity knocked. A magazine called *Architectural Forum* became available. It had begun as the *Brickbuilder* in about 1890, had changed its name along the way, and in 1928 had been sold to a new corporation, National Trade Journals, for $1 million. This organization went bankrupt in 1931 and in the receivership sale was sold for $75,000 to Gordon W. Reed, chairman of the finance committee of American Metal Climax, Inc. Reed made some desultory efforts to revive it, but was soon happy to sell Luce a 75 percent interest for $80,500 in 1931. A year later, Luce bought the remainder for a total of $111,000.

At that point, *Forum* was not too far from the abandoned *Skyline,* and it was easy for Luce's men to inject some *Time*-like features, such as a department called "Building Money" and a news section. This, however, was only the beginning. Knowing that previous architecture and building magazines had been little more than well-illustrated textbooks, *Forum* was given a picture-and-text treatment in the Luce style—the one supplementing the other. The magazine introduced such publishing innovations as doing a case study of a building by stating the problem involved, illustrating the solution with plans and photographs, summarizing the reasons for considering it at all, criticizing flaws, and relating the whole thing to general trends. Ruth Goodhue, who became managing editor, was right to say that with *Forum,* architectural journalism entered a new era.

In building the basic structure of the Luce empire, all the chief elements had been laid out and were doing well by the mid-thirties, with one exception—*Life,* the picture magazine that would revolutionize photojournalism. Its beginnings are still a matter of controversy. Elson, in his "official" history, asserts that Luce had wanted to do such a magazine for a long time, and in fact, had talked about it with Clare Boothe Brokaw before they were married; Clare herself is said to have had the same idea even earlier, in 1931, when she was a young editor at *Vanity Fair.* In this version, Luce had set his staff to preparing dummies, and half a dozen had been produced by 1934, when he temporarily gave up the idea because there appeared to be too many problems involved.

On the other hand, Roy Hoopes, an Ingersoll biographer who also wrote a history of the magazine, believes that Ingersoll deserves most of the credit. Hoopes notes that by 1935, Ingersoll was in effect running the Time Inc. organization, at thirty-four, as the general manager. Luce, in this version, asked him for a memo analyzing the photography in *Fortune*. In making his analysis, Ingersoll observed that the idea of a picture magazine was floating around in the publishing business, and he feared that it might constitute a future threat to the Luce enterprises unless Time Inc. moved first. He had seen the earlier dummies and liked them, but Luce remained unconvinced.

As a biographer of the magazine, Hoopes agrees with Loudon Wainwright, author of still another history published a year after Hoopes's volume, that Clare Brokaw had a great deal to do with it. Discussing a picture magazine was a part of their courtship, the writers agree, and by the time they were married in November 1935, she had convinced Luce to reconsider the project he had virtually abandoned. Hoopes sums up the case for Ingersoll: "After a long look at the record, it seems clear that Ralph Ingersoll was actually the one primarily responsible for bringing *Life* magazine into existence—not for the original idea or the finished product that was eventually published . . . , but as the catalytic force that kept pushing everyone in the company to publish what Ingersoll eventually came to consider 'my magazine.'"

Wainwright quotes Luce himself on the subject of origins. Luce, in a letter to his lawyer in 1941, after having been accused of stealing the idea, protested, "But there is no such thing as an 'idea.' . . . there are only a thousand ideas, feelings, theories, practical considerations, experiments, hunches, etc. & etc. . . . I didn't get the idea sitting on a mountain. Nor did I steal it. There was nothing to steal. I had dozens of ideas—on mountains and elsewhere. . . . So did a lot of other people who worked here or who talked with us. . . . Eventually came the determination to do something—or try to do it. We tried and we did."

Hoopes, relying on Ingersoll's account and on his own research, says that Ingersoll laid out the prospectus for such a magazine for Luce and his bride while they were on their honeymoon in Cuba. Luce told him to go ahead, but he doubted that the new journal would ever go above 150,000 circulation. He called it a "slick paper carriage trade idea" and remarked that *Time* could afford a *succès d'estime*. "It could be fun, go ahead with it," he added.

Ingersoll went ahead by buying up first-refusal rights to all the picture sources he could find; that is, the agencies and news organizations. In May 1936, he wrote a four-page memo to Luce summarizing his ideas for the magazine and insisting that it would be a mass-market periodical. Ingersoll also conducted a quiet campaign, with the help of Dan Longwell, to keep the new Mrs. Luce out of the editorship, according to Hoopes. She complained later that her husband had intimated the job might be hers, and when Ingersoll's move proved to be effective, she began what Hoopes calls "one of the longest feuds in journalism."

In any event, the new magazine moved toward the launching pad through most of 1936, with November as the target date. Time Inc. offered potential advertisers one of the advertising industry's greatest bargains—guaranteed rates for twelve months. A total of 235,000 charter subscriptions was obtained, mostly from a mail solicitation of 755,000 *Time* and *Fortune* subscribers. The return was phenomenal, nearly 10 percent.

A beginning staff was formed. Luce declared himself managing editor "for an unstated term of months or years," although it was not long before John Martin took over. Howard Richmond, art director at Macy's, came to take the same job at *Life*; Longwell was picture editor; there were three assistant editors, and a truly imposing staff of four photographers: Eisenstaedt, Stackpole, McAvoy, and Bourke-White. A dozen names for the new magazine were tossed in and argued about, but in the end, *Life* was chosen by Luce because that was the title Mrs. Luce had used in a 1931 memo; it had also been suggested by James A. Linen, Jr., whose son would later become president of Time Inc. Luce had to buy the *Life* name from the old Harvard-nurtured magazine that had contributed so much to humor and to criticism of the arts—another triumph for Yale. At that point, the nearly expired magazine had a circulation of only 70,000, and Luce got it for a mere $92,000. On the day before the new *Life* appeared, the last issue of the old one was taken off the newsstands.

Oddly enough, Luce was still so uncertain about the success of his new venture that almost on the eve of launching it, he invested in a possible competitor—just in case. John Cowles and his brother Gardner, publishers of the Des Moines *Register & Tribune*, were friends of Luce and stockholders in Time Inc., but they were planning a picture magazine of their own to be called *Look* and showed the dummy to Luce and other executives. The Cowles version of photojournalism was to be a monthly aimed directly at the mass market and printed in rotogravure on a cheaper stock than *Life*'s.

The Cowles brothers were not dismayed by possible competition and, in fact, offered Time Inc. a 20 percent interest. Luce bought 45,000 shares, but during the following year when both magazines were in active competition, John Cowles thought better of it and asked Luce to sell back his interest, which he did for $157,000 and a neat profit of $67,500.

Life appeared at last on November 19, 1936, and to everyone's surprise, except possibly Ingersoll's, it was the most remarkable instant success in the history of magazine publishing. On its cover was Bourke-White's now historic picture of the Fort Peck dam, then being built in Montana by the WPA. There was a symbolic frontispiece, of a newborn infant being held by the heels, with the caption "Life Begins."

There was news and information in its pages; but above all, superb pictures—the first aerial photo of Fort Knox; of Fort Belvedere, Edward VIII's country place; and of the newly opened Bay Bridge in San Francisco; along with views of a new Broadway hit, *Victoria Regina*, with career pictures of its star, Helen Hayes; and of a new motion picture, *Camille*, with Greta Garbo and Robert Taylor. The magazine began what would be a running commentary on contemporary art with three pages in color of pictures by John Steuart Curry, and it took its cameras to the first of many parties it would record, this one given for the British ambassador to France. There would be so many of these parties that the photos taken at them were collected in a book, *Life Goes to a Party*, just as several other of its specialties would be packaged and merchandised all over again.

A press run of 466,000 had been ordered, but it sold out at once, as the magazine did for the next four weeks, the presses failing to keep up with demand. Records were set immediately. No magazine had ever surpassed 500,000 in its first year, but *Life* sold more than this figure in its first few weeks.

If there was anyone who did not find joy in such great success, it was Ralph Ingersoll. As Hoopes tells us, the man who believed the whole thing was his inspiration and accomplishment came into the office on the morning after publication expecting to be congratulated. Instead, Luce called him in and told him that even though he understood *Life* had been Ingersoll's baby, he was taking it over now. Ingersoll would go back to being the publisher of *Time*. As Ingersoll said, ". . . the *Life* that I had nursed into being was Harry's, and I never had anything more to do with it." Three years later, he resigned to start his controversial and altogether remarkable New York tabloid newspaper *PM*.

Meanwhile, as Elson observes, "Success almost killed the company." No one was prepared for *Life*'s skyrocketing triumph, which was matched by skyrocketing losses until it was losing $50,000 a week with a circulation of 1,500,000. Advertising rates were doubled, but since those who had signed pre-publication contracts had been given the privilege of buying additional space for the same rates, most of them did, so that even with greater circulation and more advertising, the losses still increased. Moreover, it was competing for advertising not only with *Look*, but with the *Saturday Evening Post* and *Collier's* as well.

By the end of 1937, when the country was in the grip of a severe recession, Time Inc. had invested a total of $10.5 million, and its circulation guarantee for 1938 was 1,600,000. Ironically, *Life*'s success had hurt *Time*, which was chasing the same audience and advertising dollars. The company reported that it would pay no year-end dividend since the net profit was only $168,430 after taxes and the spectacularly successful *Life* had lost more than $3 million that year. In the magazine business there were rumors that *Life* might even be suspended. That thought, however, never crossed Luce's mind. He knew America was in love with his new periodical—later it would even be called "America's magazine"—and that his "photographic essay," as he called it, was outselling all its competitors, including a dozen or so imitators.

Roy Larsen tried some standard remedies for the situation. The circulation guarantee was raised to 2,000,000 without increasing the page rate, thus reducing the per thousand rate to $2.85. Production costs were also cut, bringing down the break-even point to 924 pages annually from a thousand. But then *Life* began getting returns from newsstands, something that had happened only minimally in its first year.

In the end, it was an editorial coup that turned *Life*'s precarious success into something more substantial. The catalyst was a documentary film called *Birth of a Baby*, for which the New York State censors had refused to grant an exhibition permit, even though it had been produced by the American Committee on Maternal Welfare. The result was a public outcry that led Joe Thorndike, *Life*'s movie editor, to suggest that the magazine show stills from the film. Luce saw the tremendous promotional opportunities at once.

Basically a conservative man himself, the founder was not at all averse to exploitation, although he might decry sensationalizing. He knew that the censorship of those days, which was rampant at state and local levels and in the Post Office Department, produced immediate public interest and sales whenever and wherever it was applied. He knew that the Post Office would not permit him to show bare-breasted African women; only the *National Geographic* had that privilege. But when he offered them the stills from *Birth of a Baby*, they surprised him by approving the pictures.

When they were displayed in *Life*, the result was all any promotion manager could have hoped for. The magazine was ordered off the stands by the governor of Pennsylvania; it was impounded by the Canadian Customs and banned in thirty-three American cities. Suits were filed in Boston and New Haven, and newsdealers in the Bronx were shut down by the assistant district attorney. In a reprise of H. L. Mencken and his sale of the "Hatrack" edition of the *American Mercury* in Boston, Luce sold a copy to a detective in the D.A.'s office and had himself cited for selling an obscene publication.

There was a rush of defenders—Eleanor Roosevelt, Dorothy Thompson, Surgeon General Thomas Parran, among other notables. The lawsuits were tried, and all of them failed except in Boston, where Joseph Welch, later the hero of the McCarthy hearings, argued vainly and brilliantly for the defense. But Luce had seen George Gallup's figures: seventeen million adults had viewed the birth pictures. He was content.

Luce solved *Time*'s problems by purchasing for $25,000 (in May 1938) the subscription list of the *Literary Digest* and what remained of its goodwill after its spectacularly wrong election prediction in 1936. Only sixteen years earlier, he had been advised that it would be hopeless for *Time* to compete against the *Digest*.

Luce shrewdly transferred James A. Linen III, the grandson of his old friend and sponsor, from the company's advertising department to the *Life* sales organization. Linen saw that the magazine's problem was that it did not fit into the recognized patterns accepted by major magazine advertisers. Other mass-market leaders like the *Post*, *Collier's*, and *Liberty* were directed to predominantly male audiences; *Life* was not. Its readership was almost equally divided, even though it offered none of the service functions of women's magazines. Readership surveys also showed that it had an enthusiastic following among young people of both sexes. Then, too, its format was not like any other periodical, and advertisers and their agencies seemed to be unsure how to assess it.

What Linen had to sell, however, was the magazine's astonishing pass-along readership—a 17,300,000 total audience, comfortably ahead of all competitors. Those were figures easily understood. The *Post*, for example, led all periodicals in both circulation and advertising, but its pass-along figure was only 12,900,000. By the end of 1939, *Life* had become a success by any measurement, and *Time* had recovered from its momentary setback.

Psychologists speak of the "fear of success," and oddly, that was what gripped both Luce and Larsen, who was quoted by Elson as saying, "The fact of success is not easily grasped. We felt there must be something wrong or about to go wrong somewhere. I fear that this was the attitude of both Luce and myself right up to World War II." They need not have worried. The war itself proved to be a substantial stimulus because of the excellent coverage given it by both *Life* and *Time*. But even at the point of American entry in 1941, Time Inc.'s revenues had reached a new record of $45,047,879, compared with a little more than $29 million ten years earlier. Before-tax profits in 1941 were $8,190,057. The only noncontributors to the Luce empire's profits in 1941 were *Fortune*, with small losses, and the *March of Time* newsreel series, which made up for its financial deficiency by its promotional value for *Time*.

At the start of the war, Luce also began an international expansion that would reach impressive proportions by launching a Latin American edition: "the world's first plane-delivered magazine," as it was advertised. It ran into political storms at once, from both right and left at home and in South America, but after the war, as the Luce international empire expanded, the foreign editions became identical in content with the one at home.

Luce's other and even more significant move in some respects was his much discussed "American Century" editorial in *Life* for February 17, 1941. Succinctly, what Luce proposed was the export of the American system, technologically and economically, to any other nation that would adopt this country's political attitudes and ideology. The response from letter writers was mostly favorable, but the editorial was roundly attacked by anti-imperialists of every persuasion, a reaction that hurt Luce because he thought of *himself* as an anti-imperialist. When this editorial appeared in book form, the volume contained a highly critical essay about it by John Chamberlain, then a *Fortune* writer, who many years later swung around to a quite different viewpoint as a book critic for the *Wall Street Journal*.

Criticism occasionally hurt but never deterred Luce. He had been accustomed to it from the beginning, when *Time*'s debut was greeted with a disrespectful jeer from another new magazine, Harold Ross's *New Yorker*. This led to a long-lasting feud between the two publishers, who could hardly have been more unlike. But there was other criticism of *Time* when it began, and one of those who took notice of it was the magazine's first foreign-news editor, Thomas S. Martyn, who saw an opportunity to start a rival newsmagazine that would neither imitate *Time* nor repeat its faults. He got up a prospectus asserting that "some people feel *Time* is too inaccurate, too superficial, too flippant and imitative." The magazine he proposed to establish would be "written in simple, unaffected English [in] a more significant format [with] a fundamentally sober attitude on all matters involving taste and ethics."

Martyn showed his prospectus to John Hay Whitney, Paul Mellon, and other prospective investors, raising enough capital from them to launch *News-Week* (the hyphen was soon dropped) on February 17, 1933. It had no distinctive writing style and no particular approach to the news except to digest it, yet it was cast in roughly the same format as *Time*, with the same kind of departmentalization. Martyn ran through several million dollars of his backers' money before it was clear that his magazine was making no headway at all against *Time*. He was forced out, and in February 1937, *Newsweek* was merged with another semi-news publication called *Today*, edited by Raymond Moley, one of Franklin Roosevelt's original Brains Trust. The money behind this merger was Vincent Astor's, who now became president and chief angel.

Moley was little help to the new publication. He was too busy telling graduate students at Columbia how different things might have been if the president had only listened to him. The magazine was being edited now mostly by former newspapermen, learning how to operate in a much different environment. Many of them later went on to become *Time* writers or executives, and one, John Osborne, the national-news editor, had a later career after a stint with *Time* as proprietor of the *New Republic*'s celebrated Washington column, signed T.R.B. no matter who wrote it.

Meanwhile, however, these talented people were wrestling with how to put out a newsmagazine that would cover the news as *Time* was doing without sounding imitative. The organization was the same; that is, a staff of writers supported by a corps of researchers. The news budget was set at the beginning of the week, on Tuesday (both magazines were published on Monday); every department was allotted so many lines of type for its stories); and then the department editors assigned them and specified the number of lines. The researchers used a small library and memos that were constantly pouring in from correspondents in Washington and other cities, plus the daily papers, particularly the *New York Times*, which lent itself to rewrite by everybody, and one wire service, Hearst's International News Service. The latter led the list of "sources not to be trusted and to be used with caution" pasted up on the bulletin board in the national news office.

But the problem was style—how to summarize the news in a readable way without what was now called Timestyle, which irritated many people but charmed a substantial number of others. There was also the matter of not simply summarizing the news but contributing some special insight, which *Time* pretended to do even when it was not doing it. This futile effort extended even to the 1937 Kentucky Derby, the two magazines having picked different winners in advance. On the day of the race, the sports editor of *Newsweek* was in a state of near hysteria for fear that *Time*'s horse would win.

Something of this hysterical atmosphere prevailed most of the time in these early days. Periodically, Astor would appear, confer at length with the executives, give the staggering magazine another infusion of capital, and go away again. It was not difficult to see what *Time* was contributing journalistically in this competition, like it or not, but *Newsweek*'s only visible departure was the signed column, Moley's, a practice *Time* would not get to for years. Otherwise, anonymity prevailed on both magazines, except in the lengthy mastheads. Politically, there was no difference. *Time* reflected Luce's solid conservative Republicanism, and as a disgruntled New Dealer, Moley had placed *Newsweek* on the same side. Both magazines, of course, would deny for some time that they were political at all. In style, *Newsweek* was much more serious and conventional.

In spite of these obstacles, *Newsweek* survived largely because there were so many readers who liked the general idea of what *Time* was doing but could not stand its brash idiosyncratic style. This distaste provided a powerful source of subscribers, who also disliked *Time*'s more obvious political biases. *Newsweek*'s were much less clear. Still, it never really established a solid identity until the 1960s, when it was acquired by the Washington *Post*, whose able publisher, Katharine Graham, gave it the capital and additional talent it needed to make a better place for itself.

At almost the same time that *Newsweek* appeared in 1933, another rival in the news and information field was launched. *United States News* was a by-product of the conservative syndicated columnist David Lawrence, and it was quite different editorially from its two competitors. Since it was not trying to compete stylistically with either of them, *United States News* could afford to write in its own style, which was mostly unadorned but clear enough English of the kind Lawrence used in his column. He was a political analyst who prided himself on his ability to forecast the future, and that set the tone of his magazine, which was both conservative and omniscient. Its intended audience was hard to define, but aside from its obvious direction toward conservatives, it seemed to be meant for people somewhat farther down the social, economic, and intellectual scale than readers of *Time* and *Newsweek*.

The self-image *United States News* had of itself was, of course, quite different, and that image became more apparent after 1948, when it merged with its unsuccessful companion publication *World Report*, begun in 1946, to make *United States News & World Report*. From that time onward, it continued to establish a place for itself as the alternative for those who could not stomach either *Time* or *Newsweek*, for whatever reasons. A substantial part of its audience was also made up of those who were devotees of Lawrence's syndicated political column, who found more of the same in his magazine. It would go on to establish a different identity after the founder's death.

15

Developing the Mass Market

If there was a King of the Mass-Market Hill in the twenties, and even for a time in the thirties, it was George Horace Lorimer. His *Saturday Evening Post* was the great popular success of that period, although the *Reader's Digest* would soon supplant it in circulation. But in several ways the *Digest* never held the unique place in the heart of middle-class America that the *Post* enjoyed.

The most enduring image of those remarkable years has been the Norman Rockwell covers that were the magazine's trademark. Rockwell's importance and the reasons for it have been analyzed by Milton Glazer, one of the best magazine designers of our time, who worked on the formats of *New York, Esquire, More,* and other magazines. "Rockwell created and synthesized and epitomized America's mythological view of itself as being a benign and generous country ennobled by work," Glazer observed in 1979. The illustrator spoke directly to the *Post*'s primarily blue-collar audience, in Glazer's view, but this seems to underestimate considerably the number of white collars among those millions of subscribers—small-town merchants and professionals, as well as those on the rise in business, Republicans all, or nearly all.

Glazer and others have characterized Rockwell's work as "charmingly illustrated clichés," portrayals of an America that never was. Those who grew up in that America, however, and were able to look back on it later with a more sophisticated perspective could see that if Rockwell drew cliché situations, then America itself was a cliché. Not one of his drawings depicted something that did not exist. They were commonplace, everyday scenes with which millions could identify, and the idealization lay in the fact that they were always upbeat; the dark, *Main Street* side of American life that Sinclair Lewis disclosed was never shown.

Obviously, that darker side did not appear on the cover of any other popular mass-market magazine. If Rockwell poked fun at his subjects, as Glazer says, it was never satiric, but more the gentle kind familiar to ordinary families whose lives were not gripped by alcohol or poverty. But Rockwell was not a static sort of artist, merely repeating himself, as all this may imply. In the thirties, his work became more complex in terms of figures and elements on the page, whereas before he had simply painted the figures against a white field. These striking silhouettes had helped the *Post* considerably on the newsstand, where they stood out from other, more solid-colored cover designs. Gradually, however, Rockwell came to use the whole surface.

It was fortunate that Lorimer did not live to see Rockwell's contributions to the covers of the Second World War period. The "Boss," as the editor was known on his magazine, had fought Franklin Roosevelt from the day he was elected, and if Lorimer had not died in 1937, he would almost certainly have taken the same anti-Roosevelt positions during the war that his right-wing Republican friends who survived him took. Rockwell, the master of idealization, caught perfectly America's wartime image of itself by creating a character named Willie Gillis, who symbolized everybody's man in uniform, doing K.P. duty, coming home on leave, doing everything except the dirty business of dying. Yet, as in the earlier work, Rockwell hit the common chord, the generalized feelings of Americans about themselves and their world.

Glazer points out that in his "Rosie the Riveter" cover of May 29, 1943, Rockwell drew on Michelangelo's figure of Isaiah. These and his other *Post* covers never particularly reflected the magazine's contents but rather the audience at whom the contents were directed. Today, most covers are designed to inform as well as attract, and most directly reflect the contents of an issue. Rockwell, however, left an indelible mark on the American consciousness that no other magazine cover artist achieved, attested to by what is still almost a Normal Rockwell industry.

Once *Post* readers moved beyond the cover during the twenties, they found Lorimer's sure touch in the contents. The list of popular fiction writers who wrote for him was formidable: Kenneth Roberts, Ben Ames Williams, Thomas Costain (who was also an editor for a time), Sinclair Lewis, John P. Marquand, Clarence Budington Kelland, Mary Roberts Rinehart, Irvin S. Cobb, and Theodore Dreiser, among many others. They constituted the foremost mass-market storytellers of the century's first four decades.

Maureen Honey, who has analyzed the images of women in the works of the *Post*'s fiction writers during Lorimer's last years—1931 to 1936— describes the magazine's prototypical heroine as "witty, athletic, self-possessed, urbane, and appears as competent secretary, aggressive businesswoman, ambitious college graduate, or adventurous aviator." Women are shown attempting to have careers and harboring ambitions beyond being wife and mother.

These images were a reflection of Lorimer's genius, since his own views of women were those of the conventional past, even though his own wife, Alma, was an active figure in Republican party politics. He was anything but a supporter of feminism, yet he understood, a little late, that a change of female roles had occurred in the twenties, and now, in the thirties, his magazine had to reflect that fact for a new generation of women readers. Lorimer, however, had not closed his eyes entirely to what was happening in the Jazz Age. His *Post* was the first popular magazine to print a story in which a woman smoked a cigarette and the first in which a woman was described as drinking liquor, although the *Post* itself accepted no liquor advertising. (Lorimer in the twenties was a recovered alcoholic, although he did not acknowledge it.)

Smoking and drinking were never glorified in the magazine, nor anything that smacked of breaching conventional sexual morality. Lorimer knew those were the standards of his audience. But to one reader who objected to the fictional portrayal of a stretched standard, Lorimer pointed out through an associate editor that "one of the purposes of fiction is to reflect existing manners and conditions, in short, life as it is."

Under that standard, Lorimer in 1931 had no hesitation in serializing Katharine Brush's novel, *Red-Headed Woman*, in which, at the end of the first installment, the secretary-heroine was having a drink with her boss at his home, with the boss's wife away and night drawing on. To the profound shock of numerous readers, the second installment began with the two having breakfast, the wife still away and a significant lapse of time unaccounted for. To the thousands of outraged readers who wrote letters of protest, Lorimer dictated a form letter: "The *Post* cannot be responsible for what the characters in its serials do between installments."

Intellectuals and the literary world in general derided the *Post* in spite of Bernard De Voto's caustic reminder that there were only two kinds of writers who did not write for the magazine, "Those who have independent means or make satisfactory incomes from their other writings, and those who can't make the grade. Many of the former and practically all of the latter try to write for the *Post*."

Critics, including serious literary writers, did not seem to notice that the *Post* printed fiction by much-admired names. Lorimer had a single criterion for acceptance, "Will the *Post* audience like it?" At lunch one day with the Scott Fitzgeralds, Isaac Marcosson, and Hugh MacNair Kahler (one of his best and most consistent popular writers), Lorimer found himself under indirect attack from Fitzgerald, who was complaining about the commercialism of American mass magazines. "They've always published the work of mediocrities and nobodies," Fitzgerald said. "They've taken no notice of real genius. It's a safe bet that nobody in this room ever heard of the most important American writer of the early 1900s. It may be fifty years before anybody, except a few of us, even knows that Frank Norris ever existed."

Lorimer nodded soberly. "If that's your considered opinion, Fitzgerald—," he said.

"It is," Scott replied.

"Then maybe I didn't go wrong after all when I bought *The Pit* and *The Octopus* from Frank Norris and serialized them both in the *Post*." With that he leaned back, and his big laugh echoed in the room.

If he had wanted to emphasize the point, Lorimer could have pointed to an impressive list of first-class writers whose works he had printed: Stephen Vincent Benét, James Branch Cabell, Willa Cather, Joseph Conrad, Stephen

Crane, Theodore Dreiser, William Faulkner, Ring Lardner, Sinclair Lewis, Rebecca West, Edith Wharton, and a good many others. Nor were they the second-drawer works of these writers, as some critics charged. He even published superior poetry, particularly the fine early works of Carl Sandburg and Edna St. Vincent Millay, as well as Alfred Noyes's bitter "A Victory Dance" in 1920.

Easily outranking any other American editor of his time, Lorimer understood that in encompassing the broadest possible spectrum of middle-class readers, he had to reach both extremes as well as the middle. Consequently, he often published stories that would appeal to only about 10 percent of the audience.

Time and again, he confounded the presumed experts. It was said, for example, that no popular magazine could print a story about miscegenation, but Lorimer printed one by Charles Brackett. In 1909, he published the Potash and Perlmutter stories by Montague Glass, satires on the cloak-and-suit trade, in the face of solemn assurances that Jews would be offended by humorous stories about their lives and non-Jewish readers would be bored. Millions of readers gave these tales possibly the largest acceptance in the *Post*'s history. The mail showed that virtually no one was offended, and circulation hit 1,250,000 that fall. Potash and Perlmutter went on to entertain millions of others on stage and screen.

When Lorimer began to run Octavus Roy Cohen's stories about blacks, he was similarly attacked by critics. But unqualified acceptance both by blacks and by whites kept the series running for years. Cohen had a black friend who acted as censor and advisor for these stories. Offensive though they might be in our day, Lorimer knew unerringly what was acceptable in his own time.

He also knew what would be unacceptable to his audience—some Kipling stories, because "they just weren't good"; and Cather's *A Lost Lady*, which he admired but knew was not right for his readers. Lewis's *Main Street* was not turned down, as one critic charged (Lewis never offered it), even though Lorimer may well have surmised when he read the published book that the frank exposition of how his readers lived might not have pleased them. Besides, he was indifferent to females like Carol Kennicott. "I'm not interested in the romantic woman in novels," he told a friend.

Lorimer's instinct for the new was nearly infallible. He bought the first *Get-Rich-Quick Wallingford* story, the first Potash and Perlmutter, the first Ring Lardner "Busher" letters, and John P. Marquand's *The Late George Apley*— all over the dissenting opinions of his editors. He also bought Stephen Vincent Benét's *The Devil and Daniel Webster* over strenuous opposition.

The magazine's articles were aimed equally strongly and squarely at middle-class interests, which were also Lorimer's. Moreover, he had an uncanny ability to forecast events, although it was passed off by others as "*Post* luck." That luck began with Samuel Blythe's "A Calm View of a Calm Man," a piece about President Harding which was on the stands when he died. While the submarine S-54 was dropping to the bottom, Commander Ellsberg's story of the salvaging of her sister ship, the S-51, was running in the *Post*. As the newspapers were having a field day with the *Morro Castle* disaster, the *Post* was printing Manuel Komroff's stories of how the press handled the *Titanic*'s sinking.

A year later, the magazine appeared one week with a lead article, the first of three, on the life of Huey Long, and a few hours later, Long was assassinated. In the same year, Jim Collins, a test pilot, told *Post* readers how it felt to power-dive at 400 miles an hour, and while they were reading his story, Collins was killed, making exactly that kind of dive, in his final test before retirement. Then there was Jack Dempsey's article, "The Next Champion," with its prophetic sentence: "There could be a new heavyweight champion and a new runner-up when this is printed."

Remarkable coincidence, the skeptics thought, but it was really the result of Lorimer's superb sense of timing, a feeling for the flow of events. As an example of his "planned luck," one could cite Gordon MacCreagh's article on Ethiopia, which forecast Mussolini's invasion the week it began. Lorimer had picked MacCreagh to do this piece because he considered him the outstanding authority on Ethiopia, and both believed that invasion was imminent, probably soon after the rainy season began. The timing was excellent. Lorimer was also prescient about the Crash. In the midst of twenties' prosperity, not yet halfway through the decade, he was sending warning signals in the *Post*'s columns. Five months before the event, an article by Garet Garrett, the magazine's veteran Washington correspondent, forecast imminent disaster. "Everybody knows it will burst—only, not yet," he wrote.

The *Post* carried articles by presidents, particularly Herbert Hoover, who was Lorimer's political best hope until he was elected. It covered developments in science, depicted the golden age of sports in the twenties, and worked hard for Republicanism in Lorimer's unsigned editorial column. The editor divided the magazine's articles into political and nonpolitical; the former were the most effective, but even the nonpolitical pieces had purpose and direction in many cases. There were few with no other purpose than to entertain. Sometimes it was hard to discern where interpreting America left off and telling America began. But the *Post* was also a teacher, leaning heavily toward autobiography and biography. "Much is to be learned between the lines of an autobiography as well as *in* the lines," a *Post* promotion piece observed sententiously. "More from an actual record of life than from deliberate and deadly attempts to tell the young man to be good and he will be happy."

Lorimer changed with the times. In the gloomy reactive days at the beginning of the twenties, with disillusionment heavy in the air, the editor, in one of his extremely rare newspaper interviews, asserted that "what this country needs is more professional Pollyannas." But five years later, he wrote, "There is in all so-called 'inspirational' literature a Victorian blinking of facts, a Pollyanna point of view that disgusts any young man of this day whose intelligence is greater than that of a child. . . . Pollyanna does not write our advertisements, neither does she contribute to our editorial columns."

Intellectuals scoffed at this assertion, but in fact, the *Post*'s articles were very much like its fiction; that is, faithful representations of what average people were thinking and talking about or of what they aspired to think about.

An examination of the *Post* for a single year, 1922, illustrates the mix that made Lorimer the chief interpreter of American middle-class taste. In that year, the magazine printed 272 short stories and 269 articles. The serials included two contemporary classics, *Merton of the Movies* and *The Covered Wagon*, as well as others by Irvin Cobb, Frances Noyes Hart, Nina Wilcox Putnam, Hal G. Evarts (the notable "Tumbleweeds"), and Clarence Budington Kelland. Fitzgerald and Ben Ames Williams were among those who contributed two-part stories. Among the articles were Augustus Thomas's autobiography, reminiscences by the famous Chicago newspaper publisher Herman Kohlsaat, Eleanor Egan's exploration of the still new Soviet Union, and Isaac Marcosson's "Europe in Transition" and "The Changing East." Lorimer also attempted to make world situations simultaneously real and palatable to the ordinary reader with Philip Gibbs's "Miss Smith of Smyrna" and "The Beggar of Berlin," short stories in form but articles in effect. Other articles dealt with transportation, immigration, Prohibition, conservation, farm prices, taxes, radicalism, the perenially sad state of the theater, motion-picture censorship, crime and the courts, town building, highways, floods, the Ku Klux Klan, the Boy Scouts, currency problems, yellow journalism, and, most particularly, articles on business and finance, the backbone of the *Post*'s nonfiction.

In that year of 1922, the *Post* also undertook to advise its readers on a variety of other subjects, including how to settle an estate, buy a home, get out of debt, bring up children, take care of babies, and collect antiques. It rehearsed the problems of traffic cops and telegraphers, explored the mysteries of radio and of balanced work, exposed disease carriers and speculation, argued about the price and distribution of milk, revealed the hotel business from the inside, praised life insurance, assailed the high cost of theater tickets, and rejoiced in the small-town home. All this encapsulated life in America as no other magazine had done. Lorimer's publishing philosophy was a simple one: "It's the business of buying and selling brains; of having ideas and finding men to carry them out." He believed there was a plebiscite on every issue of a periodical, determining its worthiness to survive, and no business, he said, could so quickly succumb if there was apathy or contentment on the part of an editor · or publisher.

While Lorimer and his magazine stood out like a mountain in Cyrus Curtis's lucrative stable of periodicals, similarities were also evident in the way all of them managed to change with the drastically shifting moods of Americans between 1921 and 1940. A content analysis by Patrick Johns-Heine and Hans H. Gerth, for example, showed that in these magazines' fiction there was a pronounced shift away from business heroes to professional heroes. In *Country Gentleman*, a similar shift of locales took place, from the farm to the small town. Presumably, these were conscious efforts being made to attract a broader range of subscribers. In the *Journal*, there was a striking tendency to make heroes and heroines younger than the readers; about 60 percent of heroines were in their twenties, while market research in the same period was disclosing that 45 percent of *Journal* readers were over thirty-five.

A sharp difference between the *Journal* and the *Post* was the former's insistence that most of its heroines be responsible for creating a happy home and family—what every woman strives for, it implied. Career women were seldom depicted as having either. The reward for virtue and appropriate middle-class behavior was love.

In these magazines, as in others, the prime factor in creating change was the shift in national mood between the expansiveness and optimism of the twenties to the traditionalism and conservatism (in a social sense) of the thirties. The rise to success was followed by the glorification of the little man, and magazines like Curtis's, directed to mass middle-class audiences, reflected this swing faithfully.

In reaching out toward the developing mass market of the twentieth century, *Collier's* had appeared to be emerging as a chief rival of the *Post*, but by the early twenties, that prospect appeared unlikely; in fact, there was a question of whether *Collier's* could survive. While the *Post* had the immense advantage of a great editor consistently at the helm, its potential competitor had gone through several editors and lost most of its editorial vitality. The issue of April 16, 1921, had only twenty-eight pages, and the pre-Christmas issue of December 10, traditionally heavy with advertising, carried only six-and-a-half pages, two of which were house ads.

Plainly, drastic measures were required, and they were taken. *Collier's* management invested more capital and at the same time shook up both the staff and the editorial format. Two new editors, Richard Walsh and Loren Palmer, began to restore some of the lost glory and circulation. By 1925, there were popular names in its columns: Octavus Roy Cohen (who did not write exclusively for the *Post*), Zona Gale, Rupert Hughes, and Sophie Kerr. There was also something new in magazine fiction, the "short-short story," which became the trademark of Collier's for many years.

But every magazine, particularly a mass-market magazine facing heavy competition, needs a superlative editor, and the Collier organization found one at last in William Ludlow Chenery, who came over from the New York *World-Telegram*. He remained as editor until he became publisher in 1943. His managing editor Charles Colebaugh completed the new team, which proceeded to give *Collier's* a long period of solid growth. The staff these men assembled remains as one of the best ever to edit (and write for) a mass-market magazine. Their names are legendary in the business: Quentin Reynolds, Walter Davenport, T. R. Ybarra, Kyle Crichton, and Kenneth Littauer. Damon Runyon, too, became a contributor, and the pages were brightened with the work of some of the best cartoonists in the business.

In contrast to Lorimer's *Post*, *Collier's* was brightly liberal. Its covers were cast somewhat in the same mold—striking figures against a white background, with the names of some contributors listed, including several of the *Post*'s best short-story writers. The accent was heavily on fiction, although there were articles as well, some of them excellent, by Reynolds, Ybarra, Crichton, and others. The cartooning was first-rate, and for a long time it appeared that *Collier's* would be one of the successful mass-market magazines, even if not in the *Post's* class. But then Colebaugh died; Chenery retired; and after the Second World War, *Collier's* fell into a decline from which it never recovered.

Oddly, it was best remembered years after its demise for its covers, as Rockwell's covers had identified the *Post*. One of the *Post's* frequent cover artists, F. X. Leyendecker, also drew for *Collier's*, along with Charles Dana Gibson (in his time), Maxfield Parrish, Arthur Szyk, and others equally famous. There were also notable cartoonists in the magazine's history, beginning with E. W. Kemble in the early days, going on to Rube Goldberg, Ted Key, Virgil Partch, Boardman Robinson, and Otto Soglow.

On the covers of *Collier's* were names that appeared in the *Post* as well, but *Collier's* developed its own stable too, including H. G. Witwer, Dana Burnet, Sax Rohmer and the redoubtable Fu Manchu, P. G. Wodehouse, Richard Connell, W. R. Burnett, and John Gunther.

As a mass-market product, *Collier's* demonstrated, as Lorimer had done, what could be accomplished with talent and ideas. It also proved once more the absolute necessity of a strong hand at the top over a long period of time, something *Collier's* enjoyed for only a relatively short period.

For all his tremendous abilities, however, Lorimer had not invented anything new with the *Post*. He had simply transformed old formulas and introduced them to a different world. Luce had been the true innovator with *Time*. DeWitt Wallace's *Reader's Digest* is often cited as another authentic revolutionary magazine development of the twenties, but it was hardly a new idea that Wallace introduced. He took a very old notion, the eclectic magazine, improved it, and made it the most successful single magazine ever published until recent times.

The Wallace story never ceases to amaze, and in fact, it amazed Wallace when he was alive. Walking with a visitor one day across the inner courtyard of the *Digest's* neo-Gothic plant in Chappaqua, New York, he looked around him and said with wonder, "Sometimes I can't believe it all really happened."

Born in St. Paul, the son of an eminent father who was president of Macalester College as well as being a Greek and Latin scholar and a Presbyterian minister, Wallace graduated from Macalester and the University of California and worked briefly for the Webb Publishing Company, a firm in the textbook and farm-magazine business, before joining Brown & Bigelow as a salesman for their line of calendars and advertising specialties. He was already thinking about his idea for the *Digest*, a modern version of the oldest magazine idea in America. The variation he had in mind, however, was significant. Instead of merely clipping and pasting from other sources, he would edit the articles he chose, shortening them for readers with less time to read and short attention spans. Shortening copy had always been a standard practice, but Wallace intended to make an art of it, preserving the essential meaning while trimming away the noncommercial ornamentation. He hoped to make as much as $5,000 a year from his idea.

While he was recuperating from wounds suffered in the Great War, Wallace practiced what he intended to do, and as soon as he came home to St. Paul, he made up a dummy that included thirty-one articles clipped from old magazines he had found in the public library and had shortened by the method he had devised. But when he showed the dummy to New York publishers and possible financial bankers, he was told the idea had no chance of success—a familiar refrain in publishing. Meanwhile, Wallace was fired from his job in the Westinghouse Company's publicity department. That proved to be a turning point, because he now turned his attention full-time to the project.

His marriage to Lila Bell Acheson provided him with not only a bride but also a full working partner in his venture. Borrowing $5,000, they put together the first issue of the *Reader's Digest* in a Greenwich Village basement. On their wedding day, Wallace sent out a mailing of several thousand circulars to solicit subscriptions. That was October 1921, and the postwar slump still lingered. Then the couple went away for a brief honeymoon. When they returned, they found 1,500 charter subscriptions waiting for them. On the basis of that response, they published the first issue of the *Digest* in February 1922.

Wallace had definite ideas about what his magazine should be. He would not select articles simply because of his personal predilections, as the old eclectics had done, but would demand that each selection meet three criteria: applicability, lasting interest, and constructiveness. These were scarcely objective standards, but it was Wallace's interpretation of the words that were the decisive factor. Like Lorimer, he had a sure instinct for the mass market, the great middle class, not yet eroded by growing extremes of wealth and poverty. "Applicability" was not a new notion; many magazines had tried to do that. "Lasting interest," by which Wallace meant articles worth reading a year later, depended solely on his judgment of the audience. But in the third criterion, constructiveness, he had hit upon a prime ingredient, one that had much to do with making the *Digest* America's favorite magazine for many years.

What did he mean by "constructiveness"? Wallace liked to answer that question because he celebrated, along with Luce, the eternal optimism of Americans. While these two men were different personalities in some ways, they did share a good many attributes, including Republicanism, a strong belief in traditionally professed American values, and a driving faith in what they were doing. Both came from deeply religious Presbyterian backgrounds, and this, too, influenced the moral postures of their magazines, particularly of the *Digest*.

It was this perennially upbeat attitude that infused the *Digest*, reflecting a dominant middle-class attitude. The *Digest* saw "sermons in stones, books in the running brooks, and good in everything." It was designed, as Wallace himself put it, "to promote a Better America, with capital letters, with a fuller life for all, and with a place for the United States of increasing influence and respect in world affairs." A prelude, it seemed, to Luce's "American Century."

It was a philosophy in which Wallace deeply believed, and it was accepted wholeheartedly by *Digest* readers.

Another element in the *Digest's* immediate success, not to be equaled until *Life* appeared, was that it offered a bargain, something people found hard to resist. Looking down the contents on the cover, readers saw articles condensed from magazines large and small, prominent and obscure. Wallace was offering his customers the "best" from the periodical press, shortened to an easy length, and cast in a pocket size that could be read easily in bedroom or bathroom, buses, trains, or airplanes—and all this for a bargain price.

The success of the *Digest* far exceeded Wallace's expectations. He and Lila soon had to move the office from their Greenwich Village basement. They relocated to Pleasantville, a small town in Westchester County north of New York City, first in a garage, with the addition of an adjacent pony shed, and then in a house the Wallaces built for themselves with early profits. In 1939, the *Digest* moved again, to nearby Chappaqua, to a hill overlooking the Saw Mill River Parkway, to a building Mrs. Wallace had designed herself. She filled it with a formidable gallery of Impressionist and post-Impressionist art that she and her husband had begun to collect.

Mrs. Wallace also supervised the decor of her husband's capacious office, changing it periodically, usually in a drastic way. Thus, a visitor might discover the founder seated at a contemporary-styled desk with a huge Chagall on the wall behind him. Returning at another time, Wallace would be ensconced in eighteenth-century splendor. Mrs. Wallace also designed the offices of top executives and editors, who were given the privilege of deciding what period they wanted to work in and the added privilege of later change. For those farther down the company ladder, however, the offices assumed a conventional sameness. Muzak was installed throughout the plant, with scientifically designed programs to stimulate workers at various times of the day. Only those in the upper echelons could turn it off.

Knowledgeable visitors were awed as they sat in a small reception room off the lobby and gazed upon the walls on which were hung genuine Utrillos, Cézannes, and Renoirs, part of a vast collection, some of which came to rest in the Metropolitan Museum of Art in New York after the death of both Wallaces. The building itself cost $1,500,000 originally, unprecedented in the magazines business, and in time it was surrounded by a colony of *Digest* personnel, living incestuous corporate lives under each other's inquiring noses. It was rural luxury.

In a relatively short time, the *Digest* became rich and powerful, which meant that it was likely to become controversial. And so it did. At some point, the initial clipping-and-shortening system was abandoned, and the *Digest* began to develop its own articles, "planting" them in selected magazines, to be "reprinted" later. The *Digest* paid the magazine for this privilege, which theoretically put the recipient in at least a partially subsidized position. To a small mazazine like the *American Mercury*, struggling along in the early forties in its post-Mencken days, the subsidy represented a substantial part of the total income.

If the *Mercury,* or any magazine under contract, refused an article sent down from Chappaqua, there was no difficulty. The *Digest* simply sent another one. Complications arose, however, if a recipient magazine shortened a *Digest*-submitted article to fit its own space. Then the new version would have to be sent in proof, or read on the telephone, to an editor in Chappaqua so that the *Digest* "reprint" coincided with whatever the recipient magazine had done.

Among those who objected to this system was Harold Ross, whose *New Yorker* had also begun in the twenties. Ross was a man of considerable moral purity in his attitudes, and he saw something essentially dishonest in the *Digest*'s practice. He denounced it publicly and declared that his magazine would have nothing to do with it. But most other periodicals did not follow his lead, either because they needed the money or because they wanted the additional publicity resulting from a *Digest* "reprint."

Those who continued to protest argued that the system gave the *Digest* power to propagandize its right-wing political views across a broad spectrum of the periodical press. This criticism grew stronger as the *Digest*'s staff came to include more and more hard-line conservatives. It was clear that the magazine used its columns and those of other periodicals to promote its highly partisan views on everything from sex to communism. (It was opposed to both.)

Individual characteristics of the magazine also came under critical scrutiny; for example, the *Digest*'s medical articles, known as "New Hope" stories because they were usually titled "New Hope for . . ." whatever human ailment was under discussion. Many times the hope lay in some drug or other treatment still in the laboratory stage. Publication of the articles produced waves of anguished telephone calls from afflicted individuals and their families to local doctors, pleading for the "new treatment" or the "new medicine." Occasionally the articles were so unsound as to be actually dangerous and eventually led to the resignation of the editor who had been in charge of them. But the series went right on for years. An office joke declared that they would not end until the final, logical title, "New Hope for the Dead."

The *Digest* had begun as a magazine without advertisements and proved the exception to the rule that such periodicals could not succeed. Having made that point, the magazine reversed itself in 1955 and announced it *would* accept advertising, though restricting the kinds it chose to carry, prohibiting liquor ads, for example. The pressures from advertisers to place ads in the magazine with the world's highest circulation were the greatest ever known in periodical history. The *Digest* could, and did, set the highest rates. As a result, it bacame even richer.

As time went on, too, the proportion of reprints—"pickups," as the editors called them then—decreased steadily. (By mid-1969 about 65 percent of the material used was original, with the percentage continuing to increase.) Thus, the *Digest* entered the postwar era, in which it would become a multifaceted corporation of record clubs, condensed book clubs, record distribution services, and other assets, with the magazine's circulation worldwide at 17 million before long. This figure was not to be challenged until the advent of *TV Guide* and later, *Modern Maturity*.

Of the other attempts to establish new beachheads in the mass market that was developing in the twenties, one of the most noteworthy was the advent of *Esquire*, the idea of a young Chicago entrepreneur named David Smart. He had the brash courage to launch his magazine in 1933, at the pit of the Depression, and to charge fifty cents for it. This was comparable to Luce's audacity in asking a dollar for *Fortune* at a time when the *Post* was selling for a nickel.

Smart was not aiming at a *Post* kind of audience, however. His perception was that not only were Americans in need of a laugh in 1933, but that there was still a younger, relatively sophisticated audience that hoped for better days. The editorial formula he devised, with the help of his talented editor, Arnold Gingrich, began with a profusion of often slightly risqué cartoons, frequently with sexual overtones. The spirit of *Esquire* was embodied in a figure known as "Esky," drawn in the form of a frisky old gentleman in a top hat.

But if sex was the underlying appeal of the cartoons and sophistication its approach to men's fashions and interests, the tone of the articles and fiction was intellectual. Gingrich knew good writing when he saw it and filled the pages of *Esquire* with the work of such authors as Ernest Hemingway, Ring Lardner, Jr., and John Dos Passos, all of whom appeared in the first issue. The articles were bright and lively explorations of contemporary culture.

After Gingrich left the magazine in the forties, it faltered for a time, having made a mistake by assuming that armed forces' readers wanted their sex cruder and more explicit, which considerably cheapened the cartoons and fiction. But it would rise again after the war was over and go through several further changes. Smart's other adventures in the magazine business were equally notable. In November 1936, he started a magazine called *Coronet*. The initial circulation of 250,000 in time would rise to surpass the figures for any other pocket journal except the *Digest*. The new publication was announced with some grandeur in *Esquire*, using black and gold facing pages. In the text, potential readers were promised "infinite riches in a little room" and "the most beautiful of magazines."

For a time, the promise was fulfilled. The five-color cover of the first number enclosed pages filled with drawings, etchings, color reproductions of paintings by Raphael and Rembrandt, fiction, articles, and photography. The first edition sold out in forty-eight hours. Subsequent issues were resplendent with classical and modern paintings, portfolios of photographs, drawings and cartoons in color, along with satire, fiction, and provocative articles.

But the recession of 1937, coupled with a country made restless by the threat of war, seemed to challenge Smart's contention that "beauty is still a very potent market," and in 1949, he was selling only about 100,000 copies. At that point, Smart called upon one of the authentic geniuses in the mass-market world, Oscar Dystel, to save his pet project.

Dystel had begun his magazine career in 1937 as circulation manager of *Sports Illustrated* and the *American Golfer* but the following year moved over to Smart's organization as circulation-promotion manager for both

Esquire and *Coronet*. He quickly became circulation manager of the latter and then, at the critical point in 1940, was called upon to assume the editor's chair. Dystel turned the magazine around in every sense, banishing great art works from the covers and replacing them with photographs of models and starlets. Inside, the content was changed to conform more with what was selling in other digests and pocket magazines. In his deft moves to create the right formula, Dystel went directly to the readers and asked for their preferences.

For the next two years, the results were sensational. Circulation figures were not made public, but it was estimated that *Coronet* reached 5,000,000 during the boom war years. Dystel left it briefly in late 1942 but returned again in 1944 to serve another four years. In 1947, he persuaded Smart to accept advertising, but at that juncture, the magazine could guarantee only 2,000,000, still a respectable figure for a publication that had been at death's door when Dystel took it over.

During the 1950s, *Coronet* was caught in the intense drive by publishers to reach ever higher circulations, and by 1961, it was losing so much money—about $600,000 a year—that in July the publisher announced the October issue would be the last. The splendid ten-color presses installed at a cost of $1,300,000 in 1958 were bought by the *Reader's Digest*, together with a subscription list divided with Curtis.

Three years later, *Coronet*'s first editor, Gingrich, wrote of it regretfully, "That lovely little art magazine, full of treasures of the world's great museums, was about twenty-five years ahead of the great art boom, on which it might have ridden like a surfboard. Instead of letting it die a natural death, we chose to make it into something that it wasn't supposed to be, with the result that it lived all the twenty-five years of its life on the crutch of family support. It became that anomaly, the world's largest midget, or the world's smallest giant, neither of which is a title with great natural box-office potentialities. In any event, it wound up in 1961 with insufficient riches in too little room. It lived to attend its twenty-fifth birthday party, reminding me of the League of Nations buildings in Geneva, which were finished just in time for World War II."

Dystel, who had given *Coronet* its one brief moment of circulation fame, moved on to the book publishing business and in 1954 helped to found Bantam, one of the most successful mass-market paperback operations in the world, and became its president.

Esquire, Inc, two years after the launching of *Coronet*, introduced a new magazine, *Ken*, whose mission, as it said, was to tell the unvarnished "insider's" truth about the world, to be a source of "unfamiliar fact and informed opinion . . . equally opposed to the development of dictatorship from either Left or Right." Its tone, however, was decidedly liberal. One of its historians, Gordon L. Cohen, calls it "one of the strangest mass-circulation publications ever to appear." It became "the magazine everyone hated," almost from the first issue, which advocated controlling venereal disease by educating women. On the other hand, in May 1938, it was also the first magazine to advocate using automobile seat belts in an article titled "Cheating Death on the Highways." *Ken* also argued for installing parachutes on passenger airliners in a piece called, chillingly, "Stand By To Crash." That cost the magazine its United Airlines advertising.

Even the illustrations were often insulting. A set of pictures showing Miriam Hopkins and her then husband Anatole Litvak at a Hollywood party had been shot from an angle that made Hopkins's face look puffy. Katharine Hepburn was shown in a picture that heavily emphasized her freckles. An unflattering chipmunk-faced picture of Claudette Colbert was explained by an assertion that studio technicians had to tone down her cheekbones before her pictures could be released.

Ken, in its eagerness to expose tyranny, corruption, and the general exploitation of people, managed to alienate most of its advertisers and stirred up controversies that not only threatened its own life but Smart's corporation as well. Gingrich, its managing editor, wrote in his 1971 memoirs, *Nothing But People,* that *Ken's* editorial approach had been "sufficient in the long run to unite the entire country against it."

Yet the intentions of Smart and Gingrich were no less than admirable. Having celebrated the good life with *Esquire* and *Coronet,* they meant *Ken* to be a fighter against fascism everywhere in the world; it was antitotalitarian, the voice of the underdog. When the Chicago *Tribune's* correspondent in Spain, Jay Cooke Allen, was fired by Colonel Robert R. McCormick, the paper's right-wing publisher, for suspected Loyalist sympathies, Smart hired him as editor and gave him a budget of $30,000 to put the first issue of *Ken* together. But Allen took the money and went back to Europe for a personal look at what was going on; this cost him his job. George Seldes held the editor's chair briefly while the magazine was in its second stage of planning, but he lasted only a few weeks. Gingrich was already busy with *Esquire* and Smart's other projects, but since no one else was immediately available, he took the job himself and brought out the first issue on April 7, 1938.

Gingrich had considered Seldes too left-wing. His intention was to make *Ken* only slightly left of center, not far enough to discourage advertisers, and at first it appeared he might be successful. The first issue carried more than fifty-three pages of advertising. It also displayed several innovations, including a four-color airbrush cover; a full-size picture magazine in the center, taking up about 30 percent of the entire contents, the photographs referring to stories on the text pages; and a synopsis (called a "Visio-graph") set in 12-point bold to introduce each article.

But even before this initial issue went to press, *Ken* had offended its potential chief contributor, Ernest Hemingway, by making an editorial insert saying that although the novelist had been contracted and announced as editor, he had had nothing to do with forming the magazine or with its policies. "If he sees eye-to-eye with us on *Ken* we would like to have him as an editor. If not, he will remain as a contributor until he is fired or quits."

Hemingway, whose sympathies had been with the Spanish Loyalists, suspected that *Ken* was going to be a red-baiting magazine and wanted no part of it. But when Gingrich said he was opposed to fascism everywhere, he made no distinction between left and right. Although Gingrich's opposition to communism was relatively mild in the first issue, he did run a color-graphic section in which Stalin was caricatured, among eight tyrants, as "his red-blooded loneliness, nobody's comrade." The third issue listed Stalin, Hitler, Mussolini, and Franco as despots who ruled through torture.

Although Hemingway stopped writing for *Ken* almost immediately, he did send some pictures from the Spanish front that were so gruesome they offended a good many readers. What he wrote to accompany them was also an offense to those who equated antifascism with communism. Hemingway also sent an article attacking fascist sympathizers in the Catholic Church and thereby touched off a boycott that seriously threatened the entire Esquire organization. It was a boycott that endured until *Ken*'s death in 1939. Parish priests advised their congregations to inform their neighborhood magazine outlets that they would buy nothing from them until they stopped selling all of Smart's magazines, and this resulted in killing the sales of the three periodicals in 1,500 cities and towns.

In the third issue, *Ken* managed to offend still more readers by running the memoirs of a Los Angeles prostitute. The piece contained no lurid details but did condemn employers who exploited women and attacked politicians who preached morality while at the same time patronizing prostitutes. The uproar over this article resulted in the temporary suspension of *Ken* from the mails.

With only two or three issues, *Ken* had managed to incur the enmity of people on both the right and left, the Catholic Church, and a wide assortment of other institutions. Its initial press run of 500,000 was cut in half, and advertisers drifted away daily. Gingrich attempted to dilute the political ingredients in the contents mix, but he did it with a Pulitzer-like concoction of sex and politics, in which sexy pictures were used to illustrate articles that were really political exposés. Thus, an article titled "Box Office Blues" and illustrated with titillating pictures of movie stars was actually an attack on the movie booking system. Similarly, an article exposing radio network hypocrisy in posing as upholders of public morality while using sex for promotional purposes was illustrated by three pages of pictures showing female stars in sultry poses and various states of undress. All these pictures had come from the networks. As *Life* would do later, Gingrich's *Ken* edged tactfully toward displaying total nudity but came no closer than a picture showing a bare-breasted South Seas beauty who could as easily have appeared in the *National Geographic*.

No matter what readers' politics might be, they would have had to acknowledge *Ken*'s investigative reporting and its frequent results. Its writers exposed the existence of Japanese and other spies in the Canal Zone, a potential coup in Mexico (that was thwarted as a result), the work of Gestapo agents in Prague, and, on a far more commonplace note, Hearst's successful attempt to suppress Walter Winchell. Long after the magazine died, its predictions about events kept on being verified.

In a final effort to save *Ken*, it was made a weekly on its first anniversary and was scaled down to a smaller size, the editors hoping by doing so to give advertisers the circulation guaranteed as a monthly. Ideologically, Gingrich

sensed that war was coming that year, 1939, and believed people would realize that he had been running a patriotic operation all the time. In June, the
price was cut, from twenty-five to ten cents. Nothing helped. It expired with
the issue of August 3.

Gingrich and Smart split up in 1946, but they renewed their relationship in
1952, when Gingrich returned as editor of *Esquire*, beginning an era when it
would be one of the most successful magazines in the country, thick with
advertising, its fiction widely read and admired, and its articles acting as the
social conscience these two remarkable men had intended from the beginning.

While the Esquire organization was experiencing its ups and downs, a
smaller rival was also exploring mass-market possibilities. Alex L. Hillman had
built Hillman Periodicals Inc. into a small empire that included confession,
factual detective, fan, and comic magazines. Observing what Smart was
doing, he began his own pocket magazine, called *Pageant*, in November 1944.
Observers in the trade joked that it looked like the offspring of a mating
between the *Reader's Digest* and *Coronet*. Like the latter, it was not taking
advertising even though it had been launched in a time of shortages and rising
production costs. Eugene Lyons, who had been editing the *American Mercury*, was the first editor but lasted only six months; it was simply not his field.
In that short time, however, *Pageant* had run through nearly $300,000 of its
original $500,000 budget, and Hillman had made a bad decision to keep it
supplied with scarce paper by suspending his profitable detective and comic
magazines.

Vernon L. Pope, who had helped to develop *Look*, succeeded Lyons, promising to bring in a million readers within six months, but when he left two
years later, circulation had fallen from 500,000 to 270,000, and $400,000 more
had vanished. The end came soon after.

With the exception of the Smart group, all of the explorations into the mass
market between the two wars employed essentially the same formula, their
model being the *Post*'s popular reflection of the American experience. They
simultaneously provided information, entertainment, escape, opinion formation, a reinforcement of moral values, and, through advertising, a vast consumer's showcase. Stephen Holder has identified these as the characteristics of
the "family magazine," which established closer relationships with large segments of the mass market.

To read these magazines today in the archival files is to survey the social
history of the times. They tell us much about what America was in those years
between the great wars. As family periodicals, they intended to conform, not
transform; to inspire, not criticize; to inform, but not to analyze—with some
exceptions. The fiction in their pages was a faithful reflection not only of the
morals but of the literary taste of the mass market. They were a constant reinforcement of what the audience already believed. That was true even of the
humor and cartoons.

As several media historians have observed, the heroes then offered by the
mass magazines were no longer the big businessmen and outsize public figures
of the nineteenth century but small businessmen, farmers, and workers. Lor-

imer may have printed Frank Norris, but the great bult of the *Post*'s fiction had a happy ending, as did that of its contemporaries. Gone, too, was the old Alger-like plot resolution through luck, achievement, or even marriage as the road to ultimate success. Now the denouements depended much more on the solving of human problems.

Family magazines in this period established a rapport with their readers which has hardly been matched since. Through editorials, contests, and various reader-participation schemes, subscribers were made to feel that it was "their" magazine. Editors were careful to establish a continuity that extended from the covers to the regular appearance of certain features and writers. If a particular issue displeased readers, at least they could be assured of finding two or three "old faithfuls" to satisfy them.

Even the advertising was keyed to this overall concept. There was continuity in much of the style and content, and often it was educational as well. Most of all, it displayed a standard of living against which readers could measure their own life-styles.

All these criteria applied to the more specialized mass-market magazines, notably those that dealt specifically with the home. Their leader was the journal E. T. Meredith began in 1922, *Better Homes and Gardens*, the beginning of a small empire that continues to thrive. Meredith, who had been secretary of agriculture in the Wilson cabinet, was already in the periodical business with *Successful Farming*, but he was prepared to stake all his resources on a magazine for home-centered American families, which would include the largest part of the American population. It would be specialization with mass-market dimensions. He launched it under the unfortunate name *Fruit, Garden and Home* but changed it two years later to the title under which it would become almost an institution.

Meredith did not make the mistake of directing his new magazine to an audience of women who, by the standards of the time, would have been considered the keepers of home and garden. He intended it for people of both sexes with moderate incomes, "average" families who would benefit from suggestions about how to care for their properties and improve them. While Chela C. Sherlock was named as first editor, Meredith took no chances; he supervised closely every step of the magazine's development, including selling advertising himself. He even used a device employed before by other publishers: urging his readers to bring in new subscriptions so he could divert more circulation promotion money to improving the product. The audience seemed to think this was a fair exchange. They brought in thousands of subscriptions and sent the circulation over the million mark in four years.

In the twenties, *Better Homes and Gardens* thus became something of a novelty: a successful magazine without fiction, fashion, or sex, the prime ingredients used to attract readers in most mass-market periodicals. Meredith also understood where the bulk of his readership lived. After Sherlock left as editor in 1927, he hired Elmer T. Peterson, a midwestern writer and editor who helped focus the magazine on the region where it was published, in Des Moines, Iowa. It was not a total concentration, however. Staff writers traveled around the country on assignments, and contributions were accepted from writers no matter where they lived. But by and large, *Better Homes and Gardens* was a midwestern magazine in the broadest sense, with midwestern editors and a confirmed midwestern publisher.

Soon after Peterson took over the reins, Meredith died, and it became the task of the new editor to nurse the magazine through the Depression. He directed it even more to providing information that readers would find useful and could apply during hard times. He also departmentalized the journal into such categories as gardens, buildings, foods and recipes, and home management. Readers were drawn into the process by being offered prizes for contributions on how they used information, either from editorial or advertising matter, to solve their own problems. The magazine also took the step of guaranteeing all the products in its advertising, and Peterson employed the old *Youth's Companion* method of offering readers prizes for bringing in subscriptions by selling to their neighbors.

Peterson was careful to let his readers know that his staff practiced what they preached. In the departments, the editors talked about their own home problems and how they were being solved. Alfred Hottes let it be known that he was a "dirt-under-the-nails" garden editor, and it was duly reported that someone had said of Fae Huttenloche, another editor, that she had put together her flower arrangements from "this and that," meaning that she had used things a housewife could conceivably find in garden or cupboard at home.

With this kind of management, *Better Homes and Gardens* survived the Depression in acceptable shape and contrived to live through the paper shortages and paucity of consumer goods during the Second World War through truly creative efforts by its staff to supply the changed needs and limitations of its readers living in wartime conditions. Momentarily, for example, remodeling and renovation took the place of new housing. The magazine entered the postwar era in better condition than most, and the period of its greatest success lay ahead.

Working in a completely different area of the mass market in the twenties, the eccentric entrepreneur Bernarr Macfadden carved out still another large segment of the mass audience by appealing to those interested in their own bodies and the bodies of others. It was Macfadden's idea to make magazines out of two primary mass-market ingredients: physical culture and sex. He made a fortune from the attempt. Sadly, it would be impossible to convey the peculiar flavor of his life and works in a few paragraphs, a life which made him eventually a familiar figure in America, with his white-maned head and sharp, aggressive features displayed in the pages of newspapers. The press happily recorded his many marriages (the last occurred when he was eighty); his bizarre exploits, like his parachute jump into the Hudson River when he was also in his eighties; his exhibitions of physical fitness decade after decade; and his views on every subject from international relations to happy marriages.

Macfadden introduced something entirely new to the business, the confession magazine, and greatly improved an old idea, the crime magazine, by introducing the true-detective periodical. His *True Story* became the proto-type of the numerous publications in that field, including a whole series of "True" magazines published by Macfadden himself. His *Physical Culture* magazine, begun in 1943, was, similarly, the prototype of all the large-muscle periodicals and was devoted to the relief of those weaklings who were weary of having beach sand kicked in their faces by muscular bullies, or so the maga-zine's advertising promised. Then there was Macfadden's notorious excursion into newspaper journalism, the New York *Graphic*, known affectionately in its day as the *Porno-Graphic* and later as "the world's zaniest newspaper." Its famous "cosmographs," which were sketches of courtroom scenes with pho-tographic heads of the participants pasted in, made newspaper history.

Perhaps some of the flavor of Macfadden's publishing life can be conveyed by recalling the days when the *Graphic* and the magazines were being pub-lished together out of the same Manhattan loft. Macfadden would stride into the *Graphic*'s city room in the morning, clad only in his familiar leopard-skin loincloth, leap on the city desk, summon his entire staff of scandal-chasers to rise to their uncertain feet, instruct other minions to open all the windows wide, even if it was the heart of winter, and lead the assembled multitude in calisthenics.

Some of Macfadden's other excursions into publishing included his purchase in 1931 of *Liberty* magazine, which he used first as a platform to advocate the election of Roosevelt. It then became a popular, cheap magazine, one of whose features was a "reading time" note preceding each article or story and guaranteeing that it would take no more than "10 minutes, 30 seconds," or whatever it might be, to absorb what was in a particular segment of type. Under another flamboyant editor, Fulton Oursler, *Liberty* prospered by exploiting sex, scandal, and politics in a time-worn pattern until Oursler expe-rienced both a financial and religious conversion, abandoned sensation, and became editor of the *Reader's Digest*, therein defending conventional mid-dle-class values with the same enthusiasm and disregard of facts.

Macfadden's final new magazine was *Babies, Just Babies*, which appeared in 1932, listing Eleanor Roosevelt and her daughter as editors, thus providing a satiric target for Roosevelt-haters but not returning much income. It was short-lived. However, there was no denying that Macfadden had the ability, as McClure and others had (perhaps beginning with Frank Leslie), to reach a mass market with material that required little cerebration.

At their peak, his magazines had a combined circulation in 1935 of more than 7 million, larger than that of any other group. Three years later, Mac-fadden Publications, Inc., was in the red, and two years after that, under the pressure of stockholders' suits, Macfadden himself had to retire as president. He died in 1955, at eighty-seven, leaving his empire in other hands. It man-aged to get itself into the black once more, but without Macfadden it could never be quite the same, even though it continued to prosper.

From both a cultural and sociological standpoint, Macfadden's most signif-icant contribution had been the confession magazine, widely copied from the *True Story* basic formula. By 1950, there were more than forty titles in this category. While they appeared to be lurid (for their time), they were actually quite moral, even though the heroines of their stories, true or otherwise, were offenders against conventional moral standards. But as their editors noted, there were two rules: sin must be followed by some kind of retribution, and morality must be recognized as such—"Slowly she came to realize," as one editor put it.

The line between "confession" and pure romance was rather carelessly drawn, and there was some overlapping. Advertisers might scorn what was between the covers of these periodicals, but they could hardly ignore the circulation figures—collectively, more than 16 million guaranteed. Among the first eighteen titles, the average sale was more than 7 million copies per issue. Even so, space buyers could not quite swallow the romance-confession category as reaching a viable market, as could be seen by the fact that twelve women's service magazines carried twice as much advertising as the sixteen leading confession journals.

George Gerbner, a sociologist who became dean of the Annenberg School of Communications, analyzed the confession magazines in 1958 and pointed out that their audience consisted mostly of people who had not been magazine readers before, people with little education and low purchasing power. The contents were written by and always edited for (said Gerbner) what the *Post* called "Macfadden's anonymous amateur illiterates." That might describe the audience, but the content of the magazines was heavily staff-written, and when contributions were accepted (for fees as low as fifty dollars), they were skillfully rewritten. Some newspapermen made extra income by "confessing" on behalf of women whose stroies they had covered in the course of their daily work. Gerbner reported that research and distribution data showed that sales were largely concentrated in small towns, the South, and the Midwest. Readers were likely to be young women from wage-earning families. Their incomes were significantly lower than the readers of the major women's magazines, their educational levels also lower, and their children more numerous.

These statistics made the magazines a particularly hard sell to advertisers, who were not much interested in audiences with little spending money. Consequently, the advertising deficiency had to be made up in circulation, which placed a double burden on the editorial side. Editors had to remain faithful to the known tastes and desires of their readers, but at the same time they hoped to find ways of altering the formula just enough to attract other readers who might persuade reluctant advertisers.

Macfadden supplied them with ammunition. Readers of confession-romance magazines, he said, lived in a mythical "Wage Town," whose inhabitants constituted 54.8 percent of all American families. They might be working-class people, but they were candidates for the middle class, and many of them made it. They were also people who were in need of direction about how to live and how to spend their money. They read these magazines for guidance. By the fifties, Macfadden's fellow publishers had found a name for this category that sounded much more respectable—Family Behavior Magazines.

Fred Sammis, editor-in-chief of *True Story* Women's Group, laid out the financial promise of this market in a booklet he called "The Women That Taxes Made: An Editor's Intimate Picture of a Large but Little Understood Market." This effort and others provide a remarkable example of how one category of magazines adhered to its basic editorial formula, knowing that was what its readers wanted, and tried hard to convince advertisers of the market's viability. They did not choose the usual alternative of shifting the formula to attract different readers, nor did they try to "weed out" unwanted readers. Sammis asserted that the formula stressed individual and family responsibility for problems, not the failure of social or economic structures. He told the advertisers that, in fact, economic grievances were minimized in these magazines, as not befitting the market status of "the woman that taxes made." Whatever social resentment existed, he said, was structured to center on "behavior problems."

Dubious though these economic arguments might have been, there was no denying that the formula had been cleverly crafted, Gerbner observed in his analysis. It focused on the insecurities of working-class life in a world where consumption centered on the middle class. Whatever risks might lie in this appeal to class divisions was minimized by making social protest appear to be irrelevant.

In the confession stories could be seen the plot formulas that would make mass-market successes in our time out of such book categories as romance novels. Readers viewed men as powerful, dominant, and sexually active, in contrast to those who read the middle-class women's journals. There was enough realism in the confession stories so that readers could identify with the characters, and use of the first person aided that effort. Critics erroneously called these stories "escapist," but in fact, they depicted real situations that readers could understand and that expressed their values. At the same time, these journals offered a satisfactory amount of advice on how bad situations could be handled and made better in the everlasting search for happiness. Heroines were shown fighting back against a brutal real world. In that world there was sin, suffering, and repentance, with suffering an essential element—just as it would be in many of the later romance novels.

Gerbner saw social class as a determining factor in the confession magazines, but others (Wilbur Schramm, for one) found it unimportant. Other researchers, taking a different approach, have argued that *True Story*'s content was not basically different from the fiction in the *Ladies' Home Journal*, since the dominant subject in both was love, with the firm implication that it is the greatest reward and satisfaction in life. *Journal* heroines were shown as preservers of values, responsible for happy homes and families, with virtue and middle-class behavior rewarded by the love of good men. *True Story* heroines, these researchers pointed out, demonstrated what happens to those who step outside the moral order: positive symbols of safety and security are changed by such an action into negative and harshly punitive symbols.

Confession magazines did not have the cheap mass market to themselves. Another substantial category was the pulp Western, a twentieth-century expression of America's love affair with the Great West that had begun in the Republic's early days. In his valuable book *The Pulp Westerns: A Popular History of the Western Fiction Magazine in America*, John A. Dinan has given a comprehensive view and analysis of this relatively neglected category that so engrossed American men for decades.

Dinan separates this field into two broad types: the super-hero magazine, and the love-story Western. Yet there was infinite variety, "a title for every possible character, subject, or mood," as he says. With few exceptions, they were all much alike. Their origins were in the dime novels of the late nineteenth century, such Street and Smith series as the Log Cabin Library, the Jesse James stories, and the Red, White and Blue Library. Competitors included *Frank Munsey's Weekly*, the *New York Detective Library*, and *Wild West Weekly*.

As Dinan observed, a starting point for this genre might well be the period from 1878, when Diamond Dick's adventures appeared, to a time just after the turn of the century, with the appearance of *Young Wild West*. Such dime-novel Westerns as *Seth Jones, or Captives of the Frontier* sold more than 400,000 copies. Clearly, a mass market for magazines was waiting, and their initial explosion occurred in the first twenty-five years of this century. Frederick Faust and Fred Glidden were the early pioneers, their Western stories, now classics, appearing in such general pulp magazines as *Everybody's* and *Popular Magazine*. The genre got a powerful stimulus from Owen Wister's *The Virginian*, a book that influenced all Western fiction for years. Its hero, as portrayed by Gary Cooper on the screen, was the prototype of his kind. His famous command, "When you say that, stranger, smile," epitomized his soft-spoken-but-tough style.

The years running roughly from the mid-1920s to the mid-1940s appear now as the golden age of Western pulp fiction. These stories were no longer sandwiched in with other adventure tales in pulp magazines directed to males but were concentrated in all-Western publications. The best were written, moreover, by men who knew from experience what the West was really like and who could capture it in exciting prose. Unfortunately, the magazines had proliferated, and the demand was so great that desperate editors had to buy from amateurs who flooded their offices with stories demonstrating only that the writers had learned the formula and were able to grind out pieces with a minimum of talent. Unlike the confession-magazine stories, they could not be rewritten nearly as easily, and besides, there was no time.

Nevertheless, two writers of formidable popular reputation emerged: H. Bedford-Jones and Ernest Haycox. Both wrote what could be called historical Westerns, which won the admiration of some historians. Haycox refined the traditional hero and made him more believable, at the same time creating stories good enough to be starring vehicles for actors like Cooper and others. Bedford-Jones painted with a broad brush, creating stories of epic proportions and going for panoramic sweep instead of character detail.

During the final period, from the mid-1940s to the ultimate death of pulp magazines and pulp fiction in the early sixties, the romance Western rose to new levels of popularity, and the general quality of the writing improved as writers who would go on to better things became contributors.

To older Americans in the closing years of the twentieth century, the names still have a peculiar magic. The oldest will remember *Western Story Magazine*, the prototype of all Western pulps, which in its thirty-year existence introduced such writers as W. C. Tuttle; Frederick Faust, whose pen name was Max Brand; and Fred Glidden, who signed his stories Luke Short. In the early 1920s, Western stories could be found in Munsey's *Argosy*, Ridgway's *Adventure*, Doubleday's *Short Stories*, and McCall's *Blue Book*, among others. Then came the all-Westerns: *West, Frontier, Cowboy Stories, Ace High, Ranch Romances, Western Adventures*, and *North West Stories*, among many others. Of them all, says Dinan, *Western Story* was no doubt the most popular. Typically in every issue were three novelettes and four or five short stories.

During the Depression, several of these magazines cut their price from fifteen to ten cents, and some made drastic changes to survive; *Wild West Weekly* introduced "Four Complete Western Stories" in a single issue, but in the form of comic strips. Such changes signaled that the end was not far off. Radio, movies, and the comic book were beginning to cut the ground from under the entire pulp-magazine industry.

In the summer of 1943, the *Weekly* failed to appear for the first time in forty-one years. In November, it was gone, ostensibly a victim of wartime shortages but in reality overwhelmed by changing social forces. By the early 1950s, the pulp era had ended for all practical purposes. The distributors as well as the public had had enough. Nearly all the Westerns were gone by the 1960s, and attempts to revive the past with brand names—*Zane Grey Magazine* was the best example—were failures. By that time, television had finished the job.

Christine Bold, who has studied the life and death of pulp Westerns, believes that the "development and decline of the storytelling voice in these magazines mirror changes in the publishing world, where authors gradually lost more and more of their authority over their fiction." This, she argues, "changed the dime author from a hired storyteller conforming to broad outlines, to a minor member of a collaboration who followed orders in the very writing of the fiction, and finally to a worker who was openly shown to hold second place to the reader in the editor's regard."

This may well have been true for the pulp industry, but it surely did not apply to other segments of publishing, where the continuing rise to power of the agent gave authors more authority than they had ever enjoyed before. It is clear, however, that pulp writers, in the Westerns and elsewhere in this broad category, were treated like house slaves. The average rate of pay was a cent per word, which at that was more than the dime-novel authors had been paid. In the gloomy thirties, however, it was still a profitable way to make a living.

One aspect of the pulp Westerns was unique. In the Letters column, a relationship was established between the magazine and its audience that was closer, perhaps, than those in any other category except women's magazines. Readers felt extremely close to the magazine; many lived in isolated places, were lonely ranch hands or real cowboys, and felt that the editors were their friends. If a woman ran the Letters department, under some such title as "The Lonely Pine Tree Mailbox," she was likely to get Christmas presents ranging all the way from saddles to live reptiles or Gila monsters from readers.

Street and Smith carried this affinity to practical lengths, telling letter writers that they could help fashion the stories themselves by informing the editors about what they would like to read or supplying real tales that authors could convert into fictional stories. Christine Bold observes that these stories in the magazines "became overt bargaining tools between publisher and public," a situation in which "the editor throws open the door of the author's study and invites all the readers in." That was admitted with admirable frankness by the editor of Street and Smith's *Popular Magazine,* the life of which extended from 1903 to 1928. "This occupation of magazine publishing," he wrote, "is, in our opinion, entirely a business proposition.... We publish the *Popular Magazine* as a money-making enterprise and we follow the line of least resistance in catering to our public."

Whatever the motive, pulp magazines in their half a century or so of glory occupied a unique place in the mass market, developing specializations beyond the West—*Love Stories,* and half a dozen other genres—and joining the confession magazines in tapping a significant portion of the mass market.

16

Intellectual Currents
in the Magazines

In a 1945 speech, Frederick Lewis Allen, then the editor of *Harper's,* outlined what he considered to be the function of a magazine in America. It should be interesting, of course, and provide news more wisely and selectively conceived than the daily press. There should be discussion and interpretation of important issues, again more completely than the newspapers but less intensely than books. Magazines, Allen said, should offer a forum for original thinkers who were not representing institutions or organizations and should also offer an outlet for "artists in literature."

Clearly, Allen was thinking of intellectual magazines like *Harper's,* since most of the mass magazines were designed for entertainment and the trade press was organized around specific areas of information. All magazines might employ parts of his formula, selectively and in their own special interest, but only those addressed specifically to an audience of "intellectuals," however defined, could hope to function as Allen had outlined their mission. Even in that category, however, there were wide differences between the intellectual character of the *Nation,* for example, and *Harper's.*

Allen recognized that his definition of a periodical's function was made up of ideals, not easily realizable realities, but he believed editors had an obligation to make the attempt to attain them. One of the obstacles he cited was pressure from advertisers, which would be extremely subtle, not direct, in an intellectual magazine like *Harper's.* Another source of pressure was the federal government, said Allen, but he was speaking in the unusual circumstances of wartime censorship and could not have foreseen the drive for secrecy and censorship by later administrations when war was not the excuse. A third source of danger to a magazine's independence, according to Allen, was its readers—better described as "pressure from the editor's own zeal for more circulation than their standards of thoroughness or of honesty or of impartiality will permit without compromise." Here Allen delivered an indirect punch at the editors of mass magazines. "There are a great many readers," he said, "people who might become readers, who want the soothing, the specious, the innocuous, the easy; and smart editors with just the right gift for reaching the popular mind can make thumping successes by diligently pleasing these people." *Harper's*, Allen implied, would never be guilty of such a thing.

By category association, all intellectual magazines could be said, ideally, to have the same goals and standards that Allen had laid out. Basically, however, they shared the same problem: how to produce a non-mass-market periodical and make it pay. The most radical ones would always have to be subsidized because they could obtain little or no advertising. The others would be offering relatively small circulations to potential advertisers, relying on the quality of their subscribers. John Fischer, perhaps the best of *Harper's* editors in its later years, observed in 1959 that the journal had gone through a great many changes in format, content, and editorial techniques, but its audience had remained much the same.

Between the wars, *Harper's* had demonstrated the strength of its own ideals by surviving the twenties and thirties, a time in which some of its old intellectual companions, notably *Scribner's*, the *Review of Reviews*, and the *Century* disappeared. Allen thought he knew why they had not survived. "These magazines," he wrote later, "were edited for ladies and gentlemen of either means or intellectual interests, or preferably both. . . . From time to time they dealt with the question of poverty in the United States, but the tone was likely to be, unconsciously, the tone of an aristocrat reminding other aristocrats of the regrettable conditions among the unfortunate if picturesque members of the lower orders." *Harper's*, on the other hand, had tried to change with the times.

Thomas Bucklin Wells was the editor who ensured *Harper's* survival in the twenties, succeeding in 1920 the man who had been editor for fifty years, Henry Mills Alden. Alden's was a hard act to follow. He had not only developed the magazine but had substantially furthered the state of American arts and letters. A staff member said of him: "I have never known, and fear I shall never know, a being so gentle, so beneficent. . . . Above everything was his unfailing wisdom. . . . It was the wisdom of a sage among men."

Nevertheless, it was Wells who brought the magazine truly into the twentieth century in a style that Alden, editor since 1869 and a man firmly rooted in the past, could probably not have managed. Assuming the editor's chair, Wells made immediate changes, shifting the magazine's emphasis to more contemporary matters, targeting his audience as the "thinking cultured reader who seeks both entertainment and an enlarged and broadened point of view."

His first change was *Harper's* hopelessly old-fashioned cover. The cherubs that had peered out vacantly for decades from its ornate design disappeared, and it was changed to a plain brick-orange. This and other changes coincided with the removal of the entire Harper publishing establishment from its original Franklin Square location in lower Manhattan to East 33rd Street.

The changes Wells made were reflected almost immediately in the circulation figures, jumping from 83,000 to 125,000 between August and December 1925, after a slower but steady rise in the preceding two or three years. *Harper's* was once more a money-maker. Wells retired in 1931, and Lee Foster Hartman became editor, guiding the magazine through both the Depression and the Second World War. He maintained the high quality that Wells had reestablished, focusing on the great events that occurred during his tenure. Particularly notable was a special issue in 1939, discussing America's role in the world. The quality of the fiction could be judged by the fact that five stories took first place in the O. Henry Fiction Awards between 1930 and 1939. Among the notable writers who contributed to the magazine in this period were William Faulkner, John Steinbeck, G. K. Chesterton, Kay Boyle, and Marjorie Kinnan Rawlings.

Allen was Hartman's successor, moving up from associate editor in 1941, just two months before Pearl Harbor. His notable reign included war coverage by Fletcher Pratt and John Dos Passos; Bernard de Voto's conducting of that venerable department "The Editor's Easy Chair"; and the introduction of such writers as J. D. Salinger, Katherine Anne Porter, John Cheever, and Mary McCarthy. Allen edited the centennial issue in October 1950 and carried on until John Fischer assumed his post in 1953, and what might be called the "modern era" of *Harper's* began.

Meanwhile, its old Boston contemporary, the *Atlantic*, was carrying on with equal success under the editorships of Ellery Sedgwick, who was at the helm from 1909 to 1938, and then Edward Weeks. Biography had never been one of the *Atlantic's* specialties, except incidentally in the historical studies of John Fiske, John Motley, and William Hickling Prescott. When such pieces ran, they were usually eulogistic. But in the 1920s, under Sedgwick's direction, they became a regular feature, and their tone changed radically to a modern style, as introduced by Lytton Strachey and others. The change had been heralded earlier, however, in the *Atlantic's* pages by both Henry Adams and Max Beerbohm.

Sedgwick carried on a tradition established early by the magazine of fearlessness in discussing religious matters; this was particularly relevant in the decade of the Scopes trial. The *Atlantic* had reviewed both of Darwin's books favorably and also had published work on this subject by the elder Henry James and his son William. Fundamentally, it was a nonsectarian journal, although it had always leaned toward Unitarianism. An undercurrent of distrust ran through its discussion of Roman Catholicism, inspired by worry over the Church's power. Sedgwick seemed to enjoy needling the Catholics, especially with articles (usually anonymous) by failed priests.

In the Big Business climate of the twenties, Sedgwick undertook examinations of industry's problems and prospects. Brooks Adams contributed an article on monopoly, F. M. Taussig analyzed free trade, and Arthur Pound talked about what he called "the iron man" in industry. It was Pound who also contributed three articles in the fall of 1922, penetrating studies of his home town—Flint, Michigan—that anticipated Robert and Helen Lynd's examination of Middletown, U.S.A., by more than ten years. In these articles and the book that evolved from it, Pound also gave readers their first glimpse of the "new automation."

At a time when everyone was either in the stock market or talking about it, Sedgwick smelled something rotten in Wall Street and hired William Z. Ripley, professor of political science at Harvard, to examine the methods by which holders of common stock were being euchred out of their control rights by dubious legal devices. When Ripley's first article appeared in the issue of January 1928, it was advertised for the first time in the history of intellectual magazines by means of a sandwich-board man who paraded through the canyons of Wall Street. This device helped to sell out an issue that would probably have done so anyway, and soon in New York anyone who wanted to sell his own copy could get a dollar for it in the sudden black market. Ripley continued his series in subsequent issues, well ahead of his time in exposing the sharp and often illegal practices in the Street which led eight years later to creation of the Securities and Exchange Commission and its regulation of the market, followed by passage of the Public Utilities Act.

Another landmark was Dr. David Bradley's book, *No Place To Hide*, which ran in the *Atlantic* in 1948, apprising Americans for the first time that the instrument which had ended the war in Japan could be the same weapon that would end all life. It was the sensational first shot in the battle over nuclear energy. As for the Second World War itself, *Atlantic* editor Edward Weeks seemingly had been content to let *Harper's* carry the burden. The magazine's contribution was negligible.

In the fiction department, interest in the long serial had begun to fade as early as 1909, when Sedgwick's tenure began. Before it ended in 1938, he had replaced long fiction with the short story, and even before the twenties, he had set out on a voyage of discovery, looking for new talent since the magazine could not afford to compete for the current brand names. It was a distinguished list of *Atlantic* firsts he compiled: Hemingway (first, at least, in a commercial magazine), Wilbur Daniel Steele, James Norman Hall, Jessamyn West, Eudora Welty, Louis Auchincloss, Richard Bissell, and many others.

The Hemingway story deserves further amplification. In the *Atlantic* version of literary history, a Hearst correspondent in Paris sent Ray Long, editor of *Cosmopolitan*, a short story by a young writer living there. It was about a prize fighter, and the correspondent thought it was good, but Long thought otherwise and rejected it, as he admitted eventually. It went next to *Scribner's*, where, ironically, it was read and rejected by Maxwell Perkins, who would soon be Hemingway's editor. Perkins did write to Hemingway and said he might buy the story if the author agreed to some cuts, but Hemingway curtly refused. Subsequently, it was passed on to the *Post* and *Collier's*, both of which turned it down, and finally appeared on Sedgwick's desk, where it found a home. It was not the first Hemingway story in print, in any case a matter of some dispute, but it appears to be the first in a magazine of general circulation.

There were other famous names in the *Atlantic*'s fiction pages: John Galsworthy, at the peak of his career; Kipling, but not his best work by any means; E. M. Forster; H. E. Bates; Mary Levin; and Lord Dunsany.

In the intellectual arena, one of Sedgwick's most controversial efforts was his printing of a thorough analysis of the Sacco-Vanzetti case in the issue of March 1927, written by Felix Frankfurter, then a Harvard Law School professor. The case had been in the Massachusetts courts for six years. Frankfurter had convinced himself that the men were innocent, and his article simply presumed that a new trial would be granted. Instead, shortly after the piece appeared, the Superior Court of Massachusetts returned a contrary verdict, and both Frankfurter and the *Atlantic* came under heavy criticism for what was deemed an impropriety. Harvard stood by its professor. When the Overseers met to discuss the subject, and President Lowell was challenged to take a position on the matter, he replied mildly, "What would you have had him do? Wait until the men were dead?"

Edward Weeks, taking over in 1939, carried on Sedgwick's good work in his own style until he left in 1966 to become editor at the Atlantic Monthly Press, but the *Atlantic*, like all magazines, intellectual or not, faced a postwar era in which profound transformations would occur.

While the *Atlantic* and *Harper's* might think of themselves as intellectual magazines, they were regarded as merely middle-brow (a term that first appeared in the *Atlantic*) by the serious magazines of opinion, the *Nation* and the *New Republic*, whose roots were nearly as deep in the nation's history and who considered themselves much more influential.

Oswald Garrison Villard had assumed command of the *Nation* in 1918, after its umbilical cord to the New York *Post* had been cut. He restored a good deal of the importance and influence it had enjoyed under E. L. Godkin. Villard himself was a man full of contradictions. He had opposed America's entry into the Great War, but he was a strong fighter for social reforms and had been a founder of both the N.A.A.C.P. and the A.C.L.U. At the same time, as a railroad builder's son and inheritor, he led the life of an upper-class gentleman, a member of both the Century and University clubs, in good standing with high society in spite of his editorship of a magazine many upper-class people regarded as radical.

This "Bolshevik," as his enemies called him, was actually a patrician, a fair-minded man who believed he could make the *Nation* financially successful and intended to solicit advertising from big business. He was the captain of an uneasy crew in the office. As Joseph Wood Krutch wrote of the staff, it "almost without exception represented the very tendencies which made him [meaning Villard] uneasy."

In fact, during the first summer of its independence, the *Nation* established a reputation as a radical magazine by attacking the wartime suspension of civil liberties, supported overwhelmingly by everyone else. But the magazine went ahead with its grim picture of a hysterical nation—a teacher in Iowa painted yellow by a mob who thought she was disloyal, Liberty Leaguers riding a woman on a rail in Illinois on the suspicion she was pro-German, the hanging of an I.W.W. member in Montana by a masked mob, and other atrocities and humiliations suffered by those who held unpopular opinions.

When the war was over, and it was the Reds instead of the Huns who were being hunted down, the magazine continued its bold upholding of civil liberties and its commitment to social justice. It argued for government regulation of monopolies, supported the right of labor to organize and bargain, and urged worldwide disarmament and free trade. Villard defended many of those accused in A. Mitchell Palmer's celebrated Red witch-hunt, and fought the national paranoia about radicals. The magazine advocated antilynching legislation, and decried government discrimination against Asians. William Hard (later to become a right-wing editor at the *Reader's Digest*) and Ernest Gruening reported on the progress of American intervention in Haiti, Santo Domingo, and Nicaragua, acts vigorously opposed from the beginning by the *Nation*. By 1930, it was already involved in Zionism and the explosive Middle Eastern situation. Villard also found himself generally in sympathy with what liberals called the "Russian experiment."

Under Villard, circulation climbed from about 9,000 in 1918 to more than 60,000 a year later, a small increase but effective because the subscribers were scattered across the United States and in many other countries. People paid attention to the *Nation,* as liberal lawyer Frank P. Walsh (oddly enough a Hearst syndicated columnist) discovered when he wrote a critical column on the railroads and at the same time published a similar article in the *Nation*. His column drew virtually no response, but his magazine piece provoked constant telephone calls on the day it appeared. Walsh said later: "People who counted, editors, assorted reformers, and lobbyists pro or con read the *Nation* and reacted to it. They had to keep up." At that particular time, the magazine's circulation was down to 27,000, but it remained influential.

In assembling his staff in the early twenties, Villard still clung to the past in his hiring of such people as Henry Mussey, who had resigned from Barnard after vainly protesting President Nicholas Murray Butler's firing of several professors opposed to the Great War; and Emily Balch, whose pacifism had similarly led to her departure from Wellesley after a long teaching career.

But there was new blood in the acquisition of Ernest Gruening, who had been fired from the *Herald Tribune* because, as managing editor, he had been responsible for juxtaposing two pictures in the paper's celebrated rotogravure section. One depicted a recent Southern lynching. The other showed black troops returning from frontline duty, marching down Fifth Avenue. The obvious discrepancy in making the world safe for democracy while denying it at home was not lost on patriots.

On the magazine's cultural side, there was a wave of Van Dorens, who helped to make the magazine's reviews among the best anywhere. Carl came first, joining the editorial board in 1920 and giving the book section a distinction it had not enjoyed since Godkin. When he left two years later to become editor of the *Century,* his sister Irita took over for him. Then, when she departed to be book editor of the *Herald Tribune* two years later, she was followed by Mark, who also bought the poetry and raised the level in that department. Dorothy, Mark's wife, had been with the magazine since 1919 as associate editor and columnist.

Still other first-rate writers and critics, at the beginning of their careers, joined in the early twenties. Lewis Gannett, who would become the *Herald Tribune*'s daily reviewer, served the *Nation* as reporter, editorial writer, and Villard's right hand. For longer or shorter terms, the masthead listed Raymond Gram Swing, Lincoln Colcord, Albert J. Nock, Suzanne La Follette, Norman Thomas, Paul Y. Anderson, and Hendrik Willem Van Loon, who contributed cartoons and drawings. Ludwig Lewisohn was drama critic for a time and also wrote book reviews and became an associate editor. Regular book reviewers included Charles Beard, H. L. Mencken, Frank Harris, Vernon Parrington, Heywood Broun, Lewis Mumford, Franz Boas, John Erskine, Zechariah Chafee, and even Anatole France, who contributed a series of "Opinions." Among the poets whose works appeared were Edwin Robinson, Robert Frost, Elinor Wylie, Alfred Kreymborg, Maxwell Bodenheim, Babette Deutsch, Witter Bynner, and Louis Untermeyer.

One of the most remarkable series of the twenties was an examination of the states. William Allen White wrote on Kansas, Mencken on Maryland, Dorothy Canfield Fisher on Vermont, Zona Gale on Wisconsin, Douglas Southall Freeman on Virginia, Theodore Dreiser on Indiana, Sinclair Lewis on Minnesota, Sherwood Anderson on Ohio, Edmund Wilson on New Jersey, Willa Cather on Nebraska, Mary Austin on Arizona, and Ludwig Lewisohn on South Carolina. Later, these essays were published in two volumes. Small wonder the *Nation* tended to overshadow its rivals in those days.

The magazine's direct competition was the *New Republic, Survey,* and the *Dial,* a Chicago journal of the arts that, in 1918, moved to New York and became a journal of opinion. How the viewpoints of these magazines varied from the kind of mass-magazine intellectualism practiced by Luce and Lorimer was evident when the first issue of *Time* appeared in 1923, admitting that Warren Harding was no superman but, in ringing Lucean phrases, asserting he was "important and successful as the embodiment of the American idea of humility exalted by homely virtues into the highest eminence. . . . [He is] the actuality of the schoolboy notion that anybody has a chance to be President." No such endorsement came from the *Nation* and its fellows. They were on to Harding from the beginning and gave him scant respect.

A new boy came on the block in December 1932 with the advent of *Common Sense,* founded by Alfred M. Bingham, surprising everyone who remembered him from four years earlier as head of the Hoover for President campaign group at Yale, where he was studying law. *Common Sense* took a position considerably to the left of Villard. The intellectuals who launched it declared forthrightly that "a system based on competition for profit can no longer serve the general welfare." It called on the nation to hold a second Constitutional Convention, at which the principles of the American Revolution would be reaffirmed by adapting them to the modern world. The contributors proved to be liberal intellectuals, some of whom years later became equally convinced right-wing conservatives. They included John Dos Passos,

Stuart Chase, A. J. Muste, V. F. Calverton, John T. Flynn, James Rorty, and George Soule.

Surprisingly, neither the *Nation* nor *Common Sense* were strong supporters of Franklin Roosevelt in 1932, even though they shuddered away from Hoover for the most part. The *Nation* called Roosevelt "Friendly Frank," and *Common Sense* went further, labeling him "the laughing boy from Hyde Park," a phrase eagerly revived by conservatives after the election.

As the Roosevelt era began, Villard put together a new cast of editors, at least one of whom would leave a lasting mark on the magazine. They included one veteran, Freda Kirchwey, Joseph Wood Krutch, Henry Hazlitt (he lasted less than a year, this conservative at heart), and Ernest Gruening. These people heard Roosevelt's inaugural address and watched with growing approval his attack on the nation's problems. They were slow in coming around, but gradually they turned the magazine's support toward the new president, although they did not consider him radical enough.

As time went on, the *Nation* remained uncomfortable with the New Deal. When the NIRA was introduced, the magazine was alarmed, considering it a turn toward right-wing fascism. But when the Supreme Court nullified the NRA in 1935 and the AAA a year later, it was equally disturbed by what it considered a right-wing assault on a liberal president.

People who read both the *Nation* and the *New Republic* found them so similar in viewpoint and coverage that the question of a possible merger arose. When a reader's letter inquired about the possibility, Kirchwey pointed out two major differences. The *Nation* at that point was almost self-supporting, while the *New Republic* had to be substantially subsidized. Then, too, she added, they did not agree on the problem of collective security and other matters. The fact was that neither magazine displayed any willingness to lose its identity.

The two magazines split sharply as the world moved toward war at the end of the thirties. In spite of its pacifist past, the *Nation* was strongly interventionist, correctly assessing Hitler's threat to the world. The *New Republic*, which had favored American entry in the Great War, had been so disillusioned by the Versailles treaty that it took a pacifist stance right up to the blitzkrieg of 1940, which convinced it of danger at last. Between wars, the two had simply switched policies.

Freda Kirchwey was the prime mover in turning the *Nation* toward strong support of the collective security concept, and she was joined by Maxwell Stewart, Max Lerner, and Robert Bendiner. Villard's detestation of Hitler was counterbalanced by his enduring disillusionment with Versailles, and as a result, he took an isolationist position, contrary to the magazine's policy and so driving an awkward wedge between him and his editors. It seemed to Villard that he had no alternative but to retire. In his final editorial on June 29, 1940, he wrote with mingled anger and sadness: "I regret . . . that my retirement has been precipitated . . . by the editors' abandonment of the *Nation's* steadfast opposition to all preparation for war, to universal military service, to a great navy, and to all war, for this in my judgment has been the chief glory of its great and honorable past."

In a move to put the magazine on a better financial footing, Kirchwey transferred ownership to a nonprofit corporation in 1943 called the Nation Association. Subscribers were asked to become members at rates varying from ten to a hundred dollars a year. This association became the magazine's primary financial support until 1955, but then the perennial financial problems of all the opinion magazines caught up with the *Nation* once more, and it had to be sold, entering then upon a new phase of its career.

The *New Republic*'s progress, meanwhile, had not been substantially different. Caught in the wash of postwar disillusionment, its editors had appeared to give up hope momentarily of having any direct influence on public opinion and turned their attention to harpooning the follies and excesses of the twenties, a tempting target for all the intellectual opinion magazines. Mencken was a contributor before he set up his own harpooning shop, and the playwright S. N. Behrman anticipated the "man in the gray flannel suit" by a generation, with a full-scale assault on the advertising business. Stark Young and Paul Rosenfeld brought their critical talents to the staff, supplemented by George Bernard Shaw and Clive Bell, both contributing from England. In brief, the *New Republic* saw itself as a beacon of civilization, sending its beams across a sea of Babbittry. Progressive idealism was no longer its cause.

As the *Nation* had done, the *New Republic* ran its own survey of the states, a series that displayed the views of Sidney Howard on California, Frank Kent (who became later the magazine's first Washington columnist to sign himself T. R. B.) on Tennessee, Bruce Bliven on Massachusetts, and Margaret Sanger on New York. It did not come up to the *Nation*'s performance, but the *New Republic* did find a chink in its rival's armor, a neglect of the arts beyond literature, and hastened to take advantage of it. Gertrude Stein's brother, Leo, wrote about Picasso; Lewis Mumford contributed excellent essays on Brancusi and Marin; Padraic Colum wrote on Joyce; while Deems Taylor did a chatty piece on Walter Damrosch.

Politically, the *New Republic* seemed as disoriented as the *Nation*. It gave support to Al Smith in 1928, but its readers were astonished in 1932 to find Croly writing about Hoover as though he might be a political messiah. Croly died in 1930, however, before the editor could suffer still another disillusionment. His place was taken by younger men who had their own ideas about what was wrong with society.

Suddenly the magazine sounded much more radical than it ever had before. Its social reporter, Edmund Wilson, wrote a telling article about the Scottsboro boys, and Dos Passos provided a report of the 1931 hunger march on Washington that sounded like a call to revolution; he followed it with a similar view of the 1932 Democratic convention. The magazine viewed the New Deal with skepticism at first, as the *Nation* had done, but as time went on, some of those involved in the administration were treated approvingly. The problem was that Progressive idealism was dead, and its one-time spokesman had no idea where to go.

As its third decade began in 1934, the *New Republic* was not quite ready to give up on social reconstruction in favor of pragmatism, but it was beginning to find itself again. The magazine's tone was noticeably less moralistic. It continued to pursue the arts and challenged the *Nation* on literary grounds, with articles by Lionel Trilling on Eugene O'Neill, Van Wyck Brooks on Amy Lowell, John O'Hara on Fitzgerald, and Max Eastman on Freud.

In the mid-thirties, the magazine abandoned whatever sympathy it had with collectivism and began to push for reforms within the system, using New Deal measures as models. In a penetrating 1936 article, Marquis Childs gave a compelling account of why the rich hated Roosevelt, and before the decade ended, the magazine had become FDR's strong supporter, notwithstanding William Allen White's sympathetic profile of Wendell Willkie. (White had also once seen Herbert Hoover as the hope of America.) Meanwhile, war was approaching and the magazine's readers were given a preview of the war's anguish by Thomas Wolfe, who was in Berlin. When Pearl Harbor came, there was no equivocating this time. The *New Republic* fully supported the war.

During the conflict, Henry Wallace arrived in the editor's chair, but it was only a short detour in his career. He did, however, print vigorous attacks on General MacArthur and Senators McCarran and McKeller by Robert S. Allen, while Vincent Sheehan wrote approvingly of Nehru. But when the war was over, this liberal magazine was compelled to define its views on communism.

A latecomer to the radical, or at least left-wing, wars was the *Partisan Review*, founded in 1934, originating as a Marxist organ of the John Reed Club. It merged in 1936 with a literary magazine called the *Anvil*, whose editor, William Phillips, then became a member of the editorial board and a guiding spirit. A year later, the magazine had to suspend for fourteen months, but it then reorganized with a new editorial board that included, besides Phillips, another founder, Phillip Rahv, F. W. Dupee, Dwight MacDonald, and, peripherally, Mary McCarthy.

Besides its political mission, the magazine was devoted to publishing the work of avant-garde writers and to introducing new European authors (in translation) to an American audience. Lionel Trilling did some of his best work for the magazine, along with other critics of note. *Partisan Review* printed Saul Bellow's first work and part of James T. Farrell's *Studs Lonigan*.

Having said all this, however, nothing has been conveyed of the magazine's radical politics, which made it a journal about which controversy has never ended. Testimony to its vigorous career continues to this day, kept alive by the memoirs of nearly all those involved with it, most notably those of Sidney Hook, who was there at the magazine's creation. Hook, the noted New York University philosophy professor, had Phillips as a student and later met both him and Rahv at meetings of the John Reed Club. At the time, Phillips was reviewing books for a Communist party publication under the pseudonym of Wallace Phelps, although Hook wrote much later that he did not think Phillips was a card-carrying member. It was Rahv and Phillips who became co-editors of the *Partisan Review*. One of its most influential editors was Dwight Mac-

Donald, who was ousted from his editorial seat in 1943 as a result of intra-mural bloodletting.

That event, which split the editorial team wide apart, was the result of a disagreement over the magazine's war policy. *Partisan Review* had seen the conflict originally as an imperialistic event, but then Phillips changed his mind and supported American involvement. He was followed a little later by Rahv. MacDonald remained the holdout, and since he was odd man out, making the editorial situation impossible, he was maneuvered off the magazine by the other two editors. Delmore Schwartz succeeded him. After the war, William Barrett joined the staff, destined to be a subject of controversy. He was also a friend of Hook, who had brought Barrett to N.Y.U. to teach philosophy.

The later literary controversies over what *Partisan Review* and its editors did or did not do were misleading in a sense. They appeared to be a protracted struggle between advocates of different varieties of radical thought, but in the case of their severest critic, Hook, they had deeper ideological roots. In a 1984 memoir of his connections with the magazine, written for *The American Scholar*, Hook dismissed any pretensions to intellectuality, political or cultural influence, or any other claim to distinction the magazine might have had. He titled his article "The Radical Comedians: Inside *Partisan Review.*" There was "something truly comic about their self-conscious role as political revo-lutionaries and cultural radicals," he wrote, "about the disparity between their profession and their performance. The truth is that they were peripheral and parasitic to the genuine movements for change in the social and political life of the country."

In the memoirs of Barrett and other participants, a different story emerges, but the core of the problem was an exaggerated form of the disputes among staff members that had developed on the other opinion magazines. Views dif-fered sharply, old loyalties were shattered, and friendships and jobs lost. In the case of *Partisan Review*, the basic quarrel was the one that enveloped the entire intellectual community; that is, the question of who had, or had not, been sympathetic to the communists (or card-carrying members) and when did they swing over to the other side. Many of the early socialist-minded intel-lectuals, writers, and editors, were bitterly disillusioned by the revelation of Stalinist horrors and became the enemies not only of the Soviet Union but also of all those who took any other position, for whatever reason. The strife at *Partisan Review* was an acute example of this division.

There were radical magazines in the twenties, however, that had nothing to do with communism, in the usual sense. One, now almost forgotten except by historians, was the *Crusader*, a radical black magazine edited by Cyril V. Briggs between September 1918 and February 1922. (Its complete file was published in book form in 1988, edited by Robert Hill.) Briggs was a native of the little island of Nevis, came to New York early in the century, and was radicalized there. He joined the Communist party in 1919, seeing in it not so much the triumph of Marxism as the means to black liberation. Briggs became a vociferous leader in African-American journalism as it pushed the cause of the "New Negro." He was opposed to the accommodationist policies of Booker

T. Washington and his journalistic mouthpieces, even though he believed in self-help and the developing of entrepreneurs.

In the continuing struggle for equal rights, one of the leaders was Charles S. Johnson, who in 1924 was editor of the Urban League's new magazine, *Opportunity*. It was Johnson who helped to organize a historic meeting on March 21, 1924, at the Civic Club in New York, where black writers and artists, invited by Frederick Lewis Allen (then editor of *Harper's*), came to meet with white magazine writers and editors. Among those who attended were W. E. B. Du Bois, James Weldon Johnson, Alain Locke, Countee Cullen, Carl Van Doren, Horace Liveright, Freda Kirchwey and Evans Clark (both from the *Nation*), and Jessie Fauset (literary editor of *Crisis*). In the days following the conference, Paul Kellogg, editor of *Survey Graphic*, told Johnson that he intended to publish a special issue of his magazine that would be devoted to subjects about which the blacks present at the Civic Club meeting had been talking and writing.

Opportunity, launched in January 1923, turned out not to be the kind of journal that the Urban League's director, Eugene Kinckle Jones, had in mind. It was less inspirational than academic, with such articles as Melville Herskovitz's statistical conclusions about racial variations in physiology and Monroe W. Work's study of rural and urban demography.

The Civic Club meeting resulted in a flow of literary contributions, changing the character of the magazine. It was a period when the artistic and literary work of black intellectuals had reached an unprecedented high point, a Harlem Renaissance, as it came to be called. Most of those in it, however, did not think of themselves as participating in a black movement. Their intention was to prove their artistic validity regardless of color. Johnson and Alain Locke helped them considerably by selecting the best of their works for the *Survey Graphic* Special Edition. They looked for the most polished material, about people untainted by racial stereotypes, and they avoided vulgarity. As David Levering Lewis wrote of this team's work years later, they meticulously rejected "too much blackness, too much streetgeist and folklore—nitty-gritty music, prose, and verse—were not welcome.

Johnson tried to persuade as many of the contributors as possible to come to New York and succeeded in attracting Arno Bontemps, Wallace Thurman, Zora Neale Hurston, and the artist Aaron Douglas. They arrived in time to witness at first hand the tremendous success of the Special Edition, whose forty thousand readers represented twice the *Graphic*'s normal circulation.

Both *Opportunity* and *Crisis* made efforts to inspire more writers by offering prize money, and the former was not reluctant to take such money from a white source, the wife of Henry Goddard Leach, editor of *Forum*. Johnson pursued these efforts in every possible way; he was no advocate of art for art's sake, particularly in the case of black writers. He talked of markets, exposure, education, intercultural exchanges.

Black radicals were not happy with either *Opportunity* or Johnson. They saw these efforts as being made possible by white capital and influence, and it was true that in the early years whites, with the best of intentions, did act as a kind of benevolent censor, setting the parameters of black writing and art. But it did not last. In time, they were no longer necessary.

In many ways, the most interesting of the radical magazines begun in the turbulent twenties was the *New Masses*, a revival of Max Eastman's earlier effort, sponsored officially or unofficially (it was not exactly clear) by the Communist party. Begun in the mid-1920s, it resembled its liberal cousins the *Nation* and the *New Republic* in format—printed on butcher-paper stock, in an approximate nine-by-twelve format. Michael Gold was its editor, and, in its early years, the magazine followed the party line faithfully but was already printing work by promising young intellectuals who were sympathizers but not always party members. Among its contributors, who were not paid, were Theodore Dreiser and Whittaker Chambers, whose article, "Can You Hear Their Voices?" attracted considerable attention.

In 1928, Gold acquired a contributing editor who would soon play an important part in the magazine's career. He was Stanley Burnshaw, later to become one of America's most distinguished poets and literary critics as well as a book publisher. Burnshaw was soon lured away temporarily by V. F. Calverton (real name: George Goetz), who was editing an unorthodox radical magazine called the *Modern Monthly*, which opposed the party; Burnshaw became contributing editor there. Until Calverton died a few years later, the *Monthly* remained a serious competitor of the *New Masses*.

Brunshaw switched allegiances in the autumn of 1933 and went to call on Herman Michelson, then editing *New Masses*. He was welcomed eagerly. Not only had he just won wide critical acclaim with a book on Andre Spire (the French poet) but he was also knowledgeable as a printing expert and was currently an advertising man. Michelson told him they were about to re-launch the *New Masses* as a weekly and urged him to give whatever time he could to help the staff design and plan the new venture, the first issue of which was to appear in January 1934. Burnshaw agreed. Besides Michelson, other full-time editors were Joseph North, Joshua Kunitz, Granville Hicks (he was book editor), and Jacob Burke, who had been the *Daily Worker*'s cartoonist and was now to be art editor. Burnshaw worked first as a half-day adjunct to the staff, but then he was persuaded to come on full-time with a salary of sixty dollars a week, as much as Earl Browder, the Communist party chief, was being paid. Browder's brother, William, was also on the staff.

Michelson's hunch had been correct; Burnshaw was the multitalented man they needed. He edited articles, wrote some of his own, reviewed plays and the dance, and served as poetry editor and book reviewer. He did the makeup for every issue of the magazine, and on Monday nights, when the journal was put to bed, he set up the covers by hand, through a special arrangement with the union printers. Then, when Michelson decided to take six months off to visit the Soviet Union, Burnshaw was asked to be managing editor. He agreed, with the stipulation that he would have to resign in a few months because he intended to re-marry and needed to earn more money. He departed a week before the Spanish Civil War broke out. Burnshaw had never joined the party and by that time had lost all faith in its program.

The talented staff that edited this magazine reflected the ability of intellectual magazines in that period to attract young writers and editors at the beginnings of ther careers, carried away ideologically by strong feelings of social protest. Michelson, for example, was a newspaperman who had been editor of the socialist magazine, the *Call*, and also Sunday editor of the New York *World*. Like so many other radical writers, he was no working stiff but lived quite comfortably in a house on Bank Street in Greenwich Village with his wife, a very rich descendant of Abraham Lincoln.

On the staff, too, was Joseph North, who specialized in politics but was also a gifted short-story writer, a first-rate reporter, and a lover of poetry. Joshua Kunitz was a Russian scholar and former professor at City College of New York, known for his book *Dawn Over Samarkand*. Orrick Johns was a well-known older poet, and William Slater Brown was not only well known as a writer but also famous for having gone to jail with e. e. cummings for their resistance during the First World War. Then there was Bruce Minton, the son of a wealthy California family, who later married Ruth McKenney, creator of *My Sister Eileen*; and Isidor Schneider, an editor and writer who started the first union in the book-publishing business and eventually became book-review editor of the magazine.

After Burnshaw's departure, *New Masses* came upon difficult times because, in the late thirties, the Communist party line became the "United Front" which meant, as Stalin himself put it after signing the pact with Hitler, it was necessary then to "love people you had formerly cursed." In addition, the appeal of the party line to many of its constituents had begun to fade. In 1936, Earl Browder ran for president on the Communist ticket, but party members were instructed to vote for Roosevelt.

For an ideological magazine, *New Masses* had a remarkable number of contributors from what could be called the political middle ground—people like Malcolm Cowley, Kenneth Burke, and Archibald MacLeish. It also printed fiction by Hemingway, Dreiser, and Horace Gregory, among others. One of its best pieces of journalism in the prewar period was a series written from inside Germany by John L.´Spivak, later published as a book.

But as the Depression faded into memory and the country swung into arms making and the war years merged into general prosperity, those who had written for *New Masses* found themselves with less reason to contribute to it. Roosevelt's social programs had done for the country what they had thought communism might do. But at its peak, the magazine had been, ideology aside, an immensely stimulating publication that outshone both the *Nation* and the *New Republic* in some respects.

In fact, *New Masses* was not a magazine for the proletariat, the downtrodden masses. It was written largely by middle-class intellectuals for others like themselves, whom the editors considered were being isolated between the proletariat and the *haute bourgeoisie*. On that audience, it made considerable impact, and its circulation was frequently equal to or above its butcher-paper rivals. Certainly its influence easily exceeded that of the socialist *New Leader*, which called for revolution by the ballot, in spite of Norman Thomas's prestige. But as Browder once observed, middle-class sympathizers are always in a hurry; they want violent action. Burnshaw wrote later that this was Browder's way of saying what one of his (Brunshaw's) Cornell professors had told him: "A radical is a liberal in a hurry, but of course you don't play the violin when the house is burning down." Burnshaw added: "All of us at the *New Masses Weekly* felt that the house *was* burning down."

Among the intellectual magazines spawned in the twenties, probably the one that drew most attention was H. L. Mencken's *American Mercury*. It was not committed to any particular ideology but had a somewhat nihilistic approach to all of them. Mencken founded it in December 1923 with his friend, the theater critic George Jean Nathan. Alfred A. Knopf, already a noted book publisher and a friend of both, was the first publisher of the magazine. On December 3, Knopf issued a certificate to Crosby Gaige, a theatrical producer and gourmet whose office was in the same building, as founder subscriber No. 1 (Gaige had provided some start-up money). Below the certificate's chaste blue border, which had been printed in Knopfian style, Mencken added in his own hand, "summa cum laude." This mocking of academia (he and Nathan often addressed each other as "doctor") fit the general tone of the *Mercury*, the centerpiece of which was the editor's attack on the "booboisie" and all its works.

There was more to the magazine than that, however. Mencken opened its doors to new writers, and they flocked in, providing an excellent sampling of the better fiction being written in that decade along with articles that attacked the status quo and the follies of the nation in less ideological terms than the other opinion magazines. Although its circulation was as small as the others, seldom approaching or surpassing 100,000, its influence was remarkable. That was because the *Mercury* addressed itself particularly to young intellectuals of the middle and upper classes, who accepted its assault on everything conventional in much the same spirit as the sixties' generation embraced national upheaval. They were against the system, as Mencken was, and it became a badge of honor to have the familiar green cover of the *Mercury* protruding from a coat pocket, for which its small size made a convenient fit.

In an era of Babbittry, the *Mercury* was the organ of iconoclasm. The work of the country's best satirists appeared in its pages, led by Mencken himself, a stylist unsurpassed, so erudite, individual, and challenging that only in retrospect can it be seen that large and now indigestible portions of what he wrote were nonsense. His targets were easy, the proverbial sitting ducks— stuffiness, pretension, political flapdoodling, Rotarianism, conventional values, organized religion, and organized-everything-else. Mencken's masterful prose obscured the fact that much of what the magazine advocated in its perverse way was essentially far-right conservative belief.

There was something authoritarian about Mencken's iconoclasm, something of the savagery of the revolutionary just before he has enough power to raise his guillotines. All this became somewhat clearer, to the vast disillusionment of many faithful followers, by that time grown older and wiser, when he appeared on Alf Landon's campaign train in 1936, wearing a large Kansas sunflower in his buttonhole and exhorting the same booboisie he had scorned only a decade before to follow him and Alf into the White House.

By the late twenties, Mencken's busy life had led him to lose interest in the *Mercury*, and he sold it to Paul Palmer, who gave the magazine an extreme right-wing tone before he went on to become an editor of the *Reader's Digest*. The new owner in the thirties was Lawrence E. Spivak, who had been Mencken's business manager and who later became a nationally known television figure as the proprietor and chief inquisitor of *Meet the Press*.

Spivak's editor was Eugene Lyons, a disenchanted wire-service correspondent who had gone to Moscow with high hopes in the twenties and returned to write the book that made him well known, *Assignment in Utopia*. Lyons assembled a stable of similarly disenchanted and disenfranchised former liberals, and the magazine became the leading anticommunist publication in the country.

It had another side, however. Although it was solidly Republican, the *Mercury* was strongly pro-intervention as the Second World War began and after Pearl Harbor revealed, among other things, the nearly disastrous condition of American air power. The magazine made itself the spearhead of a drive by General "Hap" Arnold and the then colonel, later general, Hugh J. Knerr to persuade a navy-minded president and Congress that air power would win the war, and that a massive buildup was needed. Both officers fed the *Mercury* with articles they signed, ghostwritten but based on their inside information, so that the *New York Times* and other papers sometimes ran stories that began, "The *American Mercury* will say tomorrow. . . ." Further inspiration for the drive came from Alexander de Seversky, whose best-selling *Victory Through Air Power* first appeared (in part) in the *Mercury*. In the end, the magazine was responsible for generating much of the pressure that persuaded Congress to vote appropriations for heavy bombers against the die-hard opposition of navy admirals.

Spivak's version of the magazine carried on Mencken's tradition of encouraging new writers, both of fiction and poetry, but it also printed excellent material by well-established writers. Nathan continued to write the theater column for a time. Max Eastman, now thoroughly divorced from his *Masses* beliefs, wrote the leading book-review piece, succeeding Mary Colum, wife of Padraic, who so opposed the war that she refused to review any book having to do with it.

In the late forties, Spivak's other activities—he had established lucrative sidelines in *Ellery Queen's Mystery Magazines* and a line of paperback *Mercury Mysteries*—led him to sell the *Mercury*, and this time the sale was a fatal stroke. It fell into the hands of a succession of extreme-right hate merchants, much as the *North American Review* and *Scribner's* had done, and slipped away into oblivion, leaving a notable history behind.

Another new intellectual magazine fared better, probably because it had no serious political leanings at all but devoted itself to the cause of good books and the ideas they generated. In an earlier chapter, the genesis of the *Satur-*

day Review of Literature was noted: how Henry Seidel Canby took the old *Literary Review*, the book section of the New York *Post*, and made a new magazine of it. No one has told the story of the transition better than Canby himself, when he wrote of it in 1945: "With the support of Thomas W. Lamont and the cooperation of the editors of *Time*, two of whom had been students of mine at Yale, we migrated *en masse*—editors, columnists, poets, reviewers, critics, and commentators, with a baggage of ideas, and a somewhat dubiously acquired subscription list—left Vesey Street for good and all, and in three months launched the *Saturday Review of Literature*, which was the old literary review come of age, more humorous, more literary, broader in scope, better looking, but with the same will to further the cause of good thinking, good feeling, good writing, and good books."

The quality of this new magazine was high. Its editors attempted to maintain a consistent attitude toward both past and present, trying to look beyond fads and fashions, manners and aberrations, or anything else that might lead to mere labeling or indulgence in catch phrases.

In the daily press and popular magazines, readers saw the twenties in terms of rumrunners, flappers, jazz, the stock market, Teapot Dome, F. Scott Fitzgerald's Gatsby, and Sinclair Lewis's Kennicotts. Canby and his staff saw it differently. They viewed a world that was producing notable scholars and scientists, one in which centuries-old theories were challenged every day. They wanted to examine values, not simply experiment with them. Their pages were open to analyzing the educational revolution that was occurring as a result of the work of John Dewey, William Heard Kilpatrick, and Harold Rugg. The magazine also carried the works of A. E. Housman, Yeats, Frost, Robinson Jeffers, and Millay, who were finding their audiences.

Canby's magazine addressed a readership that was aware of these matters and people, literate readers who could enjoy the juxtaposition of Shaw and H. G. Wells, who enjoyed writers like Ellen Glasgow, Edith Wharton, Willa Cather, Thomas Hardy, Joseph Conrad, and E. M. Forster, yet still welcomed such new writers as Hemingway, Lewis, and Faulkner. These readers did not take themselves so seriously, however, as to disdain writers like Robert Benchley and Ring Lardner.

As Norman Cousins, the last great editor of *SR*, wrote on the magazine's fiftieth anniversary: "All four editors . . . in its first half-century were militantly opposed to cynicism. All shared substantially the same values. All were deeply rooted on native grounds, with different centers of attraction. For Canby, it was the world of Thoreau, Emerson, and Whitman. For Bernard De Voto, the early explorers of the American West. For George Stevens, the perception of America by Steinbeck and Cather. For Cousins, it was the kinds of ideas ventilated in the Jefferson-Adams correspondence and, much later, in the Holmes-Pollock letters."

In spite of such similarities, and the undeniable advantage of this sort of intellectual continuity, there were differences among these editors, particularly in their approach to editing. Canby leaned more toward critical writing than book reviewing. Both De Voto and Stevens, following him, placed emphasis on general reviewing and on attempting to cover a reasonable number of published books. Cousins created a different magazine and led *SR* to its highest circulation in the sixties by, as he said, "connecting the world of books to the world of ideas" and making a more general magazine for those who did not separate their cultural interests from world concerns. Of the four editors, De Voto was the only one who invited controversy and delighted in it. Personal feuds were not uncommon during his tenure.

Otherwise, the magazine was vaguely liberal in its approach to life, but rarely overtly ideological. Always it was on the side of nurturing creativity and against those who would stifle it. Over the years, as Cousins observed, the *SR* acted "as a bridge between artist and audience," and he added: "Both have their needs; both have their rights. Neither exists in isolation."

In 1952, the original name of the magazine was shortened to the *Saturday Review*, and its scope was broadened to include music, science, travel, the media (in time), and other aspects of American culture that were not strictly literary. Under Cousins's direction, aided by his executive editor, Richard L. Tobin, it achieved both its largest audience and highest level of distinction.

At a different level of a literary world with multiple audiences, another major event began to take place in the twenties. The history of the so-called "little magazines" has never been told in full, nor brought up to date; they await their own historian. Only the essential details of that history can be given here. This kind of literary magazine publishing began long before the twenties—as early as 1912. Between that date and about 1950, more than six hundred little magazines were published, and since then, scores of others have appeared. Some continue to be published today; many others have disappeared. What they have all had in common and what is historically significant about them is that they enabled hundreds of unknown writers to appear in print. Some of these went on to distinguished literary careers; most disappeared. But the fact that the little magazines made it their business to embrace qualities the commercial publishers might well not take a chance on was an incalculable boon to the literary world.

These small journals, struggling to survive most of the time, considered themselves to be the front rank in the struggle for a "mature" literature. They seldom paid contributors, and they made no promises, but in time, editors in publishing houses began to read them in search of new writers who might have book-length manuscripts on hand or in mind.

In this, these journals proved strikingly successful. Between 1912 and mid-century, they published for the first time about 80 percent of the foremost novelists, critics, and poets of our time, including Hemingway, Faulkner, Erskine Caldwell, and T. S. Eliot. In doing so, they were first to introduce and support all the literary movements of consequence that occurred during that time. Of the six hundred such journals that existed at one time or another, however, less than a hundred contributed substantially to the success of this medium.

By definition, the "little magazines" were (and are) designed to print literary, artistic work by writers who are unknown and who presumably would be rejected by popular magazines, but it is also true that in at least one sense they were not always a decisive factor in the careers of writers. Once launched, some made successes in commercial publications, leaving their best work behind them. On the other hand, there were those who only learned how to fly with the little magazines and later soared to the literary peaks elsewhere.

Audiences for these journals have always been limited; it is unusual to find one with a circulation of over a thousand. A staff of one or two people often sufficed, usually putting their product out on a quarterly basis. These literary quarterlies should not be confused with such established quarterlies as the *Sewanee Review*, the *Southern Review*, *Yale Review*, and the *Virginia Quarterly Review*. Those publications have a more limited freedom to experiment or to publish new writers.

The word "little" was first applied to this category in the years before the Great War and referred to the size of their audience, not their physical dimensions (which were, in fact, almost always small). "Avant-garde" would have been a better description, but that was never widely accepted; it described, however, the character of many of the personalities involved with the magazines: Ezra Pound, William Carlos Williams, Norman Macleod, Eugene Jolas, and a good many others. These writers and editors considered themselves leaders of the avant-garde, rebelling against conventional ways of writing and advocating unorthodox literary theories.

While the twentieth-century version of the little magazines began in 1910, they had their origins in that famed organ of transcendentalism, the *Dial*, edited between 1840 and 1844 by Margaret Fuller and Ralph Waldo Emerson. It was truly "little," with only 300 subscribers in spite of the future hallowed names on its pages—Emerson, Thoreau, Channing, and Parker, among others. Then in 1858 came a further step forward with Henry Clapp's *Saturday Press*, which lasted until 1866. There followed a lengthy arid stretch until the nineties, when the *Chicago Chap Book, Lark*, and *M'lle New York* flourished briefly. Another barren interval followed, until 1910.

After a few false starts, a renaissance began in that year with the founding of Harriet Monroe's *Poetry: A Magazine of Verse*, still being issued today. In the same year, Max Eastman and Floyd Dell were trying to make the *Masses* an avant-garde literary magazine, in between administered doses of left-wing ideology. In Boston, there was the *Poetry Journal*, also begun in 1912. These forerunners printed the works of such new poets as Edward Arlington Robinson, Edgar Lee Masters, Amy Lowell, and Sara Teasdale, who had all been published but were getting little attention. Other new magazines blossomed: *Glebe, Others*, the *Little Review*. For the first time, readers were introduced to Carl Sandburg, Vachel Lindsay, Marianne Moore, John Gould Fletcher, and Maxwell Bodenheim.

Between the wars, however, the little magazines came into full bloom. New journals were devoted to all, or nearly all, poetry, including the *Fugitive, Voices*, and *Smoke*, besides those already mentioned. Many of the new publications were left wing—the *Liberator*, the *Masses, Partisan Review*, and a group of more ephemeral journals of the thirties, including the *Anvil, Blast,* and the *Little Magazine.*

Regional "little magazines" appeared for the first time in 1915 with John T. Frederick's *Midland,* published in Iowa City and enduring until 1933. Frederick's journal introduced realism (or at the least insisted on it) in writing about the vast heartland, vaguely defined as somewhere between the Appalachians and the Rockies. Other regional quarterlies appeared: the *Frontier* (1920–1939); the *Texas Review* (1915–1924); *Southwest Review,* begun in 1924; the *Prairie Schooner,* starting in 1927; and the *New Mexico Quarterly Review,* begun in 1931.

There were scores of magazines specializing in experimental writing, among them the *Little Review* (1914–1929); *Broom* (1921–1924); *Secession* (1922–1924), edited by Gorham Munson; the *Dial* (1920–1929); and *transition* (1927–1938). Even James Laughlin's New Directions Press, begun in 1936, could fall into this category, as would Dorothy Norman's semiannual *Twice A Year,* begun in 1938.

Literary criticism and high-level reviewing were the chief stock-in-trade of the *Dial, Hound and Horn* (1927–1934), and the *Symposium* (1930–1933). In their pages were the literary judgments and artistic notions of Eliot, John Crowe Ransom, and R. P. Blackmur, among others. Some of this writing was more than a little precious.

A few of the magazines took their cue from Mencken's *Smart Set,* which preceded the *American Mercury,* carrying urbane contemporary comment on a literary level. *Seven Arts* (1916–1917), and *Story,* begun in 1931, were the best of these. The latter survived past mid-century under the editorship of Whit Burnett and his wife, Martha Foley, publishing new short-story writers and contributing markedly to the stream of new literature. It was revived in 1989.

Frederick Hoffman and Charles Allen, the first scholars to examine the history of little magazines, point out that while these journals published the early work of many of America's best writers, they did not necessarily discover them. An exception would be *Poetry,* which could boast a long list of poets it introduced to the public for the first time. In addition, *Double Dealer* could be credited with introducing at least five important writers, and *Fugitive* could claim to have published the first work of Robert Penn Warren. A subcategory of little magazines, without a name, ephemeral journals on the wild side, also found new talent. *Blues,* for example, first printed the work of Farrell and Erskine Caldwell, while *Bruno's Bohemia* introduced Hart Crane.

Hemingway is usually cited as the classic case history of a major writer given a start by the little magazines. His first story in any magazine appeared in *Double Dealer* in 1922. What followed was typical of the others. His work was admired and talked about, so that many more people read his next offerings, and soon, half a dozen little magazines were printing his work and his readership was growing. At that, it took a European book publisher to recognize his extraordinary talent. A relatively unknown and noncommercial press in Paris (the French equivalent of an American little magazine) published his first book, *Three Stories and Ten Poems*, and in 1926, the talent that had been recognized in *Double Dealer* four years earlier emerged commercially as *The Sun Also Rises*, published in New York by Scribner's.

It should be noted that *transition* (the lower case was an affectation of the time), the *Little Review* and *This Quarter* were among a small group of little magazines printed in English for writers in that language but published abroad, mostly in Paris. All of them, particularly the *Little Reivew* and *transition*, offered a list of contributors whose literary accomplishments then and later were impressive.

Those in the intellectual world tended to believe that quality could only be found in small magazines with small audiences, but a brash newcomer to the New York scene, Harold Ross, began to prove they were wrong in 1925. Born in Aspen, Colorado, Ross grew up in Salt Lake City, dropped out of high school after two years, and served time on various newspapers in Sacramento, Atlanta, Panama City, New Orleans, and San Francisco. As editor of the army newpaper, *Stars and Stripes*, during the Great War, he had directed a talented staff that included Alexander Woollcott, Franklin P. Adams, John Winterich, and Grantland Rice.

While there are several versions of how his idea for a new magazine began, the most credible is that he developed it in the course of long conversations with Woollcott. After a brief postwar job as editor of *Judge*, he found a financial sponsor in Raoul Fleischmann, heir to the yeast fortune, who was one of his poker partners among the gathering of wits comprising the Algonquin Hotel's famed Round Table. Out of this partnership came the *New Yorker*, which appeared for the first time on February 19, 1925.

It was not a new idea. The *New Yorker* bore more than a superficial resemblance to the old *Life* and *Judge*, in which sophisticated cartoons were blended with bright comment on the contemporary scene. Ross, however, carried this basic formula several steps forward. The prospectus he wrote for his new magazine was no doubt one of the most accurate that any prospective publisher had ever produced. He described it as follows: "The *New Yorker* will be a reflection in word and pictures of metropolitan life. It will be human. Its general tenor will be one of gaiety, wit and satire, but it will be more than a jester. It will not be what is commonly called sophisticated, in that it will assume a reasonable degree of enlightenment on the part of the readers. It will hate bunk.

"As compared to the newspaper, the *New Yorker* will be interpretive, rather than stenographic. It will print facts that it will have to go behind the scenes to get, but it will not deal in scandal for the sake of scandal nor sensation for the sake of sensation. Its integrity will be above suspicion. It hopes to be so entertaining and informative as to be a necessity for the person who knows his way about or wants to.

"The *New Yorker* will be the magazine which is not edited for the old lady in Dubuque. It will not be concerned in what she is thinking about. This is not meant in disrespect, but the *New Yorker* is a magazine avowedly published for a metropolitan audience and thereby will escape an influence which hampers most national publications. It expects a considerable national circulation, but this will come from persons who have a metropolitan interest."

So spoke Ross, and that was very much the way his magazine turned out. This large, rough man, with perpetually rumpled hair, was himself the exact opposite of Eustace Tilley, the top-hatted gentleman peering at the world coolly (some said snobbishly) through his pince-nez (a figure drawn for the first cover by Rea Irvin, the magazine's art director, and reproduced annually thereafter). Ross even had some of the attributes of the old lady from Dubuque who had been immortalized in his prospectus. His journal was as free from naughty words and sinful suggestion as though Anthony Comstock himself had vetted it.

Most of all, Ross was a perfectionist who tortured himself and those who worked for him in an impossible effort to match his ideals. One of his noted writers, E. B. White, described him as "restless, noisy, consumed by curiosity, driven by a passion for clarity and perfection." He was never satisfied with what he or others accomplished. People either loved and admired him or hated him; there was little middle ground with Ross. Those who loved him (and no doubt hated him on occasion) formed a cult of writers and editors who worked for the *New Yorker* and deeply resented any criticism of either Ross or the magazine. They took themselves and the magazine with extreme seriousness and firmly believed that Ross had revolutionized magazine publishing, if not journalism itself.

Such total self-confidence was bound to collide with the other primary ego in the field, Henry Luce, and a feud began almost at once. It started with *Time*'s disdainful review of the *New Yorker*'s first issue, and it blazed up again after the launching of *Life*. At that point, Ross assigned one of his best writers, Wolcott Gibbs, to do a "Profile" (a word Ross added to magazine language) of Luce, and the result was a hilarious parody written in *Time*'s own style (or Timestyle, as it would have said), which has been often quoted since. Of this style, Gibbs wrote: "Backward ran sentences until reeled the mind."

About Luce himself, Gibbs was at his most wicked: "Behind this latest, most incomprehensible Time enterprise [*Life*] looms, as usual, ambitious, gimlet-eyed, Baby Tycoon Henry Robinson Luce, co-founder of *Time*, promulgator of *Fortune*, potent in associated radio and cinema ventures. Headman Luce was born in Teng-chowfu, China, on April 3, 1898. . . . Very unlike the novels of Pearl Buck were his early days. Under brows too beetling for a baby, young Luce grew up inside the compound, played with his two sisters, lisped first Chinese, dreamed much of the Occident. At 14, weary of poverty, already respecting wealth and power, he sailed alone for England, entered school at St. Albans. Restless again, he came to the United States, enrolled at Hotchkiss, met up & coming young Brooklynite Briton Hadden. Both even then were troubled with an itch to harass the public. Intoned Luce years later: 'We reached the conclusion that most people were not well informed & that something should be done. . . .'"

To people like Ross, Gibbs, and the other *New Yorker* writers who had developed their own style for the magazine, which they constantly denied having done at all (and do so to this day), *Time's* style was an abomination, and Gibbs took note of it: "Yet to suggest itself as a rational method of communication, of infuriating readers into buying the magazine, was strange inverted Timestyle. It was months before Hadden's impish contempt for his readers, his impatience with the English language, crystallized into gibberish. By the end of the first year, however, Time editors were calling people able, potent, nimble . . . so fascinated Hadden was with 'beady-eyed' that for months nobody was anything else. . . . 'Great word! Great word!' would crow Hadden coming upon 'snaggletoothed,' 'pig-faced.'"

Gibbs ended his Profile, summing up Luce and his works, with a sentence since quoted so often as to become a cliché: "Where it all will end, knows God."

Before the piece ran, Ross and Luce met to discuss it, for reasons not clear. Each brought a lieutenant—Ross, his managing editor, St. Clair McKelway; and Luce, Ralph Ingersoll. The chief combatants were more or less civil, but the lieutenants nearly came to blows and had to be separated. It was a fruitless meeting, ending in bitterness on both sides.

Later, Ross wrote a self-justifying letter to Luce, in a style as typical as *Time's*. "After our talk the other night I asked at least ten people about *Time*, and, to my amazement, found them bitter, in varying degrees, in their attitude. You are generally regarded as being as mean as hell and frequently scurrilous." To which Luce replied, also typically: "It was not 'up to you' to make any explanations as far as I was concerned, but in any case I want to thank you for the personal trouble you took with the *Time*-Luce parody. . . . I only regret that Mr. Gibbs did not publish all he knows so that I might learn at once exactly how mean and poisonous a person I am."

Luce neither forgot nor forgave; nor would the world let him forget. Years later, still haunted, he referred to "that goddamn article," with its Gibbsian phrases so easy to quote. He thought the piece extremely biased, and at least one *New Yorker* stalwart agreed with him. James Thurber wrote years later: "As parody, the Luce profile was excellent, and often superb, but it seems to me that Luce and Ingersoll were justified in resenting the tone of the piece, here and there, and some of its statements."

As it developed, the *New Yorker* could be seen as a twentieth-century version of the great literary magazines of the nineteenth century, with the addition of such features as cartoons and departments that were the products of its own times. Like all great editors, Ross left his stamp firmly and indelibly on the magazine. It was as much his as the *Post* had been Lorimer's. Not the least of his talents, by any means, was his ability to gather around him one of the most brilliant collections of artists and writers ever assembled on a modern periodical. Its list of contributors was notable even at a time when competition in this area was keenest.

E. B. White set new standards in the essay, while Thurber was admired both for his prose and for his cartoons. Gibbs wrote many other witty and incisive Profiles, besides critical writing and fiction. In that category, much of John O'Hara's work in the short story first appeared in the *New Yorker*. S. J. Perelman, among the greatest of the century's humorists, was a regular contributor, and Ogden Nash's much-quoted poetry appeared frequently. For most readers, the cartoons were even more memorable, from such masters as Helen Hokinson, Otto Soglow, Peter Arno, Whitney Darrow, Jr., Charles Saxon, George Price, and a long list of others who were among the best cartoonists of their time. Some of the captions for these cartoons became a part of the cultural language and are still quoted today by older citizens. The magazine's fiction, much of which was minimalist before minimalism was invented, was so archetypical that the phrase "a *New Yorker* short story" came to be a stereotype, although the editors irritably denied that there was any such thing—and no "*New Yorker* style" either.

There were several original ideas in Ross's magazine. The "Profile" was registered as a name, but it passed into the language as a lower-case noun. It was a modern extension of what Marcosson had done earlier in the *Post,* and many of these pieces were outstanding examples of biographical writing. The magazine's filler material—typographical and other lapses clipped from newspapers and periodicals—was not a new idea, but Ross made the fragments original by adding the magazine's own unique ironic comments below them or in headings—"Social Notes from All Over," "Letters We Never Finished Reading," "Ho-Hum Department," and "Department of Utter Confusion," among others.

For a long time, the *New Yorker* was also read for its critical departments, whose writers included Robert Benchley, Dorothy Parker, Clifton Fadiman, and Edmund Wilson. "Talk of the Town," or "Notes and Comment," short and long paragraphs reporting aspects of contemporary life, was another nineteenth-century idea, raised to a high level of excellence by the work of staff writers.

New Yorker articles frequently lengthened into books or were excerpted before publication, including Ruth McKenney's *My Sister Eileen* and John O'Hara's *Pal Joey.* In 1946, a one-issue printing of the entire text of John Hersey's *Hiroshima* was a publishing landmark, as was the serialization of Truman Capote's *In Cold Blood.*

Less apparent to the public, perhaps, the magazine also made a remarkable contribution to the art of interpretive reporting, with the works of Rebecca West, Alva Johnston, Joseph Mitchell, Richard Rovere, Lillian Ross, A. J. Liebling, E. J. Kahn, Jr., and St. Clair McKelway, among others.

There was more to Ross's little miracle than its text, however. For the first time in magazine history, its advertising came to be read with as much attention as the editorial content. At the journal's peak of that early time, advertising craftsmanship and consequent reader attention reached such a point in the magazine that in the fat issues between October and Christmas, advertisers fought for contracts, and column after column was refused. All this was accomplished with a circulation base deliberately kept at Ross's insistence at somewhere near 500,000. If the magazine went much above that figure, he thought, its distinctive character might be lost, and it would become just another magazine.

Ross died in 1951, and the magazine he created entered a new postwar life, seemingly unchanged, but already beginning to develop some of the problems that would affect its future.

A survey of intellectual currents between the wars would not be complete without at least a brief mention of religious magazines—brief because to address the subject, in all its multifaith complexities, would require far more space than can be given it here. A single case history cannot stand for all, of course, but it may illustrate how the Protestants, at least, through the ecumenical *Christian Century*, confronted the intellectual issues of the times.

The *Century* had begun in Des Moines in 1884 as a magazine by and for the Disciples of Christ denomination, but in time Charles Clayton Morrison, its first editor, noticed that the names of Congregationalists and Methodists known to him were appearing on the subscription list, and it was decided to open up its pages to a broader audience. Morrison had long been an advocate of unity. In 1917, the magazine began carrying the subtitle, "An Undenominational Journal of Religion." The news department began to cover news of other denominations, until, after a few years, it became "News of the Christian World." By that time, the subscription list justified the title.

A 1921 series titled "Do the Teachings of Jesus Fit the Times?" illustrated the dimensions achieved by the *Century*'s editorial expansion. Its contributors included Jane Addams, Joseph Ernest McAfee, Herbert Croly, Vida Scudder, and Lloyd C. Douglas, not yet a best-selling author. In general, the magazine promoted anything that might bring about church unity.

Like so many other intellectual magazines, the *Century* was not enthusiastic about American entry into the Great War until 1917, when it gave President Wilson full support. But like most of the others, it felt betrayed by what happened at Versailles and by Wilson particularly. As the twenties began, there was an added disillusionment with the administration's approach to domestic problems, particularly the treatment of dissenters and objectors. The *Century* preached hard on the subject of rehabilitating the social gospel, and it came out unequivocally for Prohibition and its subsequent enforcement.

On other social issues, the *Century* was pro-labor, but not unreservedly so, and it was sincerely anti-racial, although the tone of its articles on this subject was likely to be parternalistic. The crusade against alcohol and for Prohibition, however, dominated the magazine in the twenties. Liquor was the country's "worst menace," it said repeatedly, and, in fact, Morrison had campaigned for Prohibiton since he bought the magazine in 1908. Even after the Volstead Act was repealed in 1933, the *Century* continued its unremitting fight against alcohol. The strength of its convictions on this issue even influenced the journal's politics. Its support of Wilson first began to wane when the editors suspected that he was too friendly with the liquor interests.

As the tensions and discontents of the twenties, obscured by prosperity, moved toward the Crash, the *Century* pursued an optimistic course, telling its readers that adverse conditions represented opportunities for doing good, not causes for lamentation. The editors wanted nothing to do with Warren

Harding's return to "normalcy"; their direction was forward. A rejuvenation of the social gospel had occurred in the twenties, and the *Century* was its foremost advocate. That meant advocacy of a program that called for a constitutional amendment to end child labor, the setting of a minimum wage, and the abolition of the twelve-hour day. While the Federal Council of Churches believed that strikes were against Christian beliefs, the *Century* nonetheless supported that right, even as it severely criticized the A.F.L.'s conservatism.

It was not easy to tread the line between Christian belief and the real world. Thus the *Century* was unequivocally against Marxism and for American democracy, but at the same time it was a severe critic of laissez-faire capitalism, carrying frequent articles attacking it by Harry Ward, a professor of Christian ethics at Union Theological Seminary. Similarly, the *Century* saw no conflict of interest between devotion to religious dogma and fervently defending the right of dissent and free speech. When the Daughters of the American Revolution and the American Legion produced a blacklist of people and institutions suspected of subversion, the *Century* assailed both the list and the organizations. It also opposed the bill signed by Calvin Coolidge that excluded the Japanese, supported Philippine independence, and urged recognition of the Soviet Union. The *Century* had some doubts about the flow of migrant blacks from the South but pleaded strongly for their just treatment. Indeed, Robert Moats Miller, in his *American Protestantism and Social Justice, 1919–1939,* asserts that the magazine's sensitivity toward blacks in this period was "unmatched in any paper in the country, religious or secular."

Over and above social issues, however, the major concern of the *Century* was the search for peace and a world without war. That search was emphasized editorially, or in some other way, in nearly every issue.

The *Century* may have been the only periodical in America to regard the Crash not as a disaster but as an opportunity. It gave the public "the privilege of sobering up," an editorial said, after a "speculative debauch." The Depression, it added, was not a temporary disturbance of the system but a clear signal of fundamental flaws in it. The journal advocated "voluntary liquidation" of the old system but gave few specifics for a new one, although the ideal appeared to be clearly a socialist one. Just the same, perhaps hedging its bets, the magazine supported Hoover in 1932.

Oddly enough, Morrison (who remained editor until 1947) had come to believe that Franklin Roosevelt would go to the White House hopelessly indebted to William Randolph Hearst, and in the spring of 1932, in a series of editorials, he began preaching the virtues of a third party, but of a different kind. It would have no candidates, and its platform would consist of "disinterested political principles" rather than special interests. This party, having no candidates, would exist solely for the benefit of the platform and would endorse only those regular party candidates who accepted it.

Roosevelt's overwhelming victory seemed to dispel Morrison's doubts about his political obligations, and the magazine became an increasingly warm supporter of the New Deal, particularly such innovative social moves as creating the Tennessee Valley Authority. Except for the president's "good neighbor" policy toward Latin America, however, Morrison was uneasy about Roosevelt's foreign policy. Even as it endorsed him in 1936, the *Century* deplored the creation of a big navy and other militaristic moves. This alarm increased after the election, as Roosevelt talked about quarantining Japan, and the *Century* began to see "the folly of 1917" in danger of being repeated. It was more interested, at first, in what Ghandi was doing in India than what Hitler was doing in Germany.

This latent pacifism did not last. As the enormity of the Nazi menace became clearer, the *Century* denounced "the unspeakable brutalities" against Jews, urged the government to provide a haven for those who could leave, and expressed its additional concern about Protestant churches in Germany, particularly for those 6,000 pastors who had publicly opposed Hitler's notion of a "German church."

Just before the attack on Pearl Harbor, the *Century* acquired the *World Tomorrow*, a Christian socialist magazine, and Kirby Page, its editor and a well-known pacifist, became a contributing editor. By this time Morrison was calling for a "new liberalism," but he continued to open his pages to other viewpoints, even those of Reinhold Niebuhr, who had once called the *Century*'s opinions "pure moonshine."

During the war years, as Martin Marty has observed, a line could have been drawn clearly separating the *Century*'s contents. On one side was the magazine's coverage of the war, particularly as it affected religion, with editorials and articles on the issues of war and peace. On the other side were "reports on Christian church life and American culture." As Marty notes, "the latter had no 'business as usual' stamp, since the war colored almost everything in those years, but one senses that the editors were saying life must go on; faith needs nurture; the subtleties of life matter; there are trenches in America as well as on the front lines. . . ."

The events of 1941 had brought about a final break between Niebuhr and Morrison. Niebuhr first defected to the *Nation* and then in 1941 helped found and then edited his own magazine, *Christianity and Crisis*. The break had been precipitated by the *Century*'s delay in recognizing that the Holocaust was occurring. When Stephen Wise had presented the editors with statistics, they took the figures to the State Department, which assured them that Wise was doubtless exaggerating, and for a while that verdict was accepted. But there was little minimizing as time went on, and the *Century* duly reported on whatever evidence it could find. Nevertheless, as Marty says, the magazine's "approach through the Holocaust years was limited, blinded and the editors were benumbed; so was the U.S. government; more so were other Christian periodicals that almost or entirely ignored the subject. Even Jewish organizations saw little to do, and most did little. . . ."

The division in content represented the *Century*'s determination not to become a secular journal, even though that would mean more circulation, income, and prestige. The magazine clung with determination to its principles as a religious journal. This could be seen in its famous "How My Mind Has Changed" series, which began in the late thirties and ran for years. But the editors were not content to be a "theological cafeteria line," as Marty notes, they "argued with the authors, plundering, ransacking, resisting—and sometimes being changed themselves."

To its credit, the *Century* was also ahead of its time in its relative lack of discrimination against women on its editorial staff. There had been four women editors by mid-century, and many others held lesser jobs—a good record for the times. The magazine also furthered the cause of women in the church. In a 1941 editorial, it declared forthrightly: "The American church is still one of the most backward of all institutions in the place it accords to women and the attitude which it exhibits toward them."

With the departure of Morrison in 1947, the *Century* entered a new phase of its history.

17

Photojournalism: A New Breed

From the turn of the century, photography began to change the face of magazines, newspapers, and books in America as the camera came into its own at last. The ready availability of cameras for amateurs and the images projected by a growing movie industry fed the public's appetite for black-and-white photographs. Illustrators would never be superseded entirely, but as the craze for picture taking developed—"you press the button, we do the rest," the Kodak company advertised—whole new vistas of illustration opened to publishers in all the media.

Magazines responded to the trend cautiously at first, mixing photos and conventional illustration, but as advertisers and newspapers rushed to take advantage of public demand, publishers were caught up in the general enthusiasm. By the late twenties, every magazine staffer was said to be walking around with a plan for a picture magazine tucked away in his desk. All these plans were eclipsed by the appearance of *Life* and then *Look* in the thirties. For the first time, in their pages, the real world was depicted realistically, aided by rapid technological advances. Now there was a new way of reporting events to the public, and it was called photojournalism, a word whose origin is attributed to Erich Salomon, a German photographer, although other claimants include Henry Luce and Dan Mich, *Look*'s editor.

The success of both these major picture magazines was built on several trends which had been at work for some time in the United States. For decades, people had been enjoying themselves with their own cameras, selling themselves on the merits of photography. They had grown so accustomed to seeing events translated into visual form that they wanted more. Advertisers, too, had contributed to the increasingly visual perception of the world. They used photography in a good many inventive ways, even in their trademarks, having been among the first to see what it could do to attract readers. Because the richer advertisers had more money to spend, they could afford to experiment with new techniques. For a time, there were more pictures in the advertising than in the editorial content of most magazines.

A researcher has only to leaf through the pages of journals published in the thirties to be struck by the vividness of advertising photography, some of it in color, and the relative drabness of much of the black-and-white illustration used by editors. J. Stirling Getchell, one of the early giants of advertising, was first to use photographs of cars in automobile advertising, and later, in 1938, he started his own picture magazine, *Picture: The Photographic Digest.* Getchell and others were able to hire the best photographers for advertising work, including Edward Steichen and Margaret Bourke-White.

Newspapers used photography so successfully that they, too, provided a powerful impetus for magazine publishers to go and do likewise. It was a natural marriage. As early as 1914, the New York Times Company had begun publishing *Mid-Week Pictorial,* a magazine-like supplement, and photographically covered the war in Europe. After the war, the *Pictorial* went on for a time, providing photographic coverage of the news on a regular basis. Monte Bourjaily, of the United Press, bought it in 1936 and revamped it, grouping the pictures to tell stories, in the manner of *Life* and *Look*—picture-text stories, as they were called. Circulation went up fourfold, but unfortunately the new *Pictorial* appeared only a month before *Life* went on the stands and became the first casualty among Luce's many competitors. Bourjailly gave up after a year.

When Joseph Patterson launched the New York *Daily News* in 1919, he subtitled it "New York's Picture Magazine," and it became the forerunner of numerous other picture-text tabloids, as well as Sunday supplement magazines. In the twenties, New York had not only the *News* but also Hearst's *Mirror* and Macfadden's *Graphic.* The *Graphic* (known in the trade as the "porno-graphic," you may recall) did not stop with printing sensational pictures but improved on them if they were not sensational enough.

News photographers now became extensions of the reporters and magazine writers, poking their lenses into places that could only be described before and permitting readers to see what only the reporters—or maybe only the photographer—had witnessed. The apotheosis of the new tabloid journalism, perhaps, was the front-page picture in the *Daily News* of Ruth Snyder, a convicted murderer, getting the full voltage of the electric chair in the death house of Sing Sing Prison. A two-word headline surmounted the picture: RUTH FRIES. A reporter designated as a witness had strapped a camera to his leg and took the picture surreptitiously.

Remarkable circulations among all these media testified to the success of photojournalism. Soon, in this pre-television era, Time Inc. pushed its own advantage in the new field to create a newsreel, the *March of Time,* in which motion pictures encouraged the public to look for visual representations of the news. Conditioned by movies, advertising, and the proliferation of pictures in newspapers and tabloids, Americans were ready to accept picture magazines when they emerged explosively in the thirties. In some ways, however, the periodicals represented no more than modern improvements on the old stand-bys and were not completely new developments. *Harper's Weekly*, for example, had covered the Civil War in pictures and text, as had some of its contemporaries. Frank Leslie's various illustrated papers had pictured society, scandal, civic problems, politics, and a variety of other matters in essentially the same way the tabloids and their supplements were doing half a century or more later. The difference, of course, was photography and the technology of reproducing pictures. They had come a long way from the *Philadelphia Photographer*, the first magazine to print a photograph using half-tone reproduction.

The viewpoint of editors had changed, too. In the early twenties, before its collapse, the *Century* had agonized over the use of photographs, its editors still clinging to engravings. Even when such older, elite journals finally accepted photography, they treated the pictures primarily as art rather than as conveyors of information or reflections of reality. It took the uninhibited editors of *McClure's*, *Munsey's*, women's magazines, and other less-expensive periodicals to forge ahead with the new art. The cameras were turned on public figures, actors and actresses, inventors like Thomas Edison, writers like William Dean Howells, and ordinary people involved in the news. For once, new technology did not have a high price; the photographic process was much cheaper than woodcuts.

Photography was already prevalent in magazines by the end of the Great War. Fashion magazines seized the idea eagerly, as *Harper's Bazaar* and *Vogue* used highly stylized shots made by first-rank photographers (Steichen, for one) to show real women wearing the latest designs. When *Town and Country* and *Vanity Fair* appeared in the twenties, they demonstrated that beautiful photographs could enhance text matter without being an end in themselves. Another pioneer, the *National Geographic*, brought people and places around the globe into American living rooms. Thousands of young boys viewed female breasts for the first time in its pages, a display considered, curiously, not indecent because they belonged to women of another race. Only the rich and near rich did much traveling in those days, and the *Geographic* made various journeys possible for millions of readers. Not even the rich had viewed some of the scenes in the November 1910 issue, with its twenty-four pages of color photographs of Korea and China. The black-and-white shots had been colored by hand; it was the first time a magazine had used color photos in large numbers. Later, the *Geographic* pioneered again in photographing animals in their natural habitats, showing pictures of undersea life for the first time and publishing aerial views.

All this early use of photography, however, only illustrated texts. It would take time before both magazines and newspapers developed a concept of photojournalism in which the pictures themselves became the primary carriers of the story. Still, illustration had its own distinctive values, as journalism historian William H. Taft has noted. "*National Geographic* photographers," he observes, "have won major awards for the contributions to a better comprehension of our world. Since readers find such photographs valuable for years, it is no wonder that some persons fear the country eventually will sink under the weight of millions of copies of 'that magazine with the yellow cover.'"

Some of the earliest advances in photojournalism occurred in Germany in the late twenties, influenced most notably by the development of the small, unobtrusive Leica camera. Erich Salomon, a lawyer turned photographer, was first to use the compact Leica to take shots of the wealthy and powerful. When *Fortune* editors saw pictures taken by Salomon printed in the London *Tatler*, they tracked him down, bought some of his work, and brought him to America as a staff photographer for the magazine.

It would be hard to overstate the importance of the Leica in the development of photojournalism. It allowed photographers to take pictures unobtrusively and rapidly and could function indoors without a flash, adding to its usefulness in getting unposed shots. Margaret Bourke-White, who rose to fame on *Life* as one of the great photographers of her time, had learned her art on a large camera and was accustomed to posed shots in the beginning. Much of her early work had been industrial photography, where pictures could be taken over and over again. But she discovered the value of the Leica when she covered Bruno Hauptmann's trial for the kidnapping of the Lindbergh baby and began carrying one along with her large camera as a matter of routine. Other *Life* staff photographers—Eisenstaedt, Tom McAvoy, Peter Stackpole—all used the 35-mm. camera, which allowed spontaneous shots to be taken.

The picture essay was a new and different method of communication, one which took first place in the nation's visual life until the coming of television. Editors learned that pictures could be used in several ways. They could stand by themselves or be used with captions and text in various combinations, with the text subordinate. The point of photojournalism was to communicate, the photograph not necessarily being used for itself alone. On *Life*, according to Stephen Heller, "the picture story was described as an 'act,' a reference borrowed, no doubt, from the theater. It implied that the magazine was a stage. And so it was . . . *Life* went even further than the European journals to institutionalize the photographic essay by allowing greater page runs, designed with generous white space and economical typography . . . *Life* responded well to the needs of 20th-century men and women for concise visual information."

The immediate, resounding success of Time Inc.'s venture made it certain that a crop of imitators would appear. They faced the same problem that *Newsweek* had endured with *Time* in the early days; namely, how to do the same thing in a different way. By far the most successful attempt among a dozen or more imitations was *Look*.

Gardner "Mike" Cowles, Jr., and his brother John were working on an idea for a picture magazine while *Life* was being developed. Cowles had seen research in 1925, done by a young Princeton professor named George Gallup, which showed that readers of the Des Moines *Register & Tribune*, owned by the Cowles family, preferred pictures to written copy. As a result, Vernon Pope, then the editor, began to run more photos in the rotogravure section of the paper, some of them grouped as a series of pictures rather than standing alone. The experiment was so successful that by 1933 the *Register & Tribune* had formed a syndicate to sell their picture series to other papers.

The idea for *Look* originated that year. In his privately printed memoirs, *Mike Looks Back*, Cowles says he first thought of such a magazine when he was examining proofs of Lawrence Stallings's *The First World War*, a lavishly illustrated book. It would make a splendid feature for the *Register*'s Sunday rotogravure section, he thought, if it could be adapted. Working with Pope, who was now editing that section, a ten-part series was created, using about four hundred pictures. The circulation results were dramatic, and later the series was syndicated to fifteen other Sunday papers.

Convinced that photojournalism was a coming thing, Cowles first considered a new national Sunday supplement, one that would compete with the *American Weekly, This Week,* and *Family Weekly*, offering only picture stories. Other family members advised against it, however, on the ground that these supplements were not necessarily profitable. An obvious alternative was to start an independent magazine to be sold on newsstands for ten cents.

Cowles had no magazine experience and admitted in his memoirs that if he could have foreseen all the difficulties *Look* would bring him, he might have been more cautious. His was an optimism, he wrote, born largely of ignorance. Working with Pope, he laid out a crude dummy and took it to his friends Henry Luce and Roy Larsen at Time Inc. for scrutiny and advice. Cowles had heard rumors about *Life*, but he believed Luce and Larsen could be trusted.

Luce examined the dummy with great interest, then pulled out one of his own—for *Life*. Comparing the two ideas, Cowles wrote later, it was clear that *Life* was to be a weekly covering the news, while *Look* would be a monthly oriented toward features. Luce's entry would be upscale, aimed at affluent, educated people, readers of the *New York Times*, but *Look*'s readers were more likely to be buyers of the tabloid *Daily News*. One would be a slick-paper journal accepting advertising at once, the other would be printed on more modest stock and not take advertising until its circulation base was established.

Since the two publications did not appear to be in conflict, Luce volunteered to invest $25,000 in the Cowles venture (as noted earlier), thus hedging his bet on *Life*. Cowles's brother John advised him to delay *Look* until Luce's venture was on the stands, so that public reaction to a picture magazine could be measured. The response, as we have seen, was so immediate and overwhelming that *Life* was printing more than a million copies per issue, and even then could not keep up with the demand. But since so much money had been invested in it, well over $10 million, it was not until 1939 and a $5 million loss that the magazine was in the black.

Two months after *Life* appeared on the stands, *Look* published its first issue, dated February 1937 but on sale January 5. Hermann Goering's rotund face stared from the cover, heralding a story inside called "Will Former Dope Fiend Rule Germany?" Other articles talked of "Mishandled Paroles—America's Shame," "A Psychologist Reveals the Secret of Roosevelt's Popularity,"

"Auto Kills Woman Before Your Eyes," and "Trained Goldfish." There were profiles of Joan Crawford and Dolores Del Rio, and as an antidote to Goering, Greta Garbo's lovely head graced the back cover in a head-and-shoulders color shot.

"Graced" was not the proper word, as it turned out. "Stimulated" might have been more accurate, because someone discovered that when Garbo's picture was folded in half, the result was a reasonable facsimile of the female sex organ. News of this aberration spread across the nation with incredible speed, and the result was felt immediately on the newsstands. Even before this revelation, the initial print order of 400,000 had been sold out and more were on the presses, but then the sales shot up to 700,000 following the discovery. When Cowles learned the reason, he and the family believed they had no choice but to recall all the unsold copies, at a cost of $100,000. Not, however, before the Montreal police had seized several hundred copies.

No one could have asked for better free publicity. Between news reports and word of mouth, *Look's* inadvertent *faux pas* was known in every part of the nation. Again, the result was immediate. Circulation reached 1,203,000 on March 1 and continued to climb. A month later, the Cowles family decided to go biweekly; by October, the magazine was selling two million copies per issue. Before the end of *Look's* first year, advertising appeared in its pages, beginning November 9 with thirty-five advertisements. By May 1938, the magazine was guaranteeing 2,000,000.

Then an alarming event occurred. Cowles was on a two-month holiday in Europe near the end of the year when a cable from his brother John summoned him home. Circulation was suddenly descending with sickening speed; by July 1939, it would be little more than a million. Returns were clogging the warehouses, advertisers were cancelling. There were two reasons for the disaster. First, the economy was weak at that moment, so a magazine relying almost exclusively on newsstand sales was bound to be hurt. Then, too, the market was flooded with more than a dozen other picture magazines trying to cash in on the popularity of the two leaders.

But Cowles also blamed himself. In his memoirs, he observed that the family had begun the magazine on a shoestring, an initial mistake, compounded by the fact that there were simply not enough experienced hands involved in producing the journal, giving it an unprofessional look. The periodical was badly designed and did not flow from cover to cover. Its paper and printing, too, looked shabby when compared with *Life*. Confronting this situation, John Cowles was prepared to kill the magazine, and when Gardner asked the advice of Luce and Larsen, they agreed. Between the cheap paper and printing and the image of vulgarity created by the Garbo picture, they believed *Look* could not survive.

After this meeting, Cowles walked up to Central Park and spent an hour or two sitting on a bench, trying to decide what to do. Pride and an unwillingness to admit failure won out over all objections. Returning to Des Moines, he began to reorganize his staff, hiring two circulation experts, Les Suhler, from *Child Life*, and S. O. Shapiro, vice president for circulation at Macfadden

Publications. Between the two, the declining circulation trend was reversed. Other talents were brought in to the editorial department, most notably Dan Mich, formerly managing editor of the *Wisconsin State Journal* and eventually to become editor of *Look*, succeeding Vernon Pope. It took a year, but the magazine was turned around.

By 1940, most of the *Look* staff had been moved from their original home in Des Moines to New York, and the magazine was thriving under Harlan Logan, the general manager, charged with professionalizing the journal. In spite of Logan's success, however, he became involved in a series of policy and personality disputes. Believing that the magazine's wartime mission was to help people cope and to generate patriotism, he was making *Look* appear dull by comparison with *Life*, which was doing a superb job of covering the war. Logan's educational concept was pursued after the war, converting *Look* into a "how-to" magazine, anticipating the service magazines of nearly forty years later. Logan also created a separate department that produced books, developed a series of newsreels, and experimented with television, along with other visionary projects. This department, called Editorial Art Research (known to the staff as EAR) grew until it had more employees than *Look* and got most of Logan's attention.

These developments resulted in frequent clashes with Dan Mich, now executive editor, whose editorial ideas were much more like Pope's than Logan's. These disputes ended in a confrontation at an all-day editorial meeting and what amounted to an ultimatum from Logan. Cowles responded by taking personal charge of the magazine (he had been devoting much of his time to managing the Des Moines newspapers), abolishing EAR as one of his first acts. Logan departed and Mich was in full charge, under Cowles.

But then a new complication occurred when Cowles met Fleur Fenton, executive vice president of an advertising agency handling the *Look* account, and married her in December 1946. She joined *Look* the following year as an associate editor and member of the editorial board. In his memoirs, Cowles credits her with persuading him to include food and fashion features and with improving the magazine's appearance, mostly through the hiring of Merle Armitage, one of the best art directors of his time. By 1948, *Look* was selling nearly three million copies, surpassing *Collier's* for the first time.

In spite of its success, however, *Look* could not get itself accepted by the intellectual community. No schools or public libraries subscribed to it, nor was it indexed in the *Reader's Guide to Periodical Literature*; librarians seemed to believe it had no reference value. A direct-mail campaign corrected this situation.

A new crisis boiled up in 1950, when Mich was hired away by Otis Wiese to be editor of *McCall's*. Internal factors were involved. Mich had not gotten on well with Fleur Cowles, and their conflict was so open that when Mich took a brief leave to be examined at the Mayo Clinic, the office joke was that he suffered from "intestinal Fleur." Mostly, Mich objected to the huge amounts of time and money that Cowles Magazines, Inc. (as it was now called) had put into two new magazines, *Quick* and *Flair*. The latter was known as "Fleur's

Flair" (it had been her idea), and it ranked as perhaps the most unusual magazine of its own or any other time. Bright and breezy, a salmagundi of virtually everything on a reader's menu, it contained pages in different colors, odd designs, and several accordion pullouts. A *New Yorker* cartoon depicted a subway straphanger holding up the magazine from which fell a cascade of pullouts. Cowles termed it correctly "an art director's dream." Another *New Yorker* cartoon showed a house detective in a hotel lobby peering through a hole in the cover, one of its many innovations. The first issue of *Flair* had appeared in January 1950. A year later, it was showing a cash loss after taxes of $1,250,000, and Cowles closed it in January 1951 after an acrimonious battle with his wife. They were later divorced.

The weekly newsmagazine *Quick* was a different story with the same ending. It was a small-size periodical, inexpensive to produce. Woodrow Wirsig, assistant managing editor of *Look,* was drafted to be its first editor. *Quick* appeared for the first time in May 1949, selling for ten cents, intended exclusively for newsstand sale, and carrying no advertising. (It soon took on advertisers, when increasing the cover price became the only alternative.) The initial success of the magazine was remarkable, rising from an initial 320,000 copies per issue to 500,000 in only six months and more than 1,000,000 by the next year.

Nevertheless, *Quick* was still losing money in 1953, in spite of increased advertising and a circulation of 1,368,000, and Cowles decided to suspend it on June 1 of that year, merging it with *Look;* subscribers got that magazine to fill out their subscriptions. The reason was simple. *Quick's* unusually small page size simply could not accommodate many advertisers, and the magazine could not be produced at its established level of quality without increased advertising revenue. But there was an even more decisive reason for the suspension. *Collier's* was about to overtake *Look* in 1954, and Cowles believed something drastic had to be done. Folding *Quick* added enough subscriptions to the flagship journal so that it could hold the lead, important to advertisers.

In the early 1950s, Cowles had to contend with a number of staff problems—temperamental editors unable to get along with each other—and the solution to this situation appeared to be to woo Dan Mich back from *McCall's.* He returned in January 1954, after several editors and members of *Look's* board had pleaded with him, and as he returned, three of the troublemakers departed. Mich had had to wait three and a half years to get the control he wanted. His return, as Cowles wrote later, marked the beginning of *Look's* "golden years."

Those years were illuminated by Mich's accomplishments in improving both the content and appearance of the magazine, at the same time smoothing out its internal operations. In less than two years, *Look* was publishing some of its most noteworthy stories, including several on the rise of racial tensions in the South, and a series organized and written by Leo Rosten on religion. Foreign coverage was also greatly improved. Meanwhile, *Look's* close rivals, *Collier's* and the *Saturday Evening Post,* were expiring, leaving a clear field. Cowles credited Mich for the successes of the "golden years," and it was a severe blow when he died in 1965, at the peak of his and the magazine's success.

The decline began soon after the following year. In Cowles's view, television was the villain, slowly destroying his magazine's reason for being; the *Post* and *Collier's* had already succumbed. He embarked on a number of delaying actions, but the end was inevitable, and with the issue of October 19, 1971, *Look* was suspended. A little more than a year later, *Life* expired, too, with its last issue in December. Cowles wrote later that the day he announced the decision to suspend was the most difficult of his life.

In October 1978, Time Inc. revived *Life* as a monthly, with a much smaller circulation, concentrating on feature stories rather than the news. It survives today. *Look's* logo was bought by Daniel Filipacchi, publisher of *Paris Match*, and in February 1979 was launched anew with a reported $25 million investment. Only a few months later, in August, it closed for the last time, with losses estimated at a million dollars a month. It had never established an identity for itself in its rebirth, resembling a mixture of *Paris Match, People*, and *Rolling Stone*.

The success of both *Life* and *Look* in their earlier years spawned numerous imitations. One of the most successful was *Click*, a monthly picture magazine published by Moses Annenberg as part of his Philadelphia publishing empire. Begun in 1938, its picture sources were Annenberg's Philadelphia *Inquirer Sunday Magazine* and another journal called *Picture Parade. Click* quickly rose to a circulation of 1,600,000 with its second issue by virtue of lurid, sensational pictures, strongly seasoned with sex. Its covers displayed actresses in enticing poses.

John Cooney, a historian of the Annenberg dynasty, wrote that the first issue of *Click* "set the tone for a periodical that dealt with Peeping Toms, white slavery, and such oddities as a sexy woman lion tamer." Later articles were in the same vein, with jokes about drinking, cartoons, divorce, and immoral sexual relations. Circulation settled down at about 1,500,000, just behind its two major rivals, but problems developed. The Canadian government banned *Click*, and the Catholic Church publicly deplored it. Annenberg, involved in Pennsylvania politics, was alarmed by these actions and feared it would give ammunition to his Democratic opponents. He instructed Emile Gauvreau, the veteran Macfadden editor who was editing *Click*, to tone down its content. Gauvreau, a true professional, did so with such success—substituting art, culture, and patriotic topics for sex—that circulation dropped precipitously.

In 1942, with the senior Annenberg doing time in the penitentiary, son Walter took over the magazine and redesigned it to be a publication intended for the family, with articles on the war and profiles of celebrities. In two years, circulation had climbed back to nearly a million. But Annenberg wanted to expand another family venture, *Seventeen*, and with wartime shortages limiting the amount of paper available, *Click* had to go.

Sex and scandal were also the primary ingredients in most of the other *Life* and *Look* imitators. *Pic, See,* and *Focus* all offered similar fare and gathered circulations of about half a million each. *Pic*, a rotogravure monthly published by Street & Smith, appeared in 1937, featuring primarily seminude young women, whose photos were printed on cheap newsprint. Its editorial content could be judged by an early cover story: "Do White Men Go Berserk in the Tropics?" At the end of the Second World War, however, the publishers decided to turn *Pic* into a respectable magazine for the men coming back home. This proved to be a successful policy, as circulation passed 600,000, but like *Click*, it had to be sacrificed for the sake of its paper and printing facilities, so that Street & Smith could start a new publication, *Mademoiselle's Living*.

The first issues of *See*, appearing in 1942, were also focused on beautiful women, but the editors soon added such authors as Pearl Buck and Sumner Welles to broaden the coverage and appeal of the magazine. This new editorial mix pushed circulation over the million mark, and eventually it became a men's magazine.

Obviously, photojournalism was filling the need of Americans for pictures, and its success reflected their increasing lack of time for reading in a world moving rapidly into the fast lane. The need was to serve up information and entertainment in a form most easily grasped. Pictures had a universal appeal, speaking to all levels of society in a way that the printed word could not. Ralph Ingersoll, in the days when he was developing *Life*, had put it well: "Pictures are for rich or poor, without regard for race, class, creed or prejudice, speaking the same language. . . . You use one vernacular to a truck driver, another to a bank president. But the truck driver and bank president will stand shoulder to shoulder to watch a parade. Only in the subject of pictures will their interests draw apart."

In some ways, the career of Margaret Bourke-White paralleled the meteoric rise and tapering off of photojournalism. She exemplified all that was glamorous about it, and her work struck responsive chords among readers. Her curiosity, her determination to get to the heart of a situation with her camera, her endless wonder about machines, wars, cities, and people—all these were reflected in her pictures and found a resonance among those who saw them. She contributed much to raising the status of her profession, as her biographer, Vicki Goldberg, points out. With her staff, she set the standards at *Life* for photographic practices and insisted that all photographers be given proper credit for their pictures. They were an elite group, and they knew it. Among them, besides those already mentioned, were Eliot Elisofon, Robert Capa, W. Eugene Smith, Carl Mydans, Arthur Rothstein, and Tom McAvoy, to name only a few. It was a sign of their status that it was the reporters who assisted the photographers, carrying cameras and holding lights. Photography was the story, not the text.

As photographers gained in power and status, and the career itself promised adventure and excitement, a new breed of photojournalists appeared, willing to follow events and action anywhere in the world to get a story in pictures. They photographed celebrities, but they themselves became celebrities too, endorsing cigarettes, wine, and air travel. At least two movies were purported to be versions of Bourke-White's life.

Illustrations sometimes appeared on magazine covers and to decorate fiction, but the decline of the magazine short story, beginning in the forties and fifties, further reduced their use. Obviously, photographs could capture the essence of nonfiction far more accurately. That was particularly true, of course, in the Second World War, when the picture magazines truly came into their own, in the absence of television. Photographs had lent credibility to the devastation of the Depression in the thirties (as it was documented particularly by photographers working for the Farm Security Administration), pictures that eventually found their way into magazines and newspapers. Similarly, images from the various theaters of war brought the reality of the conflict to the millions who were far from the fighting. Risking their lives to take the pictures, photographers enhanced their own status. In 1935, Dan Longwell, one of those who had helped to found *Life,* had predicted that a war would be a boon to picture magazines, and he was right.

There was censorship of war pictures in the magazines. Until September of 1943, by government decree, no pictures of dead Americans could be published, although such shots of other nationalities were permitted. When *Life* finally printed a picture of three dead soldiers on Maggot Beach in Buna, their effect was immediate. Enlistments declined sharply the following week. After the war, photographers sent back horrifying shots of concentration camps, once again indelibly stamping the American consciousness.

In the postwar world, the trend in photography shifted to pictures representing emotional and psychological depths, moving away from the staged, symbolic or public photos so characteristic of Bourke-White's work. Photographers tried to capture internal states rather than simply to exhibit the more obvious external objects. An example was the 1948 series in *Life* photographed by Leonard McCombe called "The Private Life of Gwyned Filling." These pictures showed a day in the life of a young working woman, conveying her state of mind through such common symbols as cigarette butts. Other photographic essays soon followed, depicting people's feelings and moods and the atmosphere of situations.

The growing use of color in the fifties brought another change. Advertisers had been using color for a long time, but the lengthy processing needed for color film had discouraged its use in news photography. By the late forties, however, technology had advanced far enough so that color film could be processed much more easily than before, and the news magazines began using it with greater frequency.

There were those who thought that the demise of *Life* and *Look,* along with the rise of television, might adversely affect photojournalism, or even signal its end, but that proved to be a serious misjudgment. Photographic images went right on conveying to the public a sense of reality. All the words written about the shooting of students by National Guardsmen at Kent State during a Vietnam War protest could not surpass the picture of a distraught girl weeping over a wounded student. The range was as epic as life itself—President Jimmy Carter stumbling in a race, a Buddhist monk setting fire to himself, and the horrors of Vietnam. *Time* and *Newsweek* were chief among the potent instruments delivering such images to the public after the two major picture magazines disappeared.

Memorable images, in the opinion of some media analysts, often impressed themselves on the public consciousness more deeply than the abundant footage shown on television. In a study of Vietnam War reporting by the major newsmagazines, Oscar Patterson II found that the actual coverage in these periodicals averaged only about 7 percent, compared with almost 25 percent on the networks. Yet the dramatic pictures chosen by the magazine editors and published in the journals stayed with viewers longer than the fleeting images on the television screen.

In 1974, Time Inc. extended the range of photojournalism by bringing out a new picture magzine, *People*, focusing on celebrities and ordinary people doing extraordinary things. *Life* had done as much and more, but *People* somehow managed to project a somewhat different appeal and, like its predecessor, became an instant success. Sean Callahan, writing about this phenomenon in *American Photographer*, quoted the photographer Mary Ellen Mark as saying that *People* had "created a new form of photojournalism, one borrowed in part from the French. . . . You never have more than two days, and you learn to think on your feet."

Many of the pictures in *People* were staged, attempting to place the rich and famous in zany or embarrassing positions. The photographers created the poses. Describing the hazards in this new kind of journalism, reporter Paula Span wrote: "Bette Midler, though very anxious to promote her 'Baby Divine' book, would not climb into the Victorian baby carriage photographer Theo Westenberger had brought at considerable expense. Gloria Vanderbilt declined to pose in post-deb evening clothes, insisting she'd only wear items from her sportswear line. But *People* photo-editor Dunn had her photographed amidst a dozen jeaned buttocks. Gloria Steinem refused to allow photographer Marianne Barcellona into her home but agreed to be photographed on a book tour and would show up in a bubble-bath picture anyway."

People represented a continuity in the march of photojournalism. Richard P. Stolley, its managing editor, had been with *Life* for twenty years and so was able to draw on his experience there. But he found himself involved in a different kind of photojournalism, one that brought the reader into the most intimate moments of the famous and almost famous. The method was called "personality journalism." It did not glorify the "candid shot" but rather valued the bizarre or offbeat posed picture. The method was not without its critics. Nora Ephron called *People* the magazine "for people who don't like to read." Circulation and profits, however, continued to soar, and in ten years *People* had become the third most widely read magazine, according to Time Inc.

The impressive readership figures (21.8 million in 1984) once more bred imitations, among them, *Us*, *In the Know*, and *Celebrity*. Of these, the most successful was *Us*, published by the New York Times Company. Lavishly using pictures, it focused on celebrity but in its initial statement declared it would not be devoted simply "to people in action but to the action people—

to the dramatic and inspiring events of the day that can be captured by the eye of a camera. We believe that the growing millions of better-educated young adults will respond to superior photojournalism." But *Us* never really got off the ground. It failed to match *People* either in appearance or popularity, and its circulation of about 500,000 was well below expectations. In 1980, it was bought by Peter J. Callahan, of the Macfadden Women's Group, which then included *True Story, True Confessions, Secrets, Photoplay,* and *Cheri.*

A time when celebrity had become a national obsession, the eighties encouraged the trend toward personality journalism, in which photojournalism played a vital role. Yet at the same time, the newsmagazines were increasingly employing photographs to interpret the contemporary scene while decreasing the percentage of those devoted to human interest or entertainment. That fact only underlined the importance of photographs and photojournalism in magazine making. Whether reporting the news or illuminating aspects of American society, pictures were becoming a vital part of magazine journalism, with few exceptions.

IV

Magazines Since
the Second World War

18

The Transformation
of an Industry

Everything in the magazine industry's past was no more than a prelude to the vast changes that have occurred since the Second World War. All of American life has been transformed since then, in fact, and not much remains of our pre-1939 existence. The explosion has been both economic and cultural. In reflecting and interpreting what has happened, each of the media has had its own developmental history, each history overlapping the other in several of its aspects. But in focusing on one medium—magazines—it is possible for us at least to glimpse the interplay of forces that has wrought the dramatic postwar transformation.

It is an extremely large and unwieldy picture we have before us, this portrait of the past forty-five years. The magazine industry has changed so much, and continues to change at such an unprecedented rate, that it is almost impossible to define its constantly shifting outlines, let alone its internal boundaries. What is true today may not be true tomorrow. Even so seemingly simple a matter as gauging the dimensions of the industry lapses into a sea of qualifications. No one knows exactly how large the magazine business is because, in the first instance, it would be impossible to frame a definition of a "magazine" in terms broad enough to include all of them.

Some reference sources say there are more than 60,000 magazines in North America. Industry experts estimate between 2,000 and 3,000 consumer publications, the kind most familiar to the public. But then there are the special interest magazines, and often the line is extremely blurred between them and others. The business and professional press, by far the largest segment of the industry, is larger than the 12,000 often cited. Taking everything into account, allowing for error, and omitting newsletters of every description entirely, it is reasonably safe to say that there are about 22,000 magazines

243

deserving of the name operating in the United States in 1990. The figure constantly shifts, of course, as magazines come and go.

They fall into several broad categories: the top mass magazines (*TV Guide, Reader's Digest,* and *Modern Maturity*); news-and-opinion magazines; the business press, both general and specialized; women's magazines, led by the "Seven Sisters"; the "skin books," *Playboy, Penthouse,* and others; magazines for the young; periodicals for older readers; second-level mass magazines, such as *People;* city, state, and regional magazines; sports; science; home and shelter; those for farmers; religious publications; those of all kinds devoted to health; those directed to minorities; magazines for and about the media; culture, including travel, food, and pets; and a catchall category that would include none of the above, but those simply too difficult to classify.

One thing is certain. There is a magazine for every human being in the United States who can read, and there is no interest known to mankind which does not have at least one magazine to serve it.

A distribution network more far-reaching than any of the other media, not excepting television and radio, conveys these magazines from publisher to reader. Beyond the traditional subscription and newsstand sale, we now have a nationwide network of supermarkets acting as distributors, and more recently, magazine racks have appeared in bookstores. They are available in other stores as well and in every kind of transportation terminal. Millions of people who may read only one or two magazines at home become part of readerships as they sit waiting in doctors' and dentists' offices. Magazines are, in short, ubiquitous and a far more varied purveyor of information and entertainment than the other media.

Having said this, however, it must be added that the history of magazines since 1945 has been marked by much turbulence and upset. Always a chancy business in any event, the high-powered climate in which magazines operate today has made periodicals more of a gamble than ever in an intensely competitive struggle for survival, often for extremely high stakes. The formation of large groups through mergers and acquisitions, with capital sometimes supplied by foreign entrepreneurs and sometimes by media barons who manipulate conglomerates involving many enterprises, has transformed the magazine business (as it also has book publishing) from not much more than a cottage industry into a highly complicated pattern of enterprises.

The transformation began in the first decade after the war. In 1955, magazines totaled $725 million in advertising revenue, a figure nearly double that of 1946 and seven times higher than in the Depression years. Only forty magazines had circulated for every one hundred people in the previous generation, but in 1955, magazines examined by the Audit Bureau of Circulation alone were distributing more single-issue copies than there were people in the nation.

There were obvious reasons for this sudden advance. In booming postwar America, incomes were higher, people were better educated and enjoyed more leisure, and soon 80 percent of them would be living in towns and cities where magazine outlets were easily accessible. In fact, there were few places remaining in America where the inhabitants were not within easy driving distance of a magazine outlet. If there was any obstacle on the horizon at that point, it was the movement toward suburban living, where well-stocked newsstands were less frequent. That had already resulted in a decline of newsstand sales from 47 percent of all magazine purchases in 1947 to 38 percent in 1954. At the same time, the rapid growth of the paperback book industry, rising precipitously from 96,000,000 to 292,000,000 copies between 1947 and 1953, limited magazine growth, as did the arrival of full-scale commercial television.

Surveying this situation in *Journalism Quarterly* in 1956, Leo Bogart cited surveys showing that "the people who did the most magazine reading were the first to own television. From television's very beginnings, research indicated that reading had suffered even among these formerly heavy readers. In general, magazines were hit harder than newspapers. . . . The evidence from a variety of researchers is that television cuts down the amount of time spent with magazines."

While this knowledge had a temporarily dampening effect on the business, it was clear that although magazines might have to compete with television as a source of entertainment, when it came to conveying information of every kind, they could do what the broadcasters, focused primarily on entertainment, could not.

Still, the industry faced formidable obstacles as the postwar boom continued to develop. High circulation figures, it turned out, did not necessarily mean financial health if the readers came mostly from subscriptions. The cost of getting them and of servicing them after they were obtained could be prohibitive. Already, however, the surge had begun toward greater specialization, targeting audiences that were more and more specific; this forecast a major trend in magazine publishing.

It appeared in the early sixties that magazines were being assailed from every quarter. Marya Mannes, writing in the August 16, 1962, issue of the *Reporter,* a short-lived opinion magazine, charged that advertising was coming to look so much like editorial matter that it could scarcely be separated and that advertising and articles were often juxtaposed in an incongruous way. Investigating these charges with its own research team, the *Columbia Journalism Review* analyzed the top eight circulators at the time (*Reader's Digest, McCall's, Look, Life, Ladies' Home Journal, Good Housekeeping, Better Homes & Gardens,* and the *Saturday Evening Post*) and concluded in its fall 1962 issue that what Mannes had said was generally true, "but in degrees varying from magazine to magazine."

These giants, in spite of their circulation and advertising strengths, were feeling uneasy in the early sixties. They had recently witnessed one of the most spectacular collapses in magazine history with the decline and fall of the Crowell-Collier empire. It was, apparently, the victim of rapidly changing times, unable to keep up with shifting tastes and lacking the strong control at the top it had enjoyed earlier.

The decline had begun as early as 1953, when all the company's magazines were losing money, producing a deficit of nearly $5 million that year. Desperate measures were necessary. Paul G. Smith, who had been general manager of the San Francisco *Chronicle* for seventeen years and regarded as a boy wonder, was brought in. By tight management practices, Smith lifted the company into the black within two years, but then a new group of twenty-six investors gained control in the summer of 1955, and the handwriting was on the wall. When it came time to make a corporate decision about strengthening the company's profit position, it was clear that by eliminating the magazine division, Crowell-Collier could solve its primary problem and live on its book business. In the new climate that was now developing, there was no sentiment involved in killing off old and honorable magazines. *Collier's, Woman's Home Companion,* and the *American Magazine* were simply extinguished.

The death of *Collier's* was particularly hard. A succession of editors had searched for a new formula that might save it, but it slipped steadily to fourth in a field of four, well behind *Life, Look,* and the *Post.* Smith worked particularly hard to rescue it, putting in sixteen- and eighteen-hour days. He even succeeded in stopping the circulation sag by 1956 and adding more than half a million readers, bringing circulation to an all-time high. But advertisers (and the bankers) were holding back, waiting to see what would happen, a pause that proved fatal.

These events were chronicled in a barely disguised work of fiction, *The View from the Fortieth Floor,* by Theodore White, and by Smith himself in his memoir, *Personal File.* Credit sources were shut off, and at last, on December 14, 1956, the decision was made to discontinue *Collier's.* The last issue was January 4, 1957. On the cover was a picture of a pregnant and smiling Princess Grace and a blurb for a new complete Nero Wolfe thriller.

Blurbs on another cover that never appeared, dated January 18, 1957, indicated the kind of magazine Smith was producing: "That Shocking Françoise Sagan—how she really lives"; "Stalin's Ghost Stalks the Kremlin—by H. R. Trevor-Roper"; "Make It Yourself—$750 evening ensemble for $45." Smith, it appears, was ahead of his time.

By the early sixties, most of the mass magazines were also facing the kind of problems that had beset Crowell-Collier. Costs were soaring, competition from other magazines and television was increasing, and there was a tendency among advertisers to question how much they were getting for their dollars. "Over the years," wrote the *New York Times* advertising columnist Peter Bart, "the concept of the magazine as a marketing entity rather than as an editorial entity has gained predominance."

Problems had arisen as a result of "the rise of the magazine as merchant," Bart observed. Among them, he listed declining emphasis on circulation revenue, passing more and more costs from subscriber to advertiser, and the draining off of advertising dollars by rival media, especially television. The squeeze on profits was already pushing publishers toward diversification. Just how severe the squeeze had been was indicated by James Kobak, then of J. K. Lasser & Company, who offered a study showing that between 1950 and 1960, net profits of thirty-five leading magazine publishers fell from 0.3 percent to

1.7 percent of gross income. In the good year of 1960, nearly 40 percent of these magazines were still losing money, and in the bad year of 1961, a majority of magazines were operating in the red, according to Kobak's figures for Lasser.

To counter this trend, some magazines were changing formats, others were developing new or altered formulas. Oddly enough, in the midst of this momentary setback, a flood of new magazines appeared in 1961. During the autumn of that year, nearly two dozen periodicals either made their initial appearance or announced themselves for the near future. All of these new journals were directed to specialized audiences, a clear indication of the wind's direction. Cover prices were higher, too, another significant step. The reader, not the advertiser, was being invited to pay his way.

By the early seventies, these new approaches were paying off. Magazines directed toward specific audiences had made considerable inroads on the general-circulation journals. In short, circulation was now being defined not as how many people read a magazine but rather who they were.

The invasion of the business by investors rather than magazine people was already well underway by this time. Businessmen were buying magazines to get their hands on subscription lists, which they would then use to market merchandise by mail, the periodicals themselves becoming not much more than merchandising tools.

It was plainly evident in the early seventies that some of the old general-magazine giants were going to be victims of specialization and competition from television. *Look* had already succumbed, and *Life*, fading rapidly, had cut its circulation from 8.5 million to 7 million in 1972 without the hoped-for result and planned to cut another 1.5 million. Yet it was not so much that the magazine business as a whole was suffering; there were still freshets of new journals every year, and many of the others were flourishing. It was simply that the rules of the game had changed.

Of the several new factors now at work, the cost of obtaining new subscribers appeared to be a particular villain, and some of the frantic efforts to get them were out of control. That had been the primary cause of *Look*'s demise, industry analysts said. At both *Look* and *Life*, managements had gone by the traditional rule that large circulations meant more advertising, and so these two giants had engaged in a futile battle to boost their figures as high as possible. Advertisers, on the contrary, wanted to know who these subscribers were, how loyal they were, and whether they were likely buyers.

During the sixties, postal rates had climbed an average of 110 percent and were predicted to go even higher during the seventies. That led to a dramatic and long-lasting change in the shape of most magazines, inspired, too, by a 7 percent increase in the cost of paper, with further rises sure to follow. Large-format magazines were now dinosaurs. Led by *Esquire, Holiday, McCall's* and *Boy's Life*, pages were reduced to newsmagazine size. The old familiar *Saturday Evening Post* size that generations of readers had become accustomed to was on its way out. For these smaller versions of old favorites, subscribers would have to pay more. Analysts like Kobak were advising publishers to change them enough so that a magazine could exist without advertising revenue. That, however, did not take place on any significant scale.

Editors were conscious that their readers were different in the seventies. The Kinsey reports of 1946 and 1953 had shocked middle-class America with the revelation that the sexual behavior of their fellow citizens was not what most of them wanted to believe, and in the sixties, their children were part of an upheaval that changed the cultural life of the nation in far-reaching ways.

In the spring of 1972, *Public Opinion Quarterly* reported the results of a survey covering two decades, from 1950 to 1970, to determine how much the so-called "sexual revolution" had changed the content of mass magazines in terms of sexual references, and if so, what these references were. Using a sample of seven popular mass periodicals (*Reader's Digest, McCall's, Life, Look, Saturday Evening Post, Time,* and *Newsweek*), these researchers discovered, unsurprisingly, that there were progressively more references to sex between 1950 and 1970.

What were these enlightened readers absorbing that had been denied them before? References to sex relations aside from intercourse had decreased most, the report said, while references to abortion, birth control, pregnancy, and illegitimacy showed the greatest percentages of increase. The most frequent references were to sex organs, sexual desires, nudity, and obscenities. Homosexuality and sex crimes were being discussed, but mostly in terms of alarm. The general trend had been from conservative to liberal, the survey concluded. While the report did not say so, this swing had begun with national magazine (as well as newspaper) coverage of the Kinsey reports, stories in which it had been virtually impossible to talk about sex in anything but an explicit way, to the shock of many readers.

The seventies were a transitional point for both magazines and their readers, a decade in which the forces that had been developing since the end of the war began to coalesce, creating a climate of incredible expansion and change in the eighties. Summarizing the previous decade in the industry magazine *Folio,* Barbara Love in 1982 reported that nearly everything had been "up": number of magazines (slightly), circulation as a percent of revenue, single copies as a source of revenue, and circulation costs. On the down side were the number of multiple business publishers; and advertising, editorial, and manufacturing costs as a percentage of revenue. Profit margins were up, while advertising spending in magazines had declined relative to what was being spent in other media except spot radio. Nevertheless, magazine advertising revenue had more than doubled during the period, with advertising pages showing an increase that was slightly less.

New special advertising features had appeared, Love noted. Demographic and geographic editions were now available, along with split-run editions. Color advertising was growing rapidly. Remarkable gains were recorded in specialized business publications. Overall circulations had grown, with single-copy sales showing the greatest gains. Subscription prices had risen by 157 percent, single-copy prices by 145 percent. The seventies had been boom years, this report concluded, but now the time had come to cut costs. So it seemed at the beginning of the most expansive decade the magazine industry had ever known.

A year later, *Folio* was reporting how changes in management structure had become part of the industry's transformation. Analyzing six publishing companies (Time Inc., Forbes, Gralla Publications, CBS Publications, *Mother Jones*, and Bill Communications), Karlene Lukovitz disclosed how these companies, of varying size, had applied some of the principles of modern management to their organizations. In multipublication companies, that meant centralization of such functions as finance, production, and circulation, but decentralization of editorial and sales. Thus, Time Inc.'s Magazine Group was divided into two parts, financial/administrative and editorial, each with its own line of command.

Forbes, on the other hand, publishing just one magazine, was run directly at that point by only two men, the founder and publisher, Malcolm S. Forbes and his son, Malcolm S. Forbes, Jr., both having complete authority over the administrative and editorial sides. (That, of course, changed later with the death of the senior Forbes and further alterations in the magazine's structure.)

Mother Jones, a maverick in any case, was constructed on offbeat lines, *Folio* reported. As a nonprofit periodical published by the Foundation for National Progress, its board had the authority to hire and fire both the publisher and the executive editor, who had equal authority over their own domains.

Most magazines followed the general management patterns of the past, with numerous variations, until acquisition and conglomeration changed the corporate structure in much more complicated ways. The sub-structure, however, then and now, continued to follow traditional lines, since there were not many alternatives. Today, overall management structures are either centralized, in one way or another, or decentralized, with each magazine in a group organized as a profit center—again, with variations. In short, what had once been a highly individual industry, operating in its own way, was now a part of American corporate organization as a result of conglomeration.

What remained of the old magazine world was the editorial department, however organized, which was perenially faced with the same problems magazines had been compelled to deal with since their beginnings: the search for audiences, ideas, and writers; along with presentation formulas designed to reach and hold readers. In a highly competitive market, that meant, as always, the necessity to change with a culture that itself was changing more rapidly in the eighties, perhaps, than in any other decade of the nation's history.

A major change from the past was the decline in fiction, which had once been a staple of national consumer magazines. This collapse was a direct result of television, which told stories all day long and through the prime-time evening hours to the same people who once read serials and short stories in magazines. Another reason for the shift from fiction to fact, however, was the recasting of magazines to appeal to different audiences. *Redbook*, for example, once a prime market for fiction of all lengths, had long since fired its editor of eighteen years, Edwin Balmer, and, as a Hearst Corporation acquisition, had found an entirely different market in the baby boomers, originally defined as people from eighteen to thirty-four years old. Then the magazine shifted the scope of its audience to a different age group, twenty-five to forty-four. While this change meant that *Redbook* could publish more sophisticated fiction, it also meant reaching an audience less inclined to read stories. Since buying fiction was expensive, especially when first-serial rights for books by brand-name authors were involved, it was only sensible to curtail fiction, which most of the magazines in *Redbook*'s category proceeded to do.

Such periodicals as the *Atlantic, Harper's,* and the *New Yorker* continued to print fiction, but in reduced quantities, no more than two short stories per issue. In these journals, as in the mass-market publications, fact had all but replaced fantasy. It was a development pleasing to media directors, who saw little appeal to advertisers in fiction, whether literary or not. One agency man even advanced the novel idea that fiction should never appear next to advertising because if readers did not believe the story they might not believe the advertising.

The only mourners over the reduction of fiction in the magazines were, naturally, writers. As late as 1989, the novelist Saul Bellow complained in a *Publishers Weekly* interview that writers were having trouble placing longer fiction in periodicals, which, he charged, had "failed the public badly in this respect." In a complaint more anguished than Portnoy's, Bellow said of the journals: "They've gone political; they've gone social; they've dealt with health and with sex and with sports, and really they have no patience for fiction. . . . They run a token story or poem—if it's minimalist it's even better, because it takes up less space."

More fundamental changes were taking place, however, in the early eighties, as magazines made broad, sweeping alterations. These were not simply cosmetic, but often reformulations, involving new concepts and new designs. This often involved pushing cover prices up to as high as four dollars or more in an attempt to establish elite readerships attractive to high-ticket advertisers.

The evidence of change was everywhere in 1982. *Fortune* had emerged as a virtually new magazine after a face-lift and a doubling of its frequency. *Mademoiselle* edged closer to its rival *Cosmopolitan.* After thirteen years as a tabloid, *Rolling Stone* became a more conventional magazine, aimed at a somewhat more conventional audience. *Sports Afield,* once a blue-collar hunter's guide, was transformed into an upscale magazine for outdoorsmen with different tastes and more money. *Geo,* once a photographic excursion in violence and blood, became a model of photojournalistic surveys of kinder, gentler living.

While it was true that magazines had been engaged in such changes since the nineteenth century, they were compelled in the eighties by unprecedented economic and competitive conditions, another result of conglomeration. Sometimes the changes were made simply to survive. In other instances, the purpose was to increase growth and profits for an already successful magazine in an effort to put some distance between a thriving journal and its competitors. While growth in the industry had slowed somewhat between 1979 and

1981, it was now accelerating again, with predictable results. The necessity for change was also spurred on by the increasing affluence of the eighties and the efforts of some magazines to reach more of this monied audience and the galaxy of advertisers that might surround them.

How a magazine could lose its audience and its life entirely, no matter how well established, was illustrated by the *Saturday Review*, which had enjoyed a long life as a literary weekly and then became a more general cultural magazine in the sixties under Norman Cousins's editorship. By 1970, the *Review* had increased its circulation to nearly 700,000, undreamed of earlier. But at the point of its greatest success, Cousins sold it to financial entrepreneurs who split it into four different editions, a disastrous move. The partitioning of the weekly into four monthly editions occurred in 1971. A year later, Cousins took back his magazine and tried to re-create it. A new owner revamped it again in 1978, and two years later, still another new owner gave it a final overhaul before its death in the summer of 1982. This sequence of events showed once more how shifting ownerships could kill a magazine in a relatively short time.

There are several outstanding examples of how to turn a magazine around successfully. *Parents* revival in 1985 was the result of new capital, different management, a new editorial focus, better paper, and more aggressive selling and promotion, among other factors. It, too, was delivering a more specific market than it had offered earlier. *Life*, resurrected as a monthly, was demonstrating that a combination of old and new formats could be successful.

By the early eighties, publishers of general-interest magazines had learned that they could target specific audiences by also publishing special-interest journals, sold on newsstands from one to six times a year for premium prices. These proved to be most successful in such areas as home decorating, home building, and similar service markets. Meredith, Condé Nast, and the Hearst Corporation had become the leaders in this field by the beginning of the decade, and in 1979, they could already boast collectively of more than 2,100 advertising pages, most of them occupied by high-ticket advertisers.

Publishers were also finding that they could capitalize on the success of their media rivals. By 1981, there were already six video magazines on the market, catering to a rapidly growing audience of video buyers and renters and to makers of video machines and accessories who wanted to reach that market. These journals had already attracted $3,000,000 in advertising during 1980, and a new magazine classification had been born as the VCR boom took on new strength in the decade ahead.

Change was the order of the day in the eighties, affecting every part of the industry and not exempting the giants. *Reader's Digest*, for example, had gone along for so many years in much the same way that it seemed almost immune to alterations of any significance, although the magazine was, in fact, altered from time to time. A content analysis of the *Digest* by two researchers, Ron F. Smith and Linda Decker-Amos, reported in the *Journalism Quarterly* in 1985 that while there was truth in the frequent charge that the magazine did not change, there *had* been "a steady shift away from factual articles like biographies of world leaders, and toward articles on how to live a better life and other self-improvement topics." What had not changed was the *Digest's* consistent crusading for conservative causes and individuals.

Digest readers with long memories could recall the attack made on Dr. Alfred Kinsey, for example, after the publication of *Sexual Behavior in the Human Male* in 1946. One of those outraged by its revelations was Fulton Oursler, then a senior editor on the *Digest*. Oursler had earned his reputation as a mass-magazine editor in the twenties, making *Liberty* magazine one of the most popular of the nation's periodicals. After a successful career as an exploiter of sex and sensation, he had experienced a religious conversion and become an ardent moralist, a declared enemy of sin.

When the first Kinsey report appeared, Oursler wrote an article about it that was so vitriolic that DeWitt Wallace, a mild-mannered man despite his conservatism, refused to print it. However, he was not averse to discrediting the Kinsey study if possible. Oursler's article had advocated suppressing the book, but second thoughts prevailed and a different strategy emerged, employing a blandly clever editorial device not unknown in the magazine business.

This device was the loaded symposium. In June 1948, the *Digest* published as a lead article a collection of opinions titled "Must We Change Our Sex Standards?" subtitled "A Symposium on One of the Most Vital Questions of Our Time." To knowledgeable readers, the editorial matter preceding these opinions gave away their purpose. If Oursler had written it, as was probably the case, he misrepresented Kinsey's work without naming it, raising the spec-ter of a nation whose moral values were in deadly peril. "They [meaning Americans] have just been told," Oursler (presumably) wrote, "that practices long held in abhorrence must now be regarded as acceptable. Science, so it is said, does not recognize any expression of sex as 'abnormal'; except 'maniacal' deeds, pretty much anything is all right." The Kinsey report, of course, made no such judgments, nor did it make judgments of any kind. It reported what male Americans actually did sexually, cast in a scientific framework without reference to the values of church or state.

To justify what followed, the *Digest* proclaimed piously that it had been deluged with "pleas for guidance" from parents, young people, pastors, priests, rabbis, family doctors, editors, psychologists, and many others, beseeching the magazine to lead them out of the moral wilderness Kinsey stood accused of creating. Consequently the *Digest* had sent out a heavily loaded inquiry to a handpicked group of people who could be guaranteed to give them the right answers, from Episcopal rectors to J. Edgar Hoover. Some of the responses were so vague as to constitute not much more than a high-minded defense of virtue. Others shared Oursler's outrage.

Three months later, the *Digest* followed up with a second symposium made up of letters from readers who had read the first one, but again carefully selected to include well-known names. A few statements more or less sup-porting the Kinsey report were thrown in to give a semblance of impartiality that had not even been attempted in the first symposium. Until the day he died, Kinsey believed that Oursler had played some part in the Rockefeller

Institute's withdrawal of funds from his great project, not only through the *Digest's* attack but also through the editor's personal connections with the institute.

Kinsey, along with other people and institutions the *Digest* attacked during its relentless progress, had some reason to fear its influence, and not only because of its tremendous circulation. Given its wide use in schools, and its steadily increasing dissemination abroad (54 million in the 1980s), the *Digest's* doctrinaire right-wing views, sometimes well disguised, could plausibly result in indoctrination as well as information. In the eighties, the *Digest's* ideology, as expressed in the magazine, had not changed since its founding.

While it was all but immune to assault from its critics, the *Digest* was still vulnerable to internal disorders after the death of both Wallaces created a power vacuum. A series of upheavals began in 1984, almost simultaneously with Lila Wallace's death at ninety-four. Her husband's chosen editor-in-chief, Edward T. Thompson, was summarily fired, followed by the departure of John O'Hara (not the author, of course) as chairman of the Reader's Digest Association, the governing board. A new management team brought further changes, shutting down several unprofitable divisions and firing a substantial number of staff members in an effort to bring the magazine back to its profitability levels of 1979.

These changes sent waves of unrest through the *Digest*, which had always, under Wallace's essentially benevolent influence, been a highly regimented but secure place to work, with comforting amenities in the bargain. George V. Grune, the magazine's chairman and chief executive officer, gave assurances that there would be no major changes of any kind at the *Digest*. Examining the situation, Miriam Jacob wrote in *Folio* that three factors had precipitated the upheavals: an effort to restore peak levels of profitability, pressures inside and out to preserve editorial style and Wallace's solid Republican ideology, and a simple internal struggle for control of a billion-dollar organization. Not many had appeared to notice it, but Thompson had been told his sin was that he had departed from these policies and principles laid down by Wallace and, even more important, from past standards of profitability.

This was an object lesson in the structuring of power in a modern magazine, since the *Digest* itself accounted for only about one-sixth of the association's annual billion-dollar revenue. But it was the foundation stone, and if it were going to be seriously impaired by unprecedented change, in theory the entire structure would suffer. It was (and is) the magazine that provided thousands of new subscriber names a year, a list to which the association could then sell its other products, from condensed books to audio equipment.

Apparently, there was no substantial reason to worry. By the end of the decade, the association had gone through a new period of profitable expansion. At the end of 1986, it had, for the first time, acquired another magazine, *Travel-Holiday* (what remained of Ted Patrick's splendid magazine), with a circulation of 778,000. In January 1988, it bought another, the *Family Handyman*, with a circulation of 1.2 million, one of the largest home-improvement journals. By that time, the *Digest* itself had thirty-nine editions world-wide and published in fifteen languages, with a total circulation of 28 million, 16.5 million of that in the United States.

In addition, it was printing about 400 million condensed and general books every year, issued from two separate divisions, accounting for 46 percent of total revenue. There was also the Music Division, producing records, tapes, and video cassettes. Further acquisitions were considered quite possible. An extra added attraction was the direct-mail business, marketing items and services from compact disks to globes and automobile-club memberships.

Editors were still departing from the *Digest* in 1988, but it hardly seemed to affect the magazine, which continued to enjoy consecutive record years, with 50.3 million readers, ranking second only to Bill Cosby's television show in the size of audience delivered to advertisers.

In the spring of 1989, the association took another unprecedented step. One of the most private publishing companies in America since 1922, it announced that a small and unspecified amount of stock would be publicly offered—so to speak. At least 10 percent of nonvoting shares would be made available to stockholders who wanted to dispose of their holdings, and there would be a gradual transfer of 20 percent of the voting shares to a stock-option plan for employees. Various trusts had held all the voting shares. This disclosure also underlined the dramatic jump in total revenue for the association to $1.5 billion, including all products. About half of that money was coming from the *Digest* in 1989.

One of the most remarkable aspects of the magazine industry in the boom decade of the eighties was its seemingly unlimited ability to find new audiences, thus creating new markets for advertisers. Early in the decade it was apparent that America was building a large audience of Hispanic people, émigrés from Cuba, other Caribbean islands, Latin-America, and Mexico. While many of these people found that they had fled one dreadful slum for a slightly better one, others joined the work force, created small urban communities, and began to constitute a new and growing market.

In 1983, the De Armas Hispanic Magazine Network was launched, distributing fifteen publications in Spanish-language versions of well-known magazines: *Cosmopolitan en Español*, licensed by Hearst and circulating 17,000, published since 1973; *Harper's Bazaar en Español*, with a circulation of 6,600, available since 1980; *Mecánica Popular;* and *Buenhogar*, another Hearst licensee, with a circulation of 19,900, begun in 1967. Also on the list was *Ring en Español*, a boxing magazine, circulation 16,000, available since 1979. According to *Advertising Age*, about half the editorial content of these magazines originated in their English-language counterparts, but the biggest seller for De Armas was the entirely original *Vanidades*, with a circulation of 144,800. These magazines, said the trade journal, were selling best in Miami, New England, New York, Chicago, and on the West Coast. Most of the sales were from newsstands, but De Armas was experimenting with subscription campaigns.

The ubiquitous *Reader's Digest*, however, has the magazine with the largest Spanish circulation, *Selecciones del Reader's Digest*, with 140,000 sold in the United States alone.

In the spring of 1989, a magazine entrepreneur in San Juan, Michael A. Casiano, head of Casiano Communications, concluded that middle- and upper-income Hispanic-Americans had now increased to the point of representing an untapped market of about a million Hispanic households. *Imagen*, the glossy upscale magazine he created, was launched first in the Northeast, with plans to open offices later in Miami, Chicago, and Los Angeles. It would issue separate editions aimed at each region, anticipating a total circulation of 350,000. Casiano described his creation as a combination of *Town & Country* and *Architectural Digest*.

Another minority market emerging in the mid-eighties was the gay readership, which had become increasingly more visible during the previous decade. As early as 1967, the first attempt to reach these readers was made with the *Advocate*, starting as a twelve-page newsletter with only twenty-seven dollars' worth of advertising. By 1984, as a bimonthly, it had a circulation of 82,191, with a claimed readership of 250,000. The market was upscale, with readers having an annual household income of more than $25,000. *Advertising Age* described the *Advocate* as "a color tabloid format with slick graphics and a serious approach to its journalistic content. It has extensively covered such topics as the problems gay relationships face, anti-gay violence, the AIDS outbreak, and political issues affecting gays, along with news updates, book, movie and record reviews, and exercise and health features." Today, almost every urban area is reached by a gay magazine. Examples are the *Reporter*, in the San Francisco Bay Area; the *Blade*, in Washington, D.C.; *Gay Life*, in Chicago; the *Native*, in New York; and the politically oriented *Gay Community News*, founded in Boston in 1971. Predictably, the major problem of gay magazines is attracting advertisers, many of whom shy away from the association even though it is an affluent market.

Among magazines directed to blacks, those of the remarkable John Johnson (whose autobiography was published in 1989) continued to dominate the field—*Ebony, Jet,* and *Ebony Jr.*—but other leaders included Earl Graves's *Black Enterprise*, and Edward Lewis's *Essence*. As the 1980s ended, there were strong indications that this market was an expanding one for new magazines, with several arriving since 1984. They included *Metro*, with separate editions for Baltimore, Washington, and New York; *Odyssey West*, based in Denver, published bimonthly; *Ascent*, directed to black adults in the twenty-five areas of dominant influence; and *Seanna*, an annual for black professional women. All these publications differ only slightly in editorial outlook from the well-established journals. Their problem, once more, has been advertising support. The new black magazines, said *Advertising Age* in a survey of them, were targeted toward a more sharply defined audience of "upwardly mobile, black urban professional men and women with annual salaries of more than $25,000." The task was to convince advertisers that they were, indeed, reaching this market. Media buyers recalled the quick demise of *Elan*, a magazine that intended to be the black women's *Vogue* but failed to find its audience.

Since millions of Americans spend their time flying on airliners every year, the airline market was one of the earliest to be developed; offerings to passengers included individual in-flight magazines along with standard periodicals. The largest publisher of these journals has proved to be East/West Network, which was issuing seven of them in 1989. Jeffrey S. Butler, president of this New York company, had three major competitors: the Webb Company of St. Paul, his closest rival; Halsey Publishing of Miami, with two magazines; and the periodical published by American Airlines, the only in-house effort in its field.

Advertisers have never been entirely convinced that in-flight publications sell products, or even that they are well read in view of the fact that they are competing for attention with other magazines available on a flight. In the sixties and seventies, when they began, they were looked upon as house organs, and they still face difficult editorial problems in spite of their substantial editorial progress, resembling now the general interest magazines in content. Business journals like *Fortune* and *Business Week* are the in-flight publishers' chief competitors, however.

Children and young adults constitute another large potential market for periodicals, as they have for fifty years. But this market changes constantly and faces formidable competition from comic books (frequently and ambiguously classified as magazines themselves) and television. At the beginning of the eighties, the major entries at the upper educational level included eight magazines directed generally to college students; forty-two for high-school students or teens, including Scholastic Publishing Company's thirty-two; and, if comic books are included, five groups of publishers with multiple titles and circulations in the millions.

A brief analysis demonstrates how this field has been and is being cultivated by diverse journalistic harrows. In the college area, for example, *Ampersand,* a supplement to college newspapers, was distributing nearly a million copies in the early eighties, while *Black Collegian,* published five times a year, circulated about 250,000. *Campus Life,* published by Youth for Christ and a monthly since 1949, had 300,000 circulation. Kiwanis International, the businessmen's club, was issuing *Circle K* five times a year with 10,000 circulation. There were also the two well-known magazines read by both high-school and college students (as well as adults), *National Lampoon* and *Rolling Stone,* and at least two annuals, with circulation figures in the millions.

Unquestionably the leader in the high-school and teens field is the Scholastic group. Ten of its thirty-two magazines in the eighties accepted advertising, reaching 3.4 million students and 180,000 teachers. There was a biweekly high-school package of four journals to be bought together, including *Scholastic Update* for social studies; *Scholastic Scope* for reading skills; *Scholastic Voice,* containing fiction and student essays; and *Science World*. Circulation of the four totaled more than 2 million. *Co-ed* was being published for home-economics students, and there were four magazines for teachers. The twenty-two other journals on Scholastic's list did not take advertising only because their subject matter had no appeal for advertisers or else the audience was too young.

Scholastic magazines have a prime advantage for both readers and advertisers because costs are so low, the result of bulk distribution through the schools, which sharply reduces mailing costs. Editorially, these magazines are attractive and well written; advertisers know they are read carefully. This company was first in its field to get into the computer-magazine business with *Teaching and Computing*, for teachers in grades three to six, and *Family Computing*, directed to parents of children five to fifteen. The company also has a controlled circulation journal, *Electronic Computing*, for educational administrators and curriculum directors.

Other approaches to the teen and school market come from a variety of sources. There is the venerable *Boys' Life*, published by the Boy Scouts of America since 1912, and the later blooming *Hit Parader*, a rock music monthly. Rural teenagers (and younger) read *National 4-H News* and *National Future Farmers*. Pre-teeners have a unique journal, *New York News For Kids*, an enterprising newsmagazine with both local and national news. *Seventeen*, well established since 1944, is on this list, along with such titles as *High Times*, devoted to fashion and culture; *Teen*, mostly for girls; and *16 Entertainment* for young teens.

The comics field is dominated by the Archie Comics Group; the Charlton Comics Group, with twelve titles; and the Marvel Comics Group, with forty-two titles, by far the largest with a circulation of six million. Marvel began in 1939 as Timely Comics. A smaller enterprise, *Heavy Metal*, produces adult comics that are also read by teenagers.

A significant development in the eighties was a noticeable shift among children's magazines to more serious topics. In the days of *St. Nicholas* and the *Youth's Companion*, it would have been hard to visualize such an eighties journal as *Penny Power*, a children's version of *Consumer Reports* with 130,000 circulation, both published by Consumers Union. Like its senior version, *Penny Power* carries complaints from children about products and advertising and criticizes television commercials.

Other magazines designed to stimulate children's thinking and learning instead of simply entertaining them are the three publications of the Cobblestone Publishing Company of Peterborough, New Hampshire. One is *Cobblestone: The History Magazine for Young People*, begun in 1980 and directed to the eight-to-fourteen age group. The company also issues the *Classical Calliope: The Muses' Magazine for Youth*, begun in 1981 and devoted to the classics; and *Faces: The Magazine About People*, covering anthropological subjects, launched in 1965 and edited in cooperation with the New York Museum of Natural History. For teachers, Cobblestone also has a manual, *Cobblestone Companion*.

These ideas came from Hope M. Pettegrew, a former elementary-school teacher who was convinced that the Vietnam War had greatly diminished interest in previous history and sought to revive interest in this subject, regularly voted by children as their most disliked. Her history magazine has 40,000 subscribers who apparently find it absorbing. Published without advertising, these Cobblestone publications attempt to walk the difficult line between popularizing such disciplines as history without "dumbing down" or trivializing them. Serious scholars write for these journals, which also carry lists of books, movies, and places to visit that encourage further learning experiences.

In still another market, a late nineteenth-century institution, the Sunday newspaper magazine, showed renewed signs of life in the eighties. By 1983, there were nearly sixty of them locally edited, as well as such nationally syndicated Sunday supplements as *Parade* and *Family Weekly*. Among the locals, the leaders continued to be the *New York Times Magazine*, the Hartford *Courant's* supplement, *Northeast;* the *Sunday Magazine* of the Boston *Globe;* and that of the Philadelphia *Inquirer*. Some papers include both their own and syndicated magazines in the Sunday edition. *Parade* remains the largest Sunday supplement in America.

Several of these Sunday magazines are advertising gold mines, drawing on both local and national advertisers. There is little to distinguish the *New York Times Sunday Magazine* from a regular periodical except its format, and the eighties were filled with recurrent rumors that it might emerge as a separate national periodical published by the Times Company. On the other hand, Sunday supplements relying too heavily on local retail advertising are likely to suffer in any kind of business slump.

One of the most successful marketing diversifications of the past two decades has been the growth of regional and city magazines. Regional publications have had much to offer advertisers. They can target specific markets and transfer information about goods and services along with specialized information directed to a ready-made population group.

The boom in this market began in the sixties with the sudden rise to prominence of city magazines, from which the regionals stemmed. More than sixty city journals were introduced during that decade, most of them much superior in appearance and content to earlier versions. Designed and edited to resemble any consumer magazine, these city publications began to depart from boosterism (a few were still sponsored by Chambers of Commerce, however) and examined the warts on the faces of their cities along with beauty spots. Leaders quickly emerged: *New York, Los Angeles,* the *Washingtonian,* and *Atlanta*. A Metropolitan Magazine Association was organized with the intent to obtain national advertising on a collective basis, and the effort was immediately successful because these journals were offering audiences that were at least reasonably affluent, some of whose members were in positions of power.

The outstanding success among city magazines was *New York,* emerging in 1969 to prove that it was possible to publish an exciting city magazine. Moreover, *New York's* success underlined the fact that a city journal could cast its circulation and advertising net over a wide area, thanks to urban sprawl. On a larger scale, there was Michael Levy's *Texas Monthly,* founded in 1973, to which was added *New West,* which Levy's company bought, changing its name to *California* and supplying it with a few Texans as staff members. As the market developed, however, it was clear that, with a few exceptions such as *Alaska* and Hawaii's *Aloha,* circulation was not likely to spill over very far from a region.

In the eighties, city-regional publications were winning National Magazine Awards and building advertising lineage as well. The trade journal *Marketing & Media Decisions* reported in 1982 that advertisers were attracted by the high editorial quality and the local flavor of these journals, even though their rates were slightly higher than standard magazines. In the early eighties, at least nine city-regional magazines were giving advertisers national audiences, and at least a third of their advertising was national. (Media specialists included the *New Yorker* in this category, as they had been doing for some time, but its resemblance to the others was superficial and its circulation national.)

The success of the regionals led some national magazines to start regional editions of their own, with inserts covering a particular area. They enjoyed success; but the regionals themselves were not uniformly successful. *Rocky Mountain* was suspended in spite of winning a national award because its focus simply dissipated over the broad area it was trying to cover. The Rockies were simply too big; it was not a well-knit region with similar interests, a condition required by national advertisers. On the other hand, *Texas Monthly* reflected the interests of a very large area whose citizens were more than proud to identify themselves as Texans. By 1980, *Texas* had a circulation of 275,000, its 250-page issues replete with advertising riches, half of them national.

Population shifts contributed to the success of some regional magazines, among them *Sunset,* California-based but serving 1.4 million readers in the western states; and *Southern Living,* reaching 2 million readers from Florida to Texas. The former is the oldest regional magazine, founded in 1898 by the Southern Pacific Railroad to promote western travel. *Southern Living* took *Sunset* as its model. Today it is part of Southern Progress, the largest of regional magazine groups, along with *Southern Accents, Travel South, Progressive Farmer, Cooking Light,* and Oxmoor House books. Time Inc. bought this organization in 1985 for $480 million, but it has since (as of 1989) had relative autonomy.

Other regionals include *Shenandoah Valley, Blair & Ketchum's Country Journal, Country Magazine, Yankee,* and *New England Monthly.* The latter journal is published in an old, brick brass works on the Mill River in Haydenville, Massachusetts. Begun in 1984, it was edited by Daniel Okrent, a former book editor in Manhattan who sought to create a kind of rural *New Yorker. Folio* described his style as "a far-out, fast-paced approach." By 1987, it had won two consecutive general excellence National Magazine Awards and created a 50 percent increase in its national advertising, with 597 advertising pages and $2 million in revenues. The magazine excluded advertising by plastic surgeons, time-sharing real-estate deals, and discounters, although it did accept tobacco and alcohol accounts.

At the peak of its success in 1989, *New England Monthly,* by that time

considered one of the most prominent new consumer publications of the previous five years, was acquired by Telemedia, Inc., largest publisher of consumer magazines in Canada. It was an excellent buy, having tripled its original circulation to 110,000 and boosting the number of its advertising pages to
600 the previous year. Okrent's successful editorial formula was to combine
the features usually found in city-regional magazines with the kind of literary
content found in the *New Yorker.*

The sale of *New England* emphasized the fact that multistate regional
publications had become one of the fastest growing categories in magazine
publishing. They, too, were creating groups of conglomerates. One entrepreneur, D. Herbert Lipson, owned *Boston, Philadelphia,* and *Atlanta* magazines, as well as *Manhattan, Inc.* There were more than 200 magazines in this
category by the late eighties, many belonging to the City and Regional Magazine Association.

While the city-regional magazines had pointed the way to area diversification and marketing, one magazine in the eighties had already provided the
ultimate in spreading itself around. *Farm Journal* was printing more than a
thousand versions of its magazine, combining editorial and advertising in multiple ways, a feat made possible by the Selectronic Computer bindery system,
a product of the Chicago printing giant R. R. Donnelley. This device enabled
Farm Journal to target large numbers of regional, demographic, and farming
audiences, while providing advertisers with an equal variety of placements.

Meanwhile, the great diversification in audiences continues unabated. The
culture of the eighties has produced several categories peculiar to the times;
one of the foremost is the worship of success, recalling the idolatry of the
booming twenties. "Success" in this sense means magazines about the rich, the
good life in general, and old-fashioned business success in a new-fashioned
environment. While the baby boomers have been given credit for establishing
this category, it appears to extend well beyond them, judging by the celebrity
craze with which it is linked. Broadly speaking, it is a category that concerns
itself primarily with money, diet, sex, and fitness. Even old established magazines like *House & Garden* and *Esquire* have had to reshape themselves to
cope with the trend, while the newsmagazines have added sections on health,
fashion, and family living.

Analyzing this trend in 1985, the *Washington Journalism Review* divided
the magazines serving it into five general categories. First among them was
what the *Review* called "Fine Living"—meaning *Gourmet, Food & Wine,
Bon Appetit, House & Garden, Vanity Fair,* and *GQ (Gentlemen's Quarterly).* "Fashion and Style" constitutes another division, including (again) *GQ*
and *MGF (Men's Guide to Fashion), Working Woman,* and *Vogue* (an
incomplete list). A third category was "Fitness and Health," notably the
award-winning *American Health,* but including other specialized
publications as well as emphasis on the subject in most general magazines.
"Success," and "Celebrities and the Successful," rounded out the *Review*'s list
of categories.

To consider only one of these for a moment demonstrates the bewildering specialization that continues. A nation seemingly obsessed with fitness and health has made magazines serving these needs a small industry. The eighties witnessed the arrival of *American Health, Shape, Muscle & Fitness, New Body,* and *Exercise for Men Only,* among many others. Such well-established publishers as Rodale Press have entered this field, along with one-shot entrepreneurs and pure speculators. Most, excluding those listed above, have small circulations and minimal advertising.

Early in 1988, a new entry called *Harrowsmith* appeared in the waiting rooms of doctors, dentists, and even some veterinarians. This was a nonedible branch of General Foods, designed as a medium to display its products, specifically Grape Nuts, for the benefit of people presumably interested in health. The magazine initially delivered 25,000 copies to the offices of professionals. Waiting patients discovered not only an inserted advertisement for Grape Nuts but also articles on country living, environmental concerns, and gardening. *Harrowsmith* thus joined a growing controlled-circulation field already numbering magazines like *American Health* and *Hippocrates.* A distinct advantage of such circulation is its avoidance of increasing postal rates.

Published six times a year, *Harrowsmith* was said to be reaching 225,000 homes with in-depth articles about the hazards of nuclear poisoning, vitamin deficiencies, the danger of lead in some pottery, the National Forest Service's handling of lumber contracts, and still more on the threatening effects produced by the destruction of rain forests. The journal's name comes from a town in Ontario near the home of James M. Lawrence, its founder, editor, and publisher before General Foods acquired it. Lawrence launched the magazine from his kitchen table, in his spare time from acting as the Kingston (Ontario) *Whig-Standard*'s only investigative reporter. He hoped for a 10,000 circulation in its first year; the figure was actually 40,000. Lawrence, however, was not a Canadian, having been born in Binghamton, New York, and getting his education at Cornell and Syracuse. He later sold his Canadian company, and began a new one, Camden House Publishing, which issues the American edition of *Harrowsmith.* Circulation increased from 150,000 in 1986 to 225,000 in 1988.

While Lawrence's magazine is highly visible, others are so specialized that they draw on smaller audiences. In 1988, for example, a former compulsive gambler introduced *Recovery Now,* a monthly for people addicted not only to gambling but to alcohol and other drugs. James V. Trotta, its publisher, a veteran of twenty-four years in the magazine business, hoped to reach addicts and their families with information about available services.

Another group of health magazines is designed for patients in hospitals. They bear such encouraging titles as *RX Being Well, TLC, Expressly Yours,* and *Just For You.* That latter journal offers hospitals an opportunity to have a few pages of their own public relations materials inserted in the copies patients buy.

There are endless ways to divide the health pie. *Health Journal,* for example, a Boston publication, is a national quarterly for members of health maintenance organizations, circulating its first issue in 1988 to 690,000 readers through nineteen HMOs, with circulation forecasts of 2 million or more. Since some states require periodic communication between an HMO and its customers, *Health Journal* not only meets this requirement but offers each organization its name on the cover and four pages of its own material in the edition it buys.

One of the most unusual of the health journals in *Hippocrates,* an award-winning health and medical magazine published in Sausalito, California. In winning a National Magazine Award, it was praised for establishing a niche between "high-brow medical magazines and popular health magazines." During the same year, *Hippocrates* won eight "Haggies" from the Western Publications Association. Eric Schrier had been an award winner with his previous magazine, *Science '80.* Katherine Bishop wrote in the *New York Times* that *Hippocrates* showed the impact of nutrition and medicine on people's lives, providing readers with the information needed to act in their own best interest—"service journalism in the best sense."

Bishop placed a finger on one reason for the growth of health magazines. "Patients," she wrote, "are taking a more aggressive stance in questioning decisions by their doctors, and are also seeking alternative opinions and treatments. A huge segment of the population is moving into their 40s and their old diet and exercise routines are not doing what they used to. And their parents are getting old and they are coming face to face with their own mortality."

As a matter of history, it should be noted that *Hippocrates* began as *Novus* in 1979; its name was changed to *Science '80* and for the next five years was published by the American Association for the Advancement of Science before being sold to Time Inc., which folded it into *Discovery.* Schrier then raised $5 million in venture capital and relaunched it as *Hippocrates.* Its first anniversary number was devoted entirely to medical-ethics issues: stories on organ transplant programs, prenatal testing, abortion of fetuses of an unwanted sex or those with evidence of mental retardation, and some doctors' refusal to operate on patients with AIDS.

From a publication standpoint, mental health is a small corner of the health-magazine field. For more than two decades, its leader has been *Psychology Today,* flourishing briefly as a publication of Ziff-Davis, a primarily trade-magazine company that built its circulation by the old-fashioned Cyrus Curtis method of getting small boys to sell subscriptions in small towns for premiums like basketballs and catchers' mitts. It was an unappreciative audience then, consisting mostly of people who wanted to help out the little salesmen. The audience waiting for it was college educated and in the middle- to upper-income class.

The opportunity to break out came when William Ziff fired its editor T. George Harris in 1979, the result of an ongoing dispute about content. Harris had been with the magazine for seven years, during which time he had worked for nine publishers. They had appeared to think it should look like *Cosmopolitan* and accused him of selling out to the Ph.D.s. Harris, a former reporter for *Time, Look,* and *Life,* kept a ping-pong table in his office and sometimes held editorial conferences on the beach. He had edited *Psychology Today* as a news magazine, viewing current events through the lens of psychology.

Passing into the hands of the American Psychological Association, the journal went through four unhappy years of trying to compromise between the association's scientific outlook and the necessity to sell a magazine until Harris and a partner, Owen Lipstein, bought the journal, having already started *American Health*. They made it a news magazine, but by 1987, *Psychology Today* had a new publisher, Sherwood Katsoff, and a new editor, Wray Herbert. They hired skilled science writers to report the news from professional sources and directed the magazine toward consumer interests—people dealing with people, as they put it. At that point, it had nearly a million circulation, about 60 percent of it women, who presumably were more interested in relationships than men.

Obviously, there is no end to specialization and the targeting of new audiences. Even the broad sampling offered here is no more than a ripple on the surface. Whole categories deserve study; for example, periodicals to satisfy those Americans interested in science. Between 1979 and 1981, six new mass-market science magazines were begun, most notably *Omni*, published by *Penthouse*, with a monthly circulation of nearly a million. Looking in an entirely different direction, people outside the magazine business would be astonished to know that in a society where at least 80 percent of the population lives in urban areas, there are more than 250 farm magazines—national, state, regional, and vertical, with nearly $100 million dollars in advertising revenue. A prime example on the national scene is Meredith Corporation's *Successful Farming*. In the eighties, it was approaching a million circulation, surpassed only by the venerable Philadelphia *Farm Journal*.

If farmers are well served in an urban society, so are those involved with war in a nation at peace. Virtually unknown to the public at large, a wide-ranging military press exists, consisting of independent magazines, newspapers, and newsletters. Few people, for instance, have heard of *Military Logistics Forum*, directed to decision-makers in the defense industry, but it has shown striking independence and courage in reporting adverse news. Far better known is *Aviation Week*, clearly the leader in this field and read all over the world.

Since America has also become a highly litigious nation, it is not surprising that legal publications became a growth industry in the eighties. Where once there were only a few unexciting court reporting services, there are now, as the *New York Times* disclosed in 1988, "a growing flock of journals, newsletters, and slick magazines covering everything from arcane ramifications of law-code changes to the relative virtues of expense-account restaurants."

Another aspect of changing life in America is what is often called the "graying" of the nation—a growing number of old people in its population. Both entrepreneurs and advertisers have been quick to exploit this demographic speciality, sometimes with astonishing results. The most remarkable effort has been *Modern Maturity*, rapidly climbing the mass-market heap to rank among the top three in circulation with figures approaching 20 million and with every prospect of becoming the largest. In 1988, about 30,000 readers were being added every week to this six-a-year publication of the American Association of Retired Persons; retirees get the magazine with their membership.

Modern Maturity and its competitors have their problems, however. Advertisers have had difficulty understanding the audience, and marketing professionals have similar misconceptions about it. One of them has been a reluctance to believe that older people are willing to change brands, that they are not slaves of old habits. It is also not understood that they spend more for products than has been recognized, and that a substantial part of this spending is for quality products. *Modern Maturity*'s advertising agency has reported that young people in agencies are turned off by old people and do not want to think about them. Circulation figures, however, are extremely persuasive, and they are backed by research studies, notably one by Yankelovich, that indicate that the wants, needs, desires, and pastimes of those over fifty are more similar to than they are different from those thirty-nine to forty-nine. *Modern Maturity* has carried advertising for only ten of its more than thirty-three years (as of 1990), but its total revenues are already over $50 million, a response not only to its total circulation but also to an average growth of 250,000 every issue.

With such a market available, other publishers have made serious attempts to carve out a share, once more specializing and dividing the audience into specially targeted groups. Older women are being addressed by *Lear's* and *Golden Years*, a small regional publication in Florida, which was about to be switched back in 1989 from a more local audience to a national one by its new owners, Senior Services, Inc., of Wilton, Connecticut. *McCall's* launched in 1988 a "silver edition" for women from fifty to sixty-four. Later that year, from Westport, Connecticut, Intercontinental Publications initiated *Second Wind*, aiming for a market somewhere between baby boomers and senior citizens, that is, forty-five to sixty-nine. This audience, research showed, owned 56.3 percent of the nation's net worth and 51 percent of all its discretionary income. The content of *Second Wind* was definitely upbeat, split into "challenges," health and fitness, travel and adventure, and "pure fun"—overall, designed to broaden horizons and help "doers" get involved in new activities.

Early in 1988, *Reader's Digest* acquired *50 Plus*, a struggling entry in the field, changed its name to *New Choices for the Best Years*, and in a year increased circulation from 500,000 to 575,000. Editorial content was much like *Modern Maturity*'s relatively bland diet, with such articles as "South Carolina's Classic Charm," "Close to Nature on the Outer Banks," "Beating Inflation Through Short-Term Investments," and "Olympia Dukakis and Angela Lansbury Talk About Their Toughest Decisions."

The search for new markets may have reached a summit in late 1988, when two British monthlies devoted to following the royal family—*Majesty*, and *Royal Monthly*—appeared in the American market and found even more success than they had at home. Each was selling more than 40,000 copies per issue, significantly more than in the U.K. They were directed to two different audiences: *Majesty* as a semiofficial adjunct of Buckingham Palace for unqualified idolators; and *Royalty* for the irreverent as well as the curious, with Lady Di in a bikini on one typical cover. (Since 1980, she has appeared on more than half of *Majesty*'s covers and in more than 75 percent of *Royalty*'s.) Both British publishers reported quickly growing American sales.

19

Changing Concepts
in Women's Magazines

"Never underestimate the power of a woman," the *Ladies' Home Journal* once advised advertisers and readers in a much quoted promotional slogan. But that is exactly what most Americans did until after the Second World War, even though contrary evidence lay before their eyes in the women's magazines. When that evidence was examined, it became clear that these periodicals had been the primary factor in raising women from their social status of being "legally dead" at the close of the Civil War to a position in 1959 of control over 70 percent of the country's wealth. From having no legal rights whatever in the nineteenth century, they had moved toward the final battle for complete equality in our time.

As we have seen earlier, the women's journals had already begun the struggle before the turn of the century, talking of suffrage and other freedoms even in the midst of fashions and the "cookie-and-patterns" formulas that prevailed. In the early decades of the twentieth century, one of the most remarkable things that the women's magazines did was to give their female readers the idea, often subtly, that they were sisters, democratically bonded together, no matter what their economic or social circumstances might be.

At the same time, these periodicals gradually built up the power of the matriarchy. They gave us "mom," who remains the central figure in the family constellation as far as most Americans are concerned. Mom was also a figure of responsibility, dignity, and authority in the magazines. This was not, however, a move toward equality, since in articles, fiction, cartoons, and comics, poor Dad was usually depicted as a hapless, somewhat silly creature, manipulated by the women in his family. It was a concept also reflected in motion pictures, radio drama, and many novels. Dagwood Bumstead, in the comic strip "Blondie," became the national stereotype of the family man, an image that persists today in television sitcoms.

In time, the editors believed they had gone too far in stereotyping and began to promote family values in their pages, giving rise to *McCall's* famous slogan, "Togetherness," which unfortunately produced more jokes than changed attitudes. The idea of sharing both work and pain did not increase circulations. What *did* do the job in the postwar market was specialization, with such magazines as *Parents* and *Mademoiselle* showing the way.

Along the road from old concepts to new, magazines that could not or would not change simply disappeared. The most prominent demises were those of *Pictorial Review, Delineator*, and *Women's Home Companion*, all of them mired in variations of cookies-and-patterns. Those that did change with the times found themselves in the 1940s confronted with competition from both radio and television, forcing them into the early stages of specialization and compelling some of them to treat sex as though it actually existed, with the additional notion that it might add to newsstand sales.

In the new era of specialization that had begun, the trend was toward bigger and better magazines, on a par with or better than anything being produced in other categories. It was easy, then, not to notice a modest little journal called *Woman's Day*, being distributed for free in A. & P. stores. It was about to be a prime factor in the democratizing of women and their magazines, just as the supermarkets themselves were doing. There was also another chain-distributed journal called *Family Circle* that was already in the field and doing well, although it was not the equal of *Woman's Day*, an invention of two A. & P. executives, Frank Wheeler and Donald P. Hanson, during the Depression. Starting as a small giveaway, it had been turned into a magazine selling for two cents a copy. Later the price was raised to five cents, then seven cents and ten cents.

Woman's Day and *Family Circle* were not like the "Seven Sisters" of the women's magazine world. They were edited for women who had to be careful about how they spent their money, who had to do their own housework, and who had little cash, if any, to spend on new fashions. Yet the editors never condescended to their readers, never tried to "write down" to them.

In the competition for audiences after the Second World War, women's magazines carried on a brutal struggle that media critic Edwin Diamond was moved to call "unladylike," and nowhere was the battle more fierce than on the broad field where *Woman's Day* and *Family Circle* fought. It included not only the usual newsstand and subscription arena but also the basic supermarket checkout counter and other retail outlets.

At this stage, the two magazines were still quite unlike their older rivals among the Sisters. Far from being sleek products of an art director's talents, their graphics were more pragmatic than sophisticated. Their editorial formulas had settled into a compound of money-saving menus, dress patterns, and decorating and beauty hints, all of which varied little from issue to issue. They were Plain Janes in a crowd of models. Nevertheless, as Diamond observed in 1974, they had "managed to achieve some of the most spectacular growth in the entire magazine business."

Family Circle, now owned by the New York Times Company, sold 9.7 million copies of its December 1973 issue, a record for magazines in this field. *Woman's Day*, now the property of Fawcett Publications, at the time had a circulation almost as large and, perhaps more important, led the entire field in advertising pages. In 1973, it had grown 10 percent, while both *McCall's* and *Ladies' Home Journal* were recording drops in advertising pages. Fawcett, as a privately held company, did not publish profit figures, but its estimated 1973 income of nearly $45 million made *Woman's Day* one of the three most profitable magazine in the industry.

In the success of the two supermarket queens, Diamond saw a trend: "A fundamental change in the magazine business is now occurring. Increasingly, subscribers are becoming a burden to publishers; retail outlets are now where the action is, a complete reversal of marketing strategy." It was not the end of the subscription business, however. What happened was a broadening of the market. The old established reliables—*McCall's*, the *Journal*, *Redbook*, *Cosmopolitan*, *Good Housekeeping*—simply joined their upstart rivals at the checkout counter, where they repose today, fighting for space and sales with the garish tabloids. At the same time, sales were also pushed on the increasingly overcrowded newsstands, and the pressure for added subscriptions, or at least maintaining levels, did not diminish significantly. It was true, however, that some magazines raised subscription prices to reduce overly expensive circulations.

As a publishing phenomenon, both *Family Circle* and *Woman's Day* had their roots in the early 1930s, as we have seen. They grew with suburbs and shopping centers, but as many more women joined the work force in the 1980s, they too had to adapt. They were guided by exceptional editors from time to time. Geraldine Rhoads, beginning her editorship of *Woman's Day* in 1966, looked upon her periodical as a trade journal and had added 1.5 million subscribers to its rolls by the mid-seventies. Arthur Hettich guided *Family Circle* from 1968, at a time when Gardner Cowles owned the magazine, and kept pace with his close rival. Their counterparts on the upper-tier journals were equally able, ranking with the best. Robert Stein was editor of *McCall's* twice and of *Redbook* as well, bringing to both a serious attempt to elevate quality and so separate them from their competitors. Lenore Hershey took over from John Mack Carter in 1973 when he left the *Journal* to become editor of *American Home*, and she became one of that magazine's most distinguished editors since Bok. Helen Gurley Brown joined this distinguished company when she took over the editorship of *Cosmopolitan* and made it her magazine in a way only the best editors can do.

John Mack Carter has stood out from the others in several ways; to name one, he is the only person to have been editor of all three of the Big Three in the women's magazine field, serving successively as editor of *McCall's*, the *Journal*, and *Good Housekeeping*, where he was still in charge in 1989 at sixty-one.

He could look back on a magazine career that began when he was only twenty-five, mustered out of the navy after the Korean War, and getting his first magazine job at *McCall's* as "the least of its editors." In 1990, at an age when so many editors, even the best, begin the fade away, Carter was not only editing *Good Housekeeping*, long since part of the Hearst empire, but writing a column for *Adweek*, giving speeches, writing books, serving on boards of various women's organizations, and playing a role as elder statesman in the magazine industry.

In spite of a revolving-door record when William Randolph Hearst was actively running his publications, the company Carter left, considerably reorganized, was quite the opposite in its operations. For example, only Hearst magazines leave the final acceptance or rejection of advertising copy to the judgment of the editor-in-chief. The Hearst organization also appears to be following Cyrus Curtis's method of appointing editors and letting them run their magazines for as long as they like—or, presumably, as long as they show a profit. Carter, for example, had edited *Good Housekeeping* for fourteen years in 1989, declining other jobs along the way. Helen Brown, at *Cosmopolitan;* Frank Zachary, of *Town & Country;* and Tony Massola, *Harper's Bazaar*, have been other long-running Hearst editors.

In the late seventies, women's magazines were still, as John Peter termed them in *Folio*, "the mother lode of American magazine publishing. The women's field has the golden attractions—a great number of readers . . . who are buyers of a great number of products. . . ." The advent of *Woman's Day* and *Family Circle* had been the first major change in women's magazines' solid front since the end of the Great War. But fresh winds were blowing. The old concept that the female world was divided into housewives and mothers with common needs, on the one hand, and a relatively much smaller number of women whose interests were elsewhere was about to be challenged. Suddenly, the market did not seem so monolithic. The reason, of course, was that the women's movement had changed women's view of themselves. Women, it now appeared, were different from each other; groups of them had different needs. The result was a rash of new publications and a necessity to reposition the old favorites.

There was much discussion of who and what readers of women's magazines really were. Were the old favorite journals simply "life-style magazines" after all, about to change because life-styles themselves were being transformed? Lenore Hershey saw her *Journal* readers as "new traditionalists" with upscale interests and buying power. At that time, John Mack Carter viewed his *Good Housekeeping* audience as "young homemakers" from twenty to forty years old, married, and living in the suburbs. (Toward the end of the eighties, they, too, would become "new traditionalists.") Robert Stein, of *Redbook*, believed his readers were women "above average in income, awareness, sophistication, and education," not belonging to any particular age group. He told a *Folio* writer: "What distinguishes one woman's magazine from another is the point of view and the values of the group of people who edit it. That comes through."

By the end of the seventies, both *Redbook* and *Cosmopolitan* had been repositioned. Under Sey Chassler, *Redbook* was publishing more fiction than any other magazine in the field. At *Cosmopolitan*, Helen Gurley Brown had created a distinctive magazine easily identified by "that Cosmopolitan girl" image projected in its advertising. The Cosmopolitan girl celebrated her twenty-fifth birthday in 1990.

Ironically, Brown had come to the Hearst organization originally looking for money to back a new magazine she was developing. The company's management saw her prospectus at once as the answer to their problem at *Cosmopolitan,* which was in trouble, and made her editor. She targeted it to women between eighteen and thirty-four. In her characteristic way, Brown provided a better summary of what her magazine was about than anyone else could have done. "I figure everyone has problems," she said. "It doesn't matter how much money you make, how beautiful you are, how young, sexy, attractive, sought after, you also have on the other side of the ledger some things which are not going well because that is just simply what life is all about. . . . It's a very personally edited magazine. We don't do any research to find out what people want. I'm a very common denominator kind of person. I'm quite an average person. I know what worried me as a young career girl." No doubt all this was true, but in time *Cosmopolitan* joined the club and conducted reader preference surveys, as its rivals did (and most magazines do). The results did not appear to change the image of the magazine, however.

As the decade of the seventies ended, all the women's magazines were jockeying for position in a new race, conscious that audiences were changing and that new audiences were waiting. *Mademoiselle* and *Glamour* continued to direct their efforts mostly toward fashions for women from eighteen to thirty-five, but they were injecting increasing amounts of how-to articles into their pages. *Ms.,* which had come in with the rise of the women's movement, powered by a superlative publisher, Patricia Carbine, and an unusual editor, Gloria Steinem, concentrated on being a magazine by, for, and about women. Its ratio of 65 percent editorial to 35 percent advertising was unusually high. *Essence* continued to serve the interests of black women. The changed sexual climate had produced two new magazines, *Playgirl* and *Viva,* a product of Robert Guccione's *Penthouse* factory.

Among the newer magazines were *New Dawn,* hoping to capture the young women belonging to the post-*Cosmo* generation, and *New Woman,* done in a weekly newsmagazine style, borrowing liberally from other publications. *Woman's World* was directed toward older working women, editorially realistic, providing working women with ideas and products they could afford. *Working Woman,* obviously designed for women who work, also attempted to reach those who were "having it all," holding a job and running a home.

There were also several other lesser magazines, demonstrating the specialization now in full swing. A fashion and life-style magazine, *W,* was intended for educated, affluent older women. A new food magazine, *Sphere,* emerged under the banner of Betty Crocker. One unique journal, *PariPassu,* offered a beauty counseling service along with the magazine. Judith Daniels, formerly managing editor of Rupert Murdoch's *Village Voice,* appeared as editor of a fresh magazine approach to young readers, *Savvy.*

The field at the end of the seventies was doing well, John Peter summarized in *Folio,* adding: "If the British, who produce far more women's magazine titles than we do—a lot of them weeklies—are any clue to the future, we will be hearing more about psychographics as well as demographics in the future."

Overlooked in Peter's summary because of their relative obscurity were the feminist periodicals, some of them beginning in the late sixties, that were still surviving and reaching their own audiences. In 1975, according to the *New Woman's Survival Sourcebook,* there were 171 of them, including those in Australia, Britain, New Zealand, and Canada. They had a common problem. Since they were not easily available at newsstands or in bookstores, distribution was a primary concern. Struggling, they still managed to reach their readers.

These journals ranged from the *Spokeswoman,* a Washington, D.C., newsletter, providing information on topics of interest to feminists, to another Washington journal, a lower-case radical feminist underground publication called *off our backs.* Prisoners got free subscriptions to the latter, with the focus on minorities and the poor. *Majority Report* was a feminist tabloid aimed at working women, more in the mainstream, characterized by one researcher, Nancy Cooper, as a "feminist *National Enquirer.*"

At the opposite end of the spectrum was *Women: A Journal of Liberation,* a professionally edited quarterly, without advertising and with a circulation of about 20,000, with a militant, Marxist, exclusively feminist viewpoint. Lesser journals in the field included the *Monthly Extract,* devoted to home health remedies; *Prime Time,* for older women and not really feminist; *Herself,* another journal dealing with women's health issues; *Quash,* telling women how to resist the government; *Artemis,* for self-employed women; and *Focus: A Journal for Gay Women.*

Summing them up in the *Mass Communications Review,* Nancy Cooper wrote, "The majority of feminist publications are . . . locally oriented, general feminist newsletters put out by women's centers and college and university women's liberation groups."

All these developments in the sixties and seventies, particularly in the latter decade, were a prelude to the spectacular publishing events of the eighties. Easily the most discussed event of that decade came near its end, in 1988, when Grace Mirabella, after thirty-seven years as a *Vogue* employee and seventeen years as editor of the magazine, regarded as "the most powerful woman in fashion," was summarily fired by S. I. Newhouse, Jr., the chairman of Condé Nast Publications. She heard about her dismissal for the first time from a local television news show, *Live at Five.* It was time to "reposition *Vogue* for the '90s," Newhouse said.

Even in the new climate caused by conglomerate ownership of magazines, bringing with it the morals and manner of the corporate boardroom, Mirabella's abrupt dismissal was a shock to the industry. During her regime, she had raised circulation from 400,000 in 1971 to more than 1,245,000 in 1987. Moreover, *Vogue*'s advertising revenue of $79.5 million in 1987 was considerably higher than its competitors'. Advertising revenue for the American edition of *Elle,* its nearest rival, was only $39 million, and *Harper's Bazaar* lagged still farther behind with $32.5 million.

Newhouse admitted that his editor had reflected the spirit of the 1970s. She had exploited the rise of women in the work force and anticipated the twistings and turnings of fashion with a high degree of accuracy. Now, however, as Newhouse observed, times were changing again—as decisively in the 1990s, he predicted, as from the 1960s to the 1970s. He saw an era of more informality dawning, with guidelines blurred between art and kitsch and a consequent difficulty in distinguishing between high and casual fashion. It would be a decisive change, he said, and clearly felt that an old hand like Mirabella was not equal to it. To replace her, he appointed *Vogue's* 39-year-old art director, Anna Wintour, as the new editor-in-chief. She had formerly been editor of the British *Vogue* and of *HG.*

But Grace Mirabella was not in the usual position of a 58-year-old woman fired from her job. Her departure had occurred in July, and by November, the magazine world was told that in the spring she would become editor of a new fashion magazine in the stable of Murdoch Magazines. Moreover, it would be called *Mirabella,* aimed at women "with their own portfolios, their own careers, and their own sense of style." It also meant to go *Elle* one better in quality. It was this foreign invader that had taken over a large share of the market that *Vogue* had dominated for generations, employing bold designs and a higher-quality paper. *Mirabella* would have even heavier paper, it was promised.

The renascent editor herself announced that her new magazine (developed, it was said, by Rupert Murdoch himself and by John B. Evans, president of his magazine division) would appeal to women who had been "left behind" in the change that occurred when other magazines in the field had hurried to compete with *Elle.* They would be, she said, women like herself, "58 going on 45." This was interpreted by the trade as meaning that it would be directed to an older audience.

Mirabella was about to enter a crowded and still changing market, as N. R. Kleinfield pointed out in the *New York Times.* Even though it was already being called "the magazine born with a silver spoon in its mouth," it would be competing not only with established old favorites but, also, in one way or another, with eighteen new periodicals born in 1988, including *Model, Elva, Haircut & Style, Sheer Fashion, Swimwear,* and *Swimwear Post Girls,* bringing the total number in the field to about forty. To succeed, money was required—$30 million for the first three years.

Most of the leaders, it appeared, were already going through identity crises. In *Vogue* and *Harper's Bazaar,* the movement was toward young readers— again the influence of *Elle,* whose American version had been a joint venture of Murdoch and Hachette Publications of Paris. But *Mirabella,* as its editor implied, would be aimed toward an older audience. It would have an advertising advantage because Murdoch Magazines could sell pages in packages with its other journals, *New Woman, New York,* and *Elle.* It would also be in a position to exert more clout on the newsstands since Murdoch's recent acquisition of Triangle Publications, including *TV Guide*, the *Daily Racing Form*, and *Seventeen*.

If it meant to appeal to older women, *Mirabella* would have to compete with a new magazine, already a year old, in the same field. That was Frances Lear's entry, named, as Grace Mirabella's would be, for herself. Having divorced Norman Lear, the television producer, two years earlier and consequently possessed of a substantial settlement, Ms. Lear had announced a magazine for women over forty; she herself was a white-haired but smart woman of sixty-two. "I was tired of seeing 17-year-old models and reading stories for young marrieds," she told Geraldine Fabrikant, of the *New York Times*. "We have been shut out of the media for so long that we have not seen ourselves in positive images anywhere." Her vision was a glossy, upscale magazine for the "woman who wasn't born yesterday."

Lear's appeared for the first time on February 23, 1988, with a response from subscribers and advertisers that surprised everyone. Out of her $100 million divorce settlement, Ms. Lear expected her new venture would take at least $25 million. A novice in the business, she had a reputation for "hot temper and unpredictability"—attributes not unknown in the industry. A very large market waited for her. There would be 52,000,000 women over forty in 1990, it was predicted. Research studies showed that older consumers did not change the buying habits they had established in earlier days, and the "graying of America" had stimulated more products being made for this market, with more companies looking for ways to promote them.

Modern Maturity could appeal directly to its 17 million readers of both sexes who had acknowledged being older by becoming members of the A.A.R.P., but addressing a magazine to *women* over forty might prove to be a tricky business. Experts asserted that the only way to approach that market (older people in general, as well as women) was obliquely, but Lear meant to give her older subscribers a sense that they, too, could have a glamorous image. Models would be in their forties, fifties, even sixties, but some in their early thirties as well.

The birth pains of *Lear's* were harsh. There were problems with indecisiveness, with turning down articles by well-known writers, and with the staff. The first editor and the senior editor, both men, were fired early. Nevertheless, the six issues Lear put out in the magazine's initial year were successes. Her intention was to attain 500,000 circulation and make the journal a monthly in two years. Judging by seventy-six advertising pages in the first issues, with thirty to forty considered as standard, the future looked promising and continued to appear so at the end of the decade.

Grace Mirabella, aiming at roughly the same market, although it was not officially acknowledged, had her own formula for success. There were admitted obstacles. Famous though she might be in New York's fashion world, Mirabella was not well known in the remainder of the country. *Lear's* had already demonstrated that the problems surrounding addressing a magazine to older women could be overcome by the time *Mirabella* emerged. The editor was frank about her intentions: to reach a female audience primarily interested in fashion but also interiors and who were women with both power and money. How many of these there were in a market of fourteen million women over forty with household incomes of $40,000 or more remained to be seen.

Early in February 1989, *Mirabella* was unveiled at last in a preview. Unlike most magazines for women, no female of any kind appeared on its dummy cover. There was only a woman's lips against a bright, white background, a faceless impression. These were lips that could belong to anyone over or under forty but, in any case, were not Grace Mirabella's lips. Now the editor was telling the press that her journal would not identify its readers graphically and added cautiously that the age range would be women from thirty to forty-five plus.

Forty-eight percent of the content would be devoted to matters other than fashion and personal style. The first issue included an excerpt from Tom Wolfe's new novel, a conversation between Benazir Bhutto and Margaret Thatcher, interviews with experts in various fields, and photo essays on Georgia O'Keefe and the palace at Versailles. New twists were given to the fashion pieces; for example, science-fiction's effect on international fashion and comments by both famous and unknown women on the subject of lipstick. The circulation rate base was 225,000, but projected circulation for the first issue was 400,000, with a special newsstand price of $1, soon to be raised to $1.95.

"The intelligent woman's guide to fashion," as *Mirabella* called itself, was greeted somewhat coolly in some quarters. Bernadine Morris, reviewing it in the *New York Times*, found both "originality and banality" in the first issue. Certainly, the soft-focus fashion sketches, printed like the rest of the magazine on wide pages and heavy stock, were a breakthrough. New but more controversial were the two-page style sketches that had to be turned sideways to be viewed. Morris declared that fashion illustrations had not been "so enticingly presented" since the thirties and forties, in the time of Eric and Rene Bouché.

In a page-five letter to readers, Grace Mirabella assured them that her magazine was not about fashion but style. It would not be concerned with "glitz, clutter and terminal trendiness," she said, but with "where we travel, what we read, how we socialize, how we think about ourselves as part of a larger world." Her expressed aim was to "make sense of the culture," a formidable task.

How well this aim had been carried out in the first issue was a matter of opinion. According to her statements in the *New York Times*, Morris had found that "some features and a lot of the language are as trivial as other magazines'." Chicago, for example, was hailed as "the great American city," and there were eight pages of text and pictures displaying the home and workshop of Robert Miller, the art dealer. The cover had undergone a slight alteration, with the addition of one blue eye and a nose to the original lips. All of these belonged to the film producer Diandra Bouglas, it turned out, whose entire face could be viewed on the contents page.

Mirabella's first issue had 269 pages, 120 of them with advertising, well above *Lear's* initial offering. Its articles were considered "calmer" than

Vogue's, and they ventured into such subjects as politics, psychology, and business, considered to be of interest to well-educated women. The question, it appeared, would be how well these intellectual topics mixed with the fashions, which covered the fall shows in Europe and New York, as well as the new swimsuits and an entire album of casual all-white summer clothes. At the beginning of the nineties, it was still too early to determine how *Mirabella* would fare.

Meanwhile, back at *Vogue*, the flagship from which Grace Mirabella had been thrown overboard, the magazine's new, more relaxed look was still being debated at the end of the decade. Newhouse, apparently, was no more averse than Murdoch to spending whatever money was needed. The cover for the November 1988 issue of the new *Vogue* had cost $10,000. It depicted a jeweled Christian Lacroix T-shirt, accompanied by a $50 pair of bluejeans, without a belt, and the model's midsection on display. Anna Wintour's first issue was regarded as transitional, merely pointing toward possible new directions. *Vogue's* models had once appeared remote and aloof; now they were more informal and relaxed. The covers had once been rather formal portraits, usually shown from the waist up; now, obviously, they were unbuttoned. The feature section had been expanded to include a long discussion of fashion.

Still, the question was whether *Vogue* would continue to be the leader in its field, the dominant fashion magazine. *Elle* was already in second place, reaching out to younger readers whose average age was twenty-six. New magazines like *Details* were also making inroads. Changing *Vogue* was itself a gamble. The industry experts could point to *Harper's Bazaar*, attempting to cast off its staid image and winding up confusing its readers while *Elle* was passing it in circulation and advertising pages. The problem, as always, was how to reach new readers without alienating the old ones—and how to keep both loyal in the face of competition.

Anna Wintour had already gone through one major attempt to change a magazine. She had been made editor of another Condé Nast journal, *House & Garden*, with instructions to renovate it. The alterations were drastic. It was renamed simply *HG*, and its pages were filled with articles about personalities; these pieces were doing very well in other magazines. They did not do well in *HG*. Reactions were mostly negative, an experience likely to induce caution in altering *Vogue*.

By late 1989, there were still no definite answers on the fate of the "hipper *Vogue*." It was trendier, something *Mirabella* was avoiding, with a picture of the rock singer Madonna on the cover of the May issue, with ten more pages inside showing her house. In her first seven issues, said Geraldine Fabrikant of the *New York Times*, Wintour had been trying to move into the market dominated by *Elle* for the past two years or more—a young, hip, fashionable market. The jury, said Fabrikant, was still out. Advertising pages were up, but newsstand sales, normally accounting for about 65 percent of circulation, were mixed. Wintour herself denied that the magazine was intended for a younger audience; it was meant for readers of all ages, she said. Certainly the old *Vogue* was still present in the preoccupation with French couture, features on chic homes, tips on health and beauty, and the constant attempt to create new trends. In 1989, it was still the leader, with the number of advertising pages and circulation up slightly.

From the relentless battle of the women's fashion magazines, it was something of a relief to see how the veteran John Mack Carter was dealing with changing times at *Good Housekeeping*. At the end of 1988, with a rate base of five million, its advertising pages were declining and the revenue from them was flat. At that point, Carter, who was widely believed to know more about the women's market than anyone, could see a trend toward neo-traditionalism among the women thirty-five to fifty years old in his audience. They were, he said, yuppies who had now emerged into their forties and whose mind-set was family—in short, *Good Housekeeping*'s audience.

Discovering the new pathway to this market is credited to Alan Waxenburg, who became publisher of the magazine in May 1988, and Sandra Spaeth, director of marketing and promotion. On an airplane during a business trip, they talked of relating their journal to social trends and produced a circular chart dividing the readership into decades. The fifties were labeled "Childhood," since the median age of *Good Housekeeping*'s readers is forty-one, second youngest in its field after *Redbook*. "Protest" was the appropriate label for the sixties, "Feminist" for the seventies, "Yuppies" for the eighties, and "New Traditionalists" for the forthcoming nineties.

The implication to be drawn from this chart was that idealists and yuppies had taken women away from home and family, but the traditionalists were now ready to take them back. *Good Houskeeping*, however, had never left home. That and the family had always been its stock-in-trade. This led to an advertising campaign illuminated by the slogan, "America is coming home to *Good Housekeeping*." The copy asserted: "There's a rebirth in America. There's a renewal, a reaffirmation of values, a return of quality and quality of life." That could only be good news for an organization that had seen its advertising pages and revenues fall during the first six months of 1988, and it was not a consolation to know that twelve of the twenty-three women's magazines were also down in that period.

The advertising directed to potential advertisers that backed up this message offered a "high income edition that is designed to allow advertisers to reach the top of the New Traditionalist Market." A device was offered called "The Best in the House," a national demographic subscription-only edition, with a mailing list made up of subscribers in the highest-income zip areas with a rate base of 800,000. Women in this audience had a median age of forty and a median income of $44,211—a concentration of best customers.

Alterations were also being made to other established women's magazines as the decade of the eighties ended. In May 1989, the New York Times Company announced it would buy *McCall's* from the Working Woman/McCall's Group, a joint venture of Time Inc. and Dale W. Lang, a media investor, for a reported price of about $80 million. This move considerably strengthened the Times Company's Magazine Group in its competition with the giants in the women's service-magazine field, notably the Hearst Corporation and the

Meredith Publishing Company. *McCall's* would now be sold to advertisers as part of a package with *Family Circle*, already owned by the group. *McCall's*, at that point, had a monthly circulation of 5.1 million; *Family Circle*, with seventeen issues a year, had 5.9 million. Thus, the combined magazines would have a net, unduplicated audience of 30.4 million readers, almost a third of adult women in America.

The competition was not without its own resources. Meredith, with *Better Homes and Gardens* and the *Ladies' Home Journal*, had a combined circulation of 13.2 million, while Hearst's *Good Housekeeping* and *Redbook* had combined figures of 9.2 million. The seventh of these Seven Sisters, which had dominated the mass-market women's service field for so long, was *Woman's Day*, now owned by Diamandis Communications, with a circulation of 5.6 million.

One of the most durable magazines for women has been *Seventeen*, which was forty-five years old at the end of the decade. In 1944, the "teenager" was a relatively new concept, but *Seventeen* grew and changed with succeeding generations. Over the latter part of those four decades, however, the teenage audience shrank, and only the fact that advertisers felt it necessary to reach women as early as possible saved the market from declining, too. But young women in the latter part of the decade appeared to be feeling a need to "look good" again, and that not only stopped declining revenues for *Seventeen* but increased them. Circulation reached a 1.7 million peak in 1984. At that juncture, the magazine had only two direct competitors, *Young Miss*, with 806,348 circulation; and *Teen*, with 992,469. *Glamour* and *Mademoiselle* were targeted to slightly older readers, but *Seventeen* hoped to cut into their readership.

Trouble for everyone, however, was around the corner. It arrived in the unlikely person of Sandra Yates, a 40-year-old Australian entrepreneur who came to America in 1987 with the intention of launching an American version of her country's *Dolly*, considered the most successful teen magazine in the world.

Yates's employer was John Fairfax, Ltd., an Australian media giant, which owned *Dolly*. Carrying out her assignment, she quickly established *Dolly's* American counterpart, a magazine called *Sassy*, and it revolutionized the teenage field. When it circulated a prototype issue, industry people were astonished to see such cover lines as "Sex for the Absolute Beginner." The tone of the magazine, Yates explained, would be girls speaking to girls, friend to friend. "*Seventeen* says 'you should,' we say 'we would,'" she added. Introduced in mid-February 1988, *Sassy* was an immediate sensation. Its guaranteed rate base was 250,000, with a five-year plan to reach 1 million, and these readers found a bright, breezy collection of articles and pictures covering sex, makeup and clothes, how to keep lipstick off your teeth, advice on how to buy condoms, and when to go to the doctor for vaginal problems—all this in the jargon of the teenager. The market for *Sassy* was defined as girls between fourteen and nineteen, of whom there were about 14 million in the United States, fewer than a third of whom were reading a teenage magazine.

In the magazine's first three issues, Jane Pratt, the editor, a 25-year-old Oberlin College graduate from North Carolina, offered her audience articles on losing virginity, ways to be the best kisser, what it's like to date a rock star, AIDS, incest, and the story of a teenage stripper who committed suicide. There were profiles about celebrities and a few short stories, one about a bulimic girl who was thinking of losing her virginity. While these pieces were running, *Seventeen* was talking about a Russian poet, about young people raising money for the homeless, and about how tax laws affect teenage wage-earners—and without the vernacular.

With this kind of contrast, it was not surprising that *Sassy* grew quickly, surpassing its 250,000 base so rapidly that by the end of the year, its average paid circulation was 436,677, making it the fastest-growing teenage magazine. Direct-mail campaigns produced 7 percent responses, unusually high, and the response from insert cards was overwhelming. Readers appeared to be willing to pay the two dollar newsstand price, although it was higher than any of the magazine's competitors.

Sassy was speaking honestly and openly to its readers, but as the magazine said in an advertisement, "some are not ready for the truth." It was temporarily delisted by fifty-two chain outlets; teenagers tracked it down anyway. Under pressure from the Magazine Publishers Association, the Council of Periodical Distributors Association, and the American Society of Magazine Editors, all but eight of the chains were regained. However, the journal was also subjected to an advertising boycott organized by the Reverend Jerry Falwell's Moral Majority (an organization disbanded a year later). That led to the resignation of Helen Barr, the publisher, who had favored a more traditional, conservative approach, at least until the magazine was established. But *Sassy* went right on talking about racism, abortion, and sex, with personal asides injected into articles by its editors and writers, who thus became characters themselves. The Moral Majority boycott was a flat failure; overall, advertising pages dropped briefly from about thirty-six an issue to thirty, while circulation continued to climb. In an effort to quiet fears, however, *Sassy* did run a cover story on "Why Virgins Are Cool."

While *Sassy* was establishing itself, Sandra Yates read in *Folio*, the magazine and book trade journal, that the founders of *Ms.*, Gloria Steinem and Patricia Carbine, were looking for investors. Their magazine had recently been acquired by the Fairfax organization. Would it not be better to own than invest? she wondered. Setting up an American company, Matilda Publications, she approached Fairfax with an offer to buy both *Ms.* and *Sassy*.

The offer came at a troubled moment in the affairs of the Australian giant, the second largest media conglomerate in Australia, controlling more than fifty newspapers and nearly eighty magazines. In 1986, its revenues had been $597 million, a figure likely to attract takeover suitors. At the time Matilda Publications made its offer, Fairfax was being threatened with acquisition by the fabled Australian investor, Robert Holmes á Court. Warwick Fairfax, 28-year-old heir to the empire, took the organization private in a protective effort, but the American expansion now appeared unwise. In mid-1988, he sold both *Ms.* and *Sassy* to Sandra Yates's Matilda Publications. In just three months in America, this young Australian woman had established one successful magazine and taken over another to become an owner of publishing enterprises on her own account—certainly one of the most remarkable accomplishments in modern magazine publishing history.

The plan for *Ms.* was to convert it from a foundation, its original structure, into a money-making enterprise, with expanded coverage into topics, especially politics, previously forbidden by its tax-exempt status. There would also be added most of the ingredients of other women's magazines—humor, gardening, news, but no recipes. Both *Ms.* and *Sassy* were issued from pink and purple offices overlooking Times Square and occupied by a staff of seventy.

The struggle for the younger women's market had intensified in early 1988 when *Model*, a fashion and beauty magazine, was announced as a joint venture of Family Media and Times-Mirror Magazines, with the former as chief manager. It was aimed at *Sassy*'s particular audience, girls from sixteen to twenty, and it had the same slick, hip look, printed on extra-heavy stock. Family Media planned to make it largely a newsstand magazine, with a cover price of two-and-a-half dollars and an advertising rate base of 225,000. As a new member of the Family Media group, founded in 1974, *Model* joined another acquisition, *Taxi*, an Italian fashion magazine for women from twenty-five to forty-nine, introduced to the American market.

In October 1988, Family Media made still another acquisition, Marybeth Russell, who as president would be responsible for *Taxi*, *Model*, and the company's other magazines, *Savvy Woman* and *Health*. Russell had excellent credentials; she had been the executive responsible for *Elle*'s dramatic rise, and she was expected to make both *Taxi* and *Model* into the same kind of Europeanized high-fashion successes. Family Media owned six other magazines, including *10,001 Home Ideas*, *Homeowner*, *World Tennis*, *Golf Illustrated*, *Science Digest*, and *Discover*.

All was not fashion and home in the women's market, however, nor searching out different age groups. One of 1989's entries was *Entrepreneurial Woman*, a magazine for women owning their own businesses or contemplating starting one. Published in Irvine, California, and appearing every two months, it was a sharp departure from magazines for working women such as *Savvy Woman* (later, simply *Savvy*) and *Working Woman*. It was intended for bosses, not employees. Its articles provided advice on how to start a business, how to balance it with family management, and profiles of women role models. *Entrepreneurial Woman* had an initial rate base of 200,000.

Since health was a concern of women as well as men, there was competition for that market, too. Late in 1988, Condé Nast dramatically redesigned *Self*, its health and fitness magazine for women, even though its circulation had increased by 9 percent that year to 1,225,360, with advertising pages increasing by 4.5 percent. The newly remodeled magazine, appearing in December, represented an abrupt change of direction. Athletic models were no longer on the covers, replaced by celebrities posing outdoors; the kind of breathless prose inside that had once been described as "Selfspeak" was replaced by more conventional, more sophisticated writing.

Self's history reflected the perils of this market. Begun in 1979, it reached a circulation of a million quickly and drew the kind of advertising not ordinarily seen in women's magazines, notably for automobiles. By 1986, however, its newsstand sales were flat, and it was decided the magazine had to be radically overhauled. A new editor lasted only a year, and the job was then offered to the editor of *Parents* magazine. She accepted but took a better job as president of an international advertising agency before she could take office. Anthea Disney then came on, hired away from *Us* magazine. She offered a perceptive analysis of the market, including *Self:* "It's not a crisis magazine," she said. "In *Cosmo*, the crisis is always 'him.' In *New Woman*, there's an undercurrent of angst and neurosis. This magazine is about being in charge of your life."

Ken Kendrick, former art director of the *New York Times Magazine*, was called in to redesign *Self*. He began by revising its logo, giving it a bolder, more solid look, and carrying on its bright pastel colors inside each issue. In a radical departure from the company's way of doing things, he eliminated the lengthy service section in front and divided the magazine into three graphically distinct sections, much as Otis Wiese had done at *McCall's* years before. These sections were labeled Trends, Beauty, and Health and Fitness.

In this transformation, health and fitness emerged in a subordinate role. Only two of five articles in the initial new magazine were related to health. Its revised tone could be seen in the lead article's title, "Breast Obsessed: Is the Body Biz Throwing Women a Curve?" Some analysts thought *Self* now resembled men's magazines, raising the possibility that single men and women seemed to be buying big-ticket items in the same way; *Self* could hope for increased advertising revenues.

At the end of the decade, it appeared that John Mack Carter and *Good Housekeeping* might be right, that traditionalism was about to overcome trendiness. In January 1989, a radically new kind of women's magazine called *Victoria* appeared, and in seven months it raised its rate base to 750,000, making it the greatest success of the decade in women's magazines. Its slogan was, "A gracious new world for the non-stop woman." In that world were conventional elements, as the cover promised: "Home & Garden, Fashion & Beauty, Cooking & Entertainment, Crafts & Collectibles." Beginning in August, it began to publish monthly, delivering a total audience of more than 2.5 million "baby-boom women who are seeking a gracious life, balanced between home and business," as its advertising promised. This audience, according to the magazine, had a higher median household income than *House and Garden*, *Vanity Fair*, or *Vogue*.

Certainly nothing could be more traditional than a bride; consequently, those who believed new freedom, new jobs, and an entirely new climate for women had made marriage obsolete were surprised to see a full-page advertisement for *Bride's* magazine in the *New York Times* for January 11, 1989.

The news it contained was even more astonishing: "At 952 pages, the February-March issue is the biggest magazine ever published. With more total pages and more advertising pages than any magazine in history. . . . With an expected audience of over 5 million, it will attract the largest readership ever recorded for a bridal publication. Our readers are in the market for every product imaginable. Last year alone, the bridal market generated $278 billion in retail sales."

Clearly, in the most rapidly changing market of all, some things remained unaltered. The lesson seemed to be that this market had now grown so large and so active that it could accommodate everything from brides to sassy teenagers, with numerous possibilities in between. Would the advertising dollar stretch as well, enough to cover the seekers after multiple female audiences? Only the nineties could provide answers.

20

Developing Male Audiences

Until Edward Bok began to liberate the women's magazines from their unjustified stereotype as "cookie-and-pattern" journals, these periodicals, which were so much more, suffered from an image it took them a long time to overcome. Similarly, male readers endured a kind of reverse stereotyping in the nineteenth century.

In a profoundly patriarchal society, it was assumed that males were the chief readers of serious magazines and that their other interests, assumed to be sports and sex, were catered to more or less under the counter. The symbol was the *Police Gazette,* which combined both interests with a little scandal and became a staple of barbershops across the country. Sports flourished in periodicals, but sex remained hidden until it emerged in the period after the Second World War as a profitable part of the magazine business.

In the seventies, sex was being served by three highly competitive magazines: *Playboy, Penthouse,* and *Hustler.* While they may have seemed alike to those who deplored them, there were major differences that proved critical in the circulation race.

Playboy, the oldest, was best known because of the projection of its image through its logo, the Playboy bunny, second in worldwide recognition only to Coca-Cola, and the Playboy clubs, whose live "bunnies" in their skin-tight costumes were widely publicized. Somehow *Playboy* was frequently in the news for one reason or another. Hugh Hefner, the founder, was a Wahoo, Nebraska, boy who became the proprietor not only of a successful magazine but also of a Chicago mansion filled with the beauties who appeared in various states of undress in his journal's pages. His life-style represented the supposed fulfillment of universal male fantasies.

Hefner became as well known as his magazine. He was on television for a time with a *Penthouse Party* program. Scandalous stories ran in other magazines about his Chicago mansion, with its revolving bed flanked by video cameras and its windowless walls preventing the publisher from knowing whether it was day or night for periods of time. Abe Spectorsky, *Playboy's* editor for a time, related stories of the difficulties his boss's habits created. Since there were decisions that could be made only by the publisher, often requiring immediate attention, the familiar cry in the editorial offices was, "Is Hef awake?" If asleep or otherwise occupied, he was not to be disturbed. Sometimes he was flying in his private jet, a converted 707 equipped with another circular bed and further luxuries, traveling to exotic haunts in various parts of the world.

The characteristic of *Playboy* in its earlier days was its centerfold of a nude woman, leading one wit to speculate that a generation of American young boys was growing up thinking that women had staples in their bodies. The pages were filled with albums of pictures underlined with teasing captions, of women who were not necessarily completely undressed. The pictures were high-quality photography. Articles and fiction were sometimes raunchy but otherwise harmless and, understandably, not well read.

Into this initial monopoly of the sex market, with *Playboy* soaring toward 5 million in the seventies, came Robert Guccione's *Penthouse,* a sex horse of a different color. Using the same basic ingredients, Guccione devised a magazine that went beyond *Playboy.* For example, Hefner's models might strike erotic poses, but Guccione's were frequently seen simulating masturbation, and in time there were layouts of lesbians and others entwined in simulated intercourse, as well as further variations on conventional heterosexual sex. Articles and fiction, as well, were unabashedly erotic, although with some sophistication. Letters to the editor, running as the *Penthouse Forum,* were accounts of sexual adventures straight out of sex-store literature. Some appeared genuine, but skeptics believed that most were staff-written. In any case, they were so popular that Guccione repackaged them in a spinoff called simply *Forum,* which soon added articles about sex to its formula and became a popular magazine in its own right.

Hustler, the creation of Larry Flynt, went beyond either of its rivals, capitalizing on their success to reach a market that presumably wanted much cruder sex. There was little of sophistication in this magazine. It sold raw sex and outrageous articles that kept it in constant legal difficulties. The imitations that sprang up in this field were patterned much more on *Hustler* than the others, simply because it was a formula easier (and cheaper) to duplicate.

In the periodical world, this category was known as "skin magazines," or "skin books," using ambiguous language. Surveying them in 1977, *Time* called these publishers "merchants of raunchiness." As merchants, they were doing well at that point. Both *Playboy* and *Penthouse* were guaranteeing 4.5 million copies a month, and both were among the top ten best-selling magazines, along with *Hustler* and a new entry, *Oui.* Guccione had even started a skin magazine for women, *Viva.* The message in all these journals, *Time* observed with its characteristic self-righteousness, was self-gratification.

There were anomalies in the category, however. *Hustler,* for instance, was targeted to a blue-collar audience, but its price was a high $2.25, helpful mostly to distributors, who ordinarily took 30 to 40 percent of a magazine's cover price. Again, if the two leaders in this group were so morally depraved, as many people claimed, it was hard to understand why their readers came from higher-income middle- and upper-middle-class men (and some women). It was a market attractive to advertisers, who nevertheless had to be cautious because of the magazines' editorial content. Because its image was that of the undressed girl-next-door, *Playboy* was able to attract twice the advertising revenues of *Penthouse*, whose image was of the girl-next-door enjoying her sexuality. This was also a problem for the respective editors, who were compelled to push their formulas as far as they could to attract readers but not far enough to repel advertisers.

By the mid-seventies, the skin magazines were having to deal with the rising force of the women's movement in various way. Feminists considered them pornography, demeaning to women, and wanted them abolished or at least censored. The magazines could scarcely ignore what was happening as the result of women's liberation, but they put their own spin on it, saying in effect that the new freedom was really sexual freedom, not threatening to men's power and dominance. Lee D. Rossi, writing in the *Journal of Popular Culture*, found that *Playboy* was treating women as "healthy, clean, the type of women a man would want to marry," while *Penthouse* depicted women as the "other," experienced, jaded, perverse; a woman "who could be used and abused." Its models were also older, more often black or of other ethnic groups. Hefner, he noted, had started *Oui* because he wanted to capitalize on his rival's success without changing his own magazine. Rossi summarized the *Penthouse* appeal shrewdly: "The women in *Penthouse* and *Oui* reside outside of our normative fictions about male-female relationships. A man must exercise some restraint with a wife; there are laws and conventions he must obey. But with these nudes, as with the heroines of pornography or the phantoms of our daydreams, anything is possible."

If men were cast in the popular mind (especially in the feminist mind) as aiders and abettors of pornography because they read skin magazines, their sex role had to be viewed from a broader perspective by advertisers, who had to deal with them as readers of a variety of other magazines. Two researchers, Gerald Skelley and William J. Lundstrom, sought in 1981 to analyze the male role, presenting their case in the *Journal of Communication*. Having surveyed 660 advertisements from nine magazines—general interest, of interest to men, and of interest to women—they concluded, "Men are increasingly portrayed in decorative roles and less frequently appear in situations involving their 'manly' activities. Only in more recent years, however, do we see men in nontraditional or roles in which men and women are treated as equals."

Another review of the male role was offered by two other researchers, Barbara Ehrenreich and Deirdre English, in *Mother Jones*. They saw *Playboy*, in 1982, as a forerunner and reflector of the revolt of men from the responsibilities and constraints of the family. "The first issue of *Playboy* hit the stands in November 1953," they wrote, "The centerfold—the famous nude calendar shot of Marilyn Monroe—is already legendary. Less memorable, but no less prophetic of things to come, was the first feature article in the issue. It was a no-holds-barred attack on "the whole concept of alimony and, secondarily, the money-hungry woman in general, entitled 'Miss Gold-Digger of 1953.'" In the view of Ehrenreich and English, *Playboy* encouraged a sense of membership in a male fraternity in the ongoing battle of the sexes. Hefner's so-called "*Playboy* philosophy" was popular because it combined both hedonism and chauvinism. "The *Playboy* credo," these researchers wrote, "was to have fun, to spend money, to enjoy oneself, to shed responsibility to the family. It was rebellious in that it challenged '50's values of family and 'togetherness.'" Thus *Playboy* came into the sixties with "something approaching a coherent program for the male rebellion: a critique of marriage, a strategy for liberation (reclaiming the indoors as a realm for masculine pleasure), and a utopian vision."

Inevitably, this placed *Playboy* and the others on a collision course not only with the women's movement but also with fundamentalist opponents preaching strict Christian morality. The magazines were strong enough to withstand these pressures during the seventies, but compromises were beginning to be made. Both *Playboy* and *Penthouse* began to mix in politics with sex, carrying articles on current and often controversial affairs. It was as though someone had pulled the window shade partway down on an orgy. Readers were not fooled, and the magazines could claim their minds were not entirely on sex.

By the eighties, however, changing times began to catch up. At *Playboy*, an aging Hefner had already turned over the presidency of Playboy Enterprises to his talented daughter, Christie, and an effort was made to change and grow with the times. Most of the Playboy empire was phased out gradually: the gambling casinos, the clubs, the hotels, and much of everything Hefner had built, except the magazine. Sexually, its love affair with the girl-next-door cooled somewhat, and the magazine began to take a more relaxed, casual view of sex. In the process, circulation was lost. From a high point of 7.2 million in 1972, the magazine dropped to 4.2 million in 1985. By this time, the Playboy company had acquired its own cable television channel, on which both sex and the magazine could be promoted. It was converted to pay-per-view in 1987, and later dropped entirely, to be replaced in December 1989 by *Playboy At Night*, its new pay-per-view outlet.

What the magazine tried to do under Christie Hefner's regime was to reposition itself, shedding its "skin-book" image somewhat and offering itself as a male life-style magazine. To do so, it increased its coverage of fashion, home electronics, travel, and pop music. It also added a monthly column on sports and another intended for female readers. There was a large question mark in this turnaround. Would it do anything to solve the basic problem of relying so heavily on subscriptions, with only 42 percent of revenue coming from newsstand sales compared with *Penthouse*'s 95 percent?

There was another serious distribution problem as well. Retailers were becoming the target of attacks from vigilante groups of various kinds, intending to prevent its sale. One of the first came from the National Federation for Decency, led by the Reverend Donald Wildmon, whose members picketed stores of the 7-Eleven chain in the summer of 1984, protesting the sale of

Playboy, Penthouse, and *Forum.* The action was only partially effective. A few stores took them off the racks, but the parent company, Southland Corporation, refused to stop carrying these profitable journals.

Two years later, the so-called Meese Commission, a body organized by the Reagan administration to reverse the positive verdict of the President's Commission on Pornography in the Nixon administration, struck what could have been a far more serious blow. Early in 1986, Attorney General Edwin Meese sent a threatening letter to retailers, informing them that they were involved in "the sale and distribution of pornography" by stocking such magazines. This time the Southland Corporation listened. *Playboy* and *Penthouse,* along with others, were banned from 7-Eleven stores, the most important outlet for *Playboy,* although some franchises continued to sell them. As a result, *Playboy* almost overnight lost 12,000 of its 110,000 retail outlets. A suit was brought against the commission, which drew in its horns a little and sent out a "corrective" letter. It took three months, however, for the magazine to get back 3,000 outlets.

These events climaxed seven years of declining advertising pages. But at this low point in its career, *Playboy* began to turn around. An imaginative advertisement, the opening gun in a campaign begun in February 1988, heralded what was to come. It showed the *Playboy* bunny of the logo (not the centerfold or the waitress) at the controls of a spaceship, with the headline, "The Empire Strikes Back." In the campaign copy, the magazine pointed out that it still had a readership of 13 million, the largest magazine for men in the world. It had, in fact, never been unprofitable. The business staff of *Playboy* made it clear to advertisers that nudity (they avoided the word by calling it "pictorial") was by now only 13 percent of the editorial content and that the magazine was now being accepted in homes where it had been forbidden. Three million women were among its readers.

At thirty-five in 1989, *Playboy* appeared likely to stay at the top of its field, notwithstanding *Penthouse*'s strong and steady progress. Pressure from the outside had virtually disappeared, and the magazine had added about 5,000 retail stores in 1988, including more of Southland's franchise operators. *Penthouse,* which had deviated from its original formula by adding political and socially oriented articles, was also doing well. In the first nine months of 1988, it totaled 390 pages of advertising, an increase of 7 percent from a year earlier, and circulation was slightly above 2 million, a somewhat discouraging drop of 8 percent. But then, *Playboy*'s circulation was not in much better health.

Patrick Reilly, analyzing *Playboy*'s recent difficult situation in *Advertising Age,* blamed it on "the editors' penchant for blockbuster celebrity covers, which can send monthly circulation on a rollercoaster." For example, the May 1987 issue, with a picture of Vanna White, television's *Wheel of Fortune* star, on the cover sold about 5.4 million, substantially above the base rate of 3.4 million. Consequently, circulation failed to hold up in the first half of the year. On the other hand, *Playboy*'s average paid subscriptions, on which it depended so greatly, had grown 36 percent since 1978, reflecting its greater acceptance in homes. The thirty-fifth anniversary edition of January 1989, with a five-dollar cover price, had more than 300 advertising and editorial pages.

The magazine had also begun to increase production of what the industry calls "flats," that is, one-shot magazines. A large seller was *The Girls of Summer*, consisting of unused photographs from the magazine, running without advertising and selling for $4.98 on newsstands. The company planned to turn out fourteen flats in 1989.

By the end of the eighties, the able Christie Hefner, as chief executive of Playboy Enterprises, had fairly well stabilized the empire and was looking to renewed expansion. A new publisher, 36-year-old Michael Perlis, arrived in 1989 and predicted that, in the next five years, *Playboy* would still be the company's major publication, but he expected to buy or create new magazines "aimed at men's leisure time." At the moment, *Playboy* itself was contributing 60 percent of the company's total revenues of $166 million. Other publishing ventures added 15 percent, video entertainment enterprises contributed 13 percent, licensing and merchandising accounted for 11 percent.

In the chief executive's office, Christie Hefner was still chopping away at overhead, moving the company from its seven floors of offices on Michigan Avenue in Chicago to a custom-designed duplex on the top floor of the American Furniture Mart. That the old days were truly gone, however, could be seen in the entry into conventional married life of the founder. Hefner married a former "Playmate" in mid-1989, and the event had all the earmarks of a tradition he had formerly scorned. Some thought it symbolic of the company's new status: the revolving bed had stopped rotating at last.

Penthouse, meanwhile, had been broadening its scope for some time, as *Playboy* had done, but in a different direction. A full-page advertisement in the *New York Times* of June 21, 1989, cleverly summarized its new appeal. "What *Penthouse* Bares Best Are Secrets, Plots and Crimes," the bold headline proclaimed, and the signature read, *"Penthouse*, Where Nothing Is More Naked Than the Truth."

There was truth in this advertising. Guccione's erotic "other" women were still there, and so were the uncensored and explicit letters to the editor, but there were now also articles which could be called investigative reporting—"just as probing as our cameras," the advertisement asserted. Was this enough to impel readers to part with four-and-a-half or five dollars per copy on the newsstand and so attract advertisers? Apparently that was the case, because the magazine's advertising pages increased 15 percent between 1986 and 1988, with higher recognition for the advertisers, according to Starch survey reports, than *Playboy, Esquire,* or even *Time.*

Esquire had begun as a slightly risqué magazine, never in competition with the "skin books," but operating in both pictures and articles on a level of sophisticated sex, as we have seen in an earlier chapter. Its later years of glory, after a nearly disastrous slide during the Second World War, centered on the career of Harold Hayes, who died in April 1989. Hugh Hefner had also once worked for *Esquire*, writing promotional copy, a fact on a par with Gloria Steinem's brief career as a *Playboy* "bunny" in the process of getting a story for *Show* magazine, Huntington Hartford III's ill-fated effort.

Hayes joined *Esquire* in 1956, became managing editor in 1960, and was made editor in 1964, just in time to introduce what came to be known as the New Journalism, which some believe was invented by Tom Wolfe, who appeared in the magazine's pages, along with Gay Talese, who contributed thirty articles. "*Esquire* was the red-hot center of magazine journalism," Wolfe recalled at the time of Hayes's death. "There was such excitement about experimenting in nonfiction, it made people want to extend themselves for Harold." There were many other literary bylines in the magazine, Truman Capote and James Baldwin among them. By some accounts, the New Journalism began when Norman Mailer covered the Democratic National Convention in 1960 and returned with a highly impressionistic account. Mailer covered the 1964 political conventions as well, at a time when *Esquire* was carrying the work of Jean Genet and William Burroughs. Later, in 1968, Allen Ginsburg, to considerable astonishment, covered the 1968 conventions.

Hayes knew how important covers were, and with the aid of George Lois, an exceptional art director, he produced some that were memorable, including Andy Warhol drowning in a can of Campbell's soup. Inside, there was a great deal of American life to report on in Hayes's regime: the turbulence of the sixties, civil rights, Vietnam, flower children, drugs, and the women's movement.

In piloting the magazine through these white-water rapids, Hayes demonstrated that he was among the great editors. Consequently, it was a surprise when he resigned in 1973 after a dispute with Arnold Gingrich, the publisher, whom he was expected to succeed. As usually happens with such a departure, *Esquire* floundered. Clay Felker, who had been with *New York* at its beginnings, became majority owner of *Esquire* in 1977. It was his intent to redirect the magazine toward the professional man, and he began publishing business stories that seemed oddly out of place to many readers. Phillip Moffitt, the new editor, printed mainstream articles for the most part—on vacations, relationships, careers, and consumption. It was Moffitt, together with Christopher Whittle, in the early stage of his career as a magazine entrepreneur, who bought *Esquire* in 1979.

Since such an experienced and well-regarded editor as Clay Felker had been unable to resuscitate *Esquire*, advertisers, readers, writers, and competitors waited to see what Moffitt and Whittle would do with it. What they did in the next two years was to spend $10 million without making a profit, although they projected that would occur in two more years. At the time this team acquired it, *Esquire*'s circulation was about 92 percent subscriptions, many low-profit because they had been sold through discount agencies. Moffitt and Whittle continued to let them expire, as Felker had done, meanwhile trying to build up newsstand sales. In this, they had some success, maintaining a rate base of 650,000 while increasing single-copy sales from 60,000 to 120,000, according to a report in *Folio*. Whittle had put together an unusually large single-copy marketing team to carry out this effort. The group of twenty focused on such outlets as convenience stores, supermarkets, and drugstores.

With some uncertainty about what *Esquire*'s new market was going to be, advertisers were reluctant to commit themselves, and lineage dropped; it had been declining since 1979. The new owners were clear enough about who was in their market—affluent, well-educated males from twenty-five to forty-five. The magazine was redesigned for Whittle and Moffitt, retaining some of its traditional elements while adding such new ones as innovative typography, in an attempt to make the journal easier to read. These and other measures were in vain. Whittle moved on to new adventures, far more successful ones, and sold *Esquire* to the Hearst Corporation, where it began an uncertain new life but showed no signs of collapse at the end of the decade.

Publishers looking for markets in the eighties, and not wanting to get involved with sex magazines or faltering older magazines, could and did turn to a field that was rapidly expanding, the male's presumed other interest—sports. It was a decade in which more Americans than ever before were watching sports, either in stadiums and other arenas or at home on television. Americans were taken with the craze for fitness; it was a time of intense physical activity. By 1980, there had been a 36 percent increase since 1970 in participation in active sports, according to a Harris poll. That meant a rising market for sports magazines in the eighties. There were obvious areas for further specialization and exploitation. Tennis, to take a single sport, was played by 25 million Americans, and another 30 million were joggers, representing 11 percent and 13 percent respectively of the total population, as *Marketing & Media Decisions* pointed out to advertisers.

Besides such general magazines as *Sports Illustrated*, there were a growing number of specialized journals. The *Runner*, founded in 1977 by George Hirach, who distributed it to those participating in the New York Marathon that year, had a 1980 circulation of 120,000. An earlier entry, *Runner's World*, begun in 1966, was by 1980 up to 410,000. *Jogger* and *Running Times* had circulations of about 100,000. The competition among these magazines was intense, sometimes resulting in legal actions.

The leading magazine for tennis was, appropriately, simply *Tennis*, with a circulation of 440,000 in 1980. It represented an attractive market to advertisers, who were told by Howard Gill, the publisher, that his research studies showed every part of the magazine was read by most readers, making any advertising page a good one. While interest in tennis had declined slightly, affecting both *Tennis* and its chief rival *Tennis World*, the audience increased again later in the decade.

One noted tennis champion, Billie Jean King, also a leader in the women's movement, started her own magazine, *Women Sports*, in the late sixties. It was bought later by Charter Communications, which in turn relinquished the journal to the Women's Sports Foundation; it had achieved a circulation of 75,000 in 1980.

In the sports market, there was something for everyone, from *Racquetball Illustrated* to *Skin Diver*, most of these journals growing and most of them seeking what the trade calls horizontal advertising; that is, advertising not related directly to editorial content. Some also tried to reach beyond their obvious audience. *Runner's World*, for example, carried articles on fitness, massage, vitamins, yoga, and stretching. Advertisers were attracted because it was reaching an audience with a median household income of $30,000. Not all these magazines were horizontalists, by any means. *Surfer* and dozens of others elected to stay with related products on the theory that these advertisers would be more loyal to the magazine.

In a field that was rapidly getting crowded, some of the old established magazines were having difficulty keeping pace with the newcomers. *Sports Afield*, for years faced with no serious competition, was compelled to upgrade quality and reach out toward a better-educated, more-affluent audience. Improving quality, however, meant increasing the cover price, which had a depressing effect on circulation. *Camping Journal* had also encountered serious problems in the seventies because of the shortage of gasoline, as well as unavoidable price increases. Faced with the necessity to broaden its audience in order to survive, the magazine began to target women, hitherto neglected, and the growing number of readers interested in ecology and the environment. By taking these measures, it was able to keep a more or less stable circulation of about 270,000.

In the early eighties, Americans were spending nearly $200 billion annually on sports equipment and accessories, making sports magazines a lucrative field for advertisers and providing a steadily increasing audience for the circulation departments. Turning an analytical eye on this market in 1981, *Advertising Age* made some observations that introduced at least one snake into the publishers' Paradise. "First," it said, "even the most established magazines reach a tiny percentage of those who participate. It has been suggested that active participation may be incompatible with devotion to a periodical, but whatever the reason, greater outreach remains a standing challenge. Women represent the largely untapped frontier in most sports, and they are taking up new activities at a faster clip than men. . . . Finally, what may be termed the 'professionalization' of amateur sports—a situation in which participants seek increasing similarity in their athletic pursuits to the style and experiences of professional athletes—is already widespread and would continue to expand."

In the eighties, the sports-magazine field could be divided into half a dozen categories, according the *Advertising Age*, which proceeded to do the carving. First were the outdoors magazines: *Field and Stream*, with 2,000,000 circulation; *Outdoor Life*, 1,750,000; and *Sports Afield*, risen to 542,000 in 1981. All of these were long-established publications, founded before the turn of the century. They were now, however, competing with such strong regional publications as *Fins and Feathers*, in the Midwest; *Southwest Angler; Southern Outdoors;* and *Western Outdoors*. There were also new magazines for campers, *Camping Journal* and *Backpacker*.

A second field, continuing to grow, was what the trade journal called "racquet sports," with 14 million players by 1981. Besides the tennis magazines already mentioned, there was also the newer *Tennis USA*, with 125,000 circulation. In addition to *Racquetball Illustrated*, with 105,000 subscribers, there was the official publication of the National Racquetball Association, *National Racquetball*.

"High-ticket sports" constituted a third category. There were magazines for skiers, an audience attractive to advertisers because of their relative affluence. For them, there was *Ski*, 421,000; and *Skiing*, with the same figure. Supplementing these leaders were *Powder*, for truly serious skiiers, and a new magazine *Skier's World*, published by Anderson World, Inc., owners of *Runner's World*. Golfers could choose between *Golf Digest*, with a million circulation, and the newer *Golf*, with 740,000—again, extremely attractive to advertisers since golf was assumed to be the sport of executives. Another high-ticket sport with an undeniably affluent audience was sailing, served by *Sail*, 77,590; *Yachting*, 148,000; *Cruising World*, 93,355; and *Yacht Racing/Cruising*, 37,000—not to mention the motorboaters, with *Boating* and *Boating and Sailing*.

Running, a fourth category, was well served by *Runner's World*, which had begun as a single typewritten sheet issued from the editor's house in Kansas, as well as by *Running Times*, 52,000; the *Runner*, 136,000; and *Running*, published by the Nike Company, makers of running shoes.

A fifth category was "road and off-road sports," which would include *Bicycling*, 127,000; *Cycle*, for motorbikes and snowmobiles, 447,000; *Cycle World*, 333,000; *Dirt Bike*, 164,000; *Cycle Guide*, 115,000; *American Motorcyclist*, 112,000; and *Now Goer*, 100,000.

Finally, there were "individual standouts," including *Outside*, with 200,000 circulation, an exceptionally well-written journal covering the people, activities, literature, and politics of the outdoors; *Adventure/Travel*, with a somewhat smaller circulation; and *Young Athlete*, 112,000.

The new sports magazines continued to arrive as the decade ended. One of the most innovative, launched in December 1988, was *TV Sports*, published by ESPN, the television cable company specializing exclusively in sports. A monthly, its chief purpose was the listing of all the network's sports programs broadcast nationally, as well as those of the major networks and the superstations, carried by many cable systems. Sports fanatics would find listings organized by sport and hour-by-hour programming. *TV Sports* was being distributed as a paid insert in Sunday newspapers, beginning with those in ten major cities. To ensure reaching the market, however, these supplements would only be included in home-delivered newspapers where cable television existed and where the average household income was above $35,000. Total initial circulation was 2.1 million. Analysts expected ESPN might be walking a tightwire between broadcasting and publishing a broadcasting magazine.

By 1990, the sports magazine field, like every other, was becoming increasingly a collection of conglomerates, some independent, others offshoots of largely unrelated enterprises. A prime example was the New York Times Magazine Group, which in December 1988 announced the purchase of its fifth golfing magazine, *Golf World*, with a paid circulation of 120,000, perhaps intending to corner the market. This magazine was added to a list that already included *Golf Digest; Golf Shop Operations;* and two British magazines, another *Golf World* and *Golf World Industry News*. The Times stable also included four other sports and leisure magazines: *Tennis, Tennis Buyer's Guide, Cruising World,* and *Sailing World,* besides such nonsports enterprises as *Family Circle, Family Circle Great Ideas Series, Snow Country, Decorating & Remodeling,* and *Child*.

In all this ferment, one might ask, what had happened to the red-blooded adventure magazines that so many American boys had grown up with in the earlier years of the century? There were six of them surviving and doing reasonably well: *Argosy, True, Climax, Male, Saga,* and *Stag*. Their continued existence, according to researcher David Pugh, writing in the *Journal of American Culture,* provided evidence "of the masculinity cult in the United States." Such magazines, he argued, "do not imply or merely suggest the existence of a masculinity cult in American life; they are some of the chief documents of that cult, bought and read by men whose values and ideas are shaped by an obsessive virility. The resulting super-male point of view is rooted in the past, especially the Old West, and certain rituals of manhood reenacting those traditions inherited from our mythologized frontier experience."

The magazines themselves would agree. In a typical advertisement for itself, *True* declared: "One word describes the new *True* magazine: MACHO. The Honest-to-god American MAN deserves a magazine sans naked cuties, Dr. Spock philosophies, foppish, gutless, unisex pap, and platform shoes. It's time for a refreshing change. . . . A hardy slice of adventure, challenge, action, competition, controversy, including informative features that bring the American man and American values back from the shadows. Back from the sterile couches of pedantic psychiatrists. Back from behind the frivolous skirts of libbers. . . . If you're a man, you'll like it."

In this pronouncement, there was enough to keep sociologists, psychologists, and historians of popular culture busy for some time. As Pugh observed, the thrust of these magazines included "anxiety over female intrusion, enjoyment of hunting, ambivalence about civilization, and self-imposed aloneness. . . . The he-man magazine descends from another form of popular literature in the nineteenth century, the dime novel, and the heroes of each share much in common. . . . Yet the he-man concept of the masculine male—a blend of hunter, cowboy and self-made man—creates a dilemma for those who can neither rejoin his mythologized past nor fully accept his own modernism in rejection of that past."

Pugh and others have seen the he-man magazine as an effort to retrieve the old-fashioned hero from the past and so escape changing concepts of masculinity. In some instances, these magazines are heavily masculine, to the point of sublimating sex. In the pages of *Climax,* for example, there is no heterosexual love. One of its typical stories in the eighties was "Stranger in Town," a narrative about an aging gunfighter going blind, who realizes the world he once dominated is slipping away. While these magazines provide vicarious he-man experiences, they encourage in every way the cult of nineteenth-century masculinity. Women, the city, and civilization are viewed as threats.

For men, as for women, the magazine industry reaches out to many interests, but obviously, male audiences are fewer and less specialized. Today there is also some overlapping. While few men would read fashion magazines for women, unless they had some professional interest in the audience, both sex and sports magazines aimed primarily at male audiences have attracted substantial percentages of women readers. Men, of course, have their own fashion magazine, *GQ*. All this is good news for advertisers, bad news for publishers and editors trying to pinpoint and define their audiences.

21

The Way We Live: Magazines Reflecting American Life

All magazines reflect some aspect of American life, and have from the beginning, as the preceding chapters have shown, but we have become more conscious of it perhaps in the closing years of this century as periodicals have clustered into large groups. Each of these groups represents an interest that, for the moment at least, has become an absorbing interest of Americans.

In the eighties, for example, public interest in celebrities dominated the pages of numerous magazines. This preoccupation was even more evident in television and in the book business, where volumes by or about celebrities dominated the best-seller lists—"celebrating celebrity," as the media critic Edwin Diamond put it.

The aberration began in 1974 with the emergence of *People* magazine. In its early years, it was designed to imitate television; that is, it displayed large pictures and used very little type. As Diamond reported, its managing editor, Richard Stolley, employed "the now famous Stolley's Law of Famous Covers—young is better than old, pretty is better than ugly, TV is better than music, music is better than movies, movies are better than sports, anything is better than politics."

People has been one of the great successes of its time, earning substantial profits for Time Inc. and establishing itself as a reading habit among millions of people. In its wake trail a number of periodicals that, if not imitators, are toiling in the same vineyard. One is *ET* magazine, the initials standing for a popular syndicated television show about celebrities, *Entertainment Tonight*. Others include *Us* and *Picture Week*, also a Time Inc. project for which Stolley was taken from his post at *People* to edit.

In these journals, news is fun, told primarily in pictures, concentrated on celebrities, and all of it offered in small "print bites," so to speak, like the sound bites on television, the whole designed for people with short attention spans or who read in brief spurts. The immediate success of the formula could be seen in *People*'s circulation figures: 2.5 million in two years. Even a comparatively uninspired periodical like *Us* was selling more than a million by 1985.

In that year, Jann Wenner, founder and editor of *Rolling Stone*, became a part-owner and the editor of *Us*, which had been owned previously by the New York Times Co., Macfadden Holdings, Inc., and Warner Communications. Wenner's associate in the venture was a former associate editor of *Rolling Stone*, Don Welsh, who had created such popular children's magazines as *Muppet* and *Barbie*.

To further complicate the multiple possession of magazine properties in this era of interlocking conglomerates, *ET* magazine is partly owned by the book publishers Simon and Schuster, who in turn are owned by Paramount Communications, which also owns the *Entertainment Tonight* show. Celebrity magazines rely heavily on single-copy sales at newsstands, checkout counters, and other retail outlets. This accounts for half of *Us*'s figures and two-thirds of *People*'s. The others, *ET* and *Picture Week*, are designed, as Diamond has pointed out, to "fit somewhere between *People*-reading and TV-watching, further blurring the lines that separate print and television. For example, [*ET*, the magazine] may use the show's logo and some of its graphic styles and departments, and Fier [Harriet Fier, the editor] may herself appear on *ET* to talk about pieces in the magazine. *ET* the TV show tends to attract a female audience (as do *People* and *Us*)." Citing magazine marketing experience, Diamond predicted in 1985 that "the full business potential of a journalism that celebrates celebrity has not yet been reached," and the subsequent five years have demonstrated the correctness of that estimate.

The soap-opera world is its own celebrity universe, so it was natural that in January 1988 a glossy new magazine, *Soap Opera Update*, would appear and in less than a year accumulate a circulation of nearly 300,000. This was the newest entry in a field which began with *Daytime TV*, started in 1969, published by Sterling's Magazines, Inc., also owners of *Daytime TV Presents* and *Soap Opera Stars*. None of these magazines equals the circulation of the leader, *Soap Opera Digest*, with 1.1 million.

Daytime serials, in fact, are the source of a small industry, including magazines other than the above, newsletters, syndicated columns, even books. All the periodicals are more or less alike, concentrated on interviews with the stars of the soaps but including other features. *Update* and *Digest*, however, go beyond this formula and offer synopses of the serials as a primary attraction. Both these journals were founded by Jerome and Angela Shapiro, although they sold the *Digest* in 1980 to the Network Publishing Company, acquired in turn by Gary M. Bitterman in 1986.

"Soap-opera" magazines have given at least some respectability to a phrase used only in a derogatory way until the mid-seventies. Sometime during the decade, the shows themselves acquired a new popularity, with better ratings and more publicity (often about their sexual freedom), giving the magazines a lift, too. Celebrity played a large part in this development, since a major reason for the rise in the soaps' popularity was their use of big-name stars to supplement the work of regular contract players. The result was that such daytime soaps as *General Hospital* became known to people who had never heard of it before. Recently, however, "soaps" have gone into a new decline.

These magazines rely heavily on single-copy sales, which account for about 80 percent of revenues for both *Digest* and *Update*, with only 20 percent coming from advertising, mostly for cigarettes and beauty products. Their primary outlet is the supermarket checkout counter. They are published on black-and-white newsprint, a handicap in one sense because such paper cannot accommodate color advertising. That places these journals at a disadvantage to *Update*, which comes full-sized and in full color, abundantly supplied with flashy advertising.

An audience profile shows that *Update*'s readers are mostly women from nineteen to forty-nine years old, 60 percent of them in the regular work force. About 31 percent earn more than $35,000, and another 40 percent earn more than $25,000. *Digest* readers have about the same kind of income, an average of $25,000 to $27,000, lower than those reading the Seven Sisters.

One thing separates the soap-opera magazines from their companions at the checkout counter. They scrupulously avoid gossip and scandal, no matter how lurid the soaps themselves may be. Readers do not want to hear anything bad about the real lives of the actors and actresses, even if they may be playing characters involved in outrageous behavior.

By 1990, it appeared that there would be no foreseeable end to the exploitation of celebrity for star-obsessed Americans. A late entry was *Premiere* magazine, appearing in November 1988, its initial cover promising "Tom Cruise: Hollywood's Top Gun Pours Out His Heart in *Cocktail* [his new movie]." There was, of course, a large picture of Tom. Inside were stories about "John Hughes: Who Loves Him, Who Hates Him, and Why"; "Shot By Shot: A Chinatown Case in *The Presidio*"; "Home Video: Five Top Directors Pick Their Favorite Films"; and similar pieces. For the January 1989 issue, the circulation rate base was 375,000. An advertisement for *Premiere* told potential advertisers what the magazine was doing: "For just over a year, *Premiere* has been taking a quarter of a million primary readers behind the scenes of the silver screen. Now, these active, affluent, acquisitive movie lovers are putting us way out in front. With 3.5 readers per copy, more than 1.2 million *Premiere* fans in all." At that point, *Premiere* could successfully proclaim itself to be America's number one movie magazine.

Another unusual entry in the celebrity field during 1988 was *Fame*, "today's most entertaining and enlightening magazine," its advertising asserted. No money had been spared to fulfill that promise. The magazine carried celebrity to new heights, its pages filled with articles about the stars by best-selling, name-brand authors, and illustrated with the work of noted photographers, all printed on the same paper as Thomas Hoving's *Connoisseur*. The magazine's advertising promised a garden of delights: "Learn what high achievers like Paloma Picasso, Irving 'Swifty' Lazar and Lee Iacocca think about their own celebrity"; a history of fame itself, by the author of a biography of Truman Capote; "fascinating coverage of the powers who are the behind-the-scenes creators—the people who influence business, fashion, entertainment, media, politics—but whose names you probably don't yet know." There was also coverage of theater, books, film, and music, and a collection of experts giving advice on love and the law, surviving success, and business and investments.

Fame did not consider itself in competition with the other celebrity magazines. Its competitors were *Connoisseur*, with 320,000 circulation; *HG*, 500,000; *Town & Country*, 400,000; and *Vanity Fair*, 500,000. *Fame* had a distance to go in 1989, with 200,000, but it was climbing. Its list of advertisers, filling about eighty of its 200 pages, promised instant success: Bentley, Chrysler, Estée Lauder. Tiffany, Cadillac, Dior, Credit Suisse, and Cartier. It, too, was available at checkout counters, but not in supermarkets, in the two leading bookstore chains (Waldenbooks and Dalton), in leading airports, and, of course, on newsstands.

Demographers and marketers define "rich" in figures quite different from those in the popular mind. To be a millionaire is far more common than it was even fifty years ago, but is not considered rich by the truly rich; however, it remains the aspiration of the multitudes. To media planners and to publishers planning to deploy risk capital, "rich" means readers with incomes of at least $35,000 a year, few of whom at this base figure would consider themselves wealthy. Of the 538 notable consumer magazines extant in 1987 (up from 336 in 1975), a large majority of the newer entries were aiming at this affluent market. In the *New York Times*, Fabrikant quoted an agency executive as saying: "Magazines used to be positioned for everybody, but not any more. All the growth is at the top end. You really can't be so vulgar as to publish a magazine called *Rich*, but there *is* one called *Millionaires*."

Affluent America, in fact, had made it easier to start a new magazine in the late eighties, since start-up costs were a bargain by prevailing standards in business. While major magazine companies might be willing to spend tens of millions to start a new publication, individuals could launch one for as little as $5 million, an inconsiderable sum by marketplace accounting—in comparison, the costs of buying a small daily newspaper would be at least $10 million, $100 million for a television station. On the other hand, simply to reintroduce a revived *Vanity Fair* in 1983, Condé Nast was willing to spend more than $30 million. By contrast, Herbert Lipson needed only $8 million to start *Manhattan, Inc.* in 1984. In her analysis of the new market, Fabrikant recalled that more than ten years had passed since Judy Price founded *Avenue* in 1979, perhaps the first successful magazine for the super-rich. At that time, there were only three major journals directed to this audience: the *New Yorker*, *Gourmet*, and *Town & Country* (keeping in mind, of course, that these readers were not necessarily super-rich in real terms but were capable of buying the expensive products advertised in their pages). By 1987, "upscale" had become the buzzword in the magazine business.

While it is necessary to qualify "rich," as we have done, it is also true that in the late eighties there were 36 million Americans (one in five adults) with annual incomes larger than $50,000, a segment that had doubled in size since 1983. Advertising executives believe, correctly, that there are a great many other people who spend money as though they really have it, as Americans' enormous credit-card debt testifies. But the members of this market have more in common than money, in whatever quantity. They are also united by education, value systems, and roughly comparable life-styles; they also care about status in the products they buy. Publishing for the affluent has more than the usual pitfalls. In a market so competitive, it becomes necessary to invest more and more money in editorial content, paper quality, and printing, making for considerable nervousness during the five years it takes for a typical publisher to earn back his investment.

Perhaps the most successful positioning of a magazine to take full advantage of affluence was that undertaken by *Vanity Fair*. It was accomplished by a remarkable editor, Tina Brown, who was brought over from the London *Tatler*. In America, she said later, after making some initial mistakes, it was necessary to "go for the big themes—love, death, and where America is going." Doing so brought the magazine back to life.

Condé Nast (Newhouse) was willing to invest $10 million in this effort and another $10 million to revitalize *Esquire* because the competing Hearst Corporation was willing to invest $4.5 million to promote *Connoisseur*, which it had brought over from Britain as a journal to accommodate advertisers who could not afford *Town & Country*'s rates. *Vanity Fair*, however, had to overcome another handicap—the remembrance of what it had once been in its heyday, "America's most memorable magazine," in Cleveland Amory's words. Reviving it with vol. 46, no. 1, in 1983 did not please everyone who remembered. Henry Fairlie, the professional iconoclast, greeted it in the *New Republic* with a broadside calling it "a magazine whose publisher and editors (and so its contributors) have no idea why it is needed, what they have to tell, or even in what time or place they have their being." It was, he added, an enterprise created only to make money.

To understand the task that confronted Tina Brown, it is worth recalling *Vanity Fair*'s history briefly. It appeared first in September 1913, a merger of two journals, calling itself *Dress & Vanity Fair* and promising in a subtitle to cover "Fashions, Stage, Society, Sports, the Fine Arts." The *Dress* part of the title was dropped after the fifth issue. In March 1914, the magazine entered the days of its glory when Frank Crowninshield became editor. He was already well known in cultural circles, having supported the 1913 Armory Show; later, in 1930, he helped found the Museum of Modern Art in New York City.

Crowninshield, who would preside as editor until the magazine died in 1936, announced at once that his journal would cover "entertainment, happiness and pleasure." Naturally, it would support modern art, but it would also do the same for feminism. Fairlie wrote that more than any other comparable magazine, *Vanity Fair* "captured at the time and still captures for us their spirit," meaning the journal's early contemporaries, Mencken and Nathan's *Smart Set* and *American Mercury*.

Vanity Fair in its early days was heavily laced with members of the Algon-quin Round Table—Dorothy Parker (writing first as Dorothy Rothschild) and Robert Benchley, among others—but it also offered a literary smorgasbord that included Aldous Huxley, D. H. Lawrence, Thomas Wolfe, e. e. cum-mings, Colette, T. S. Eliot, Walter Lippmann, and André Gide. These names helped substantially to sell the magazine, but Crowninshield also gave it a flair, a feeling of walking through endless delights, that charmed its audience.

Taking over the revived magazine as editor, Tina Brown (and the publisher) did not make the mistake of trying to resuscitate what had gone by. Crown-inshield's magazine could not have been duplicated, for many and obvious reasons. In her first five years, Brown created a new *Vanity Fair,* attuned to a different time and appealing to a different audience. The answer lay not only in superior design and photography but also in her ability to attract some of the best-known writers, as well as sought-after new ones.

In doing so, she improved the magazine market for contributors by elevat-ing their pay scales. Before her advent, *Playboy* had paid more than any other periodical, a dollar a word, and most of the others paid less than half that figure. Ten to twelve cents a word was at the low end. Brown paid two dollars per published word, more if a brand name required it. For 12,000 words, a writer might get as much as $25,000. Moreover, she signed many of her con-tributors to annual contracts, following *New York*'s lead. With a contract, a writer could feel himself part of the magazine. Brown also wooed writers, even if they were already signed, by doing what book publishers had long done for their best-selling authors: she sent flowers, took them to the theater and dinner, made them part of the New York social scene—her own scene.

The effect on other magazines' pay scales was beneficial but limited. The low end of the scale remained about the same for most journals, but new arrivals—*Premiere, Fame, Smart,* and *Lear's,* all making their debuts in 1988—were paying $7,500 or more for 5,000-word articles. A writer might even get as much as $15,000 for a piece that would have brought him only $5,000 five years earlier. In Tina Brown's own bailiwick, Condé Nast had to raise rates on its other magazines—*Vogue, GQ,* and *Traveler.* There was little change at such other magazines as the *New Yorker,* which had its own pecu-liar system of payment, the *Atlantic,* and *National Geographic.* But these and other periodicals were now compelled to pay more for the rights to important articles, since agents could auction them as they would any other subsidiary right.

If Americans were fascinated by celebrity, money, and the lives of the rich and famous in general, their national way of life also included a preoccupation with food, both what to eat and what not to eat. Food was not only a part of the affluent life but also an interest of middle-class readers who did their own cooking. Floods of cookbooks and magazine articles testified to this, and a whole genre of magazines about food and wine continued to grow in the eight-ies. Marketers believed that this was spurred by heavy travel to Europe when the dollar was weak, but it appeared to have other components as well.

Five magazines dominated the field in the early eighties: *Food & Wine,*

Bon Appetit, Cuisine, Cook's Magazine, and *Gourmet.* Their ownership was diverse. The largest, *Bon Appetit,* with a circulation of 1.3 million, was owned by Knapp Communications, publishers of *Architectural Digest. Cuisine* was the property of CBS, *Food & Wine* of the American Express publishing division, and *Cook's Magazine* had recently sold a 50 percent interest to the *New Yorker.*

Gourmet, the oldest of these journals, was about to pass into other hands after many years as the personal expression of its founder, Earle MacAusland, who died in 1980. Under his close direction, beginning in 1941, it had been issued every month for years from the penthouse of the Plaza Hotel in New York. MacAusland lived in this splendor with his dog, retaining a chef and the chef's wife, who together presided over the kitchen. The magazine itself occupied three or four small rooms, no doubt once used as guest rooms. Adjoining them was a two-story-high drawing room with floor-to-ceiling windows looking out over Central Park. It opened into a similar room, with views the other way, where MacAusland was the host at gourmet meals devised by the chef and the chef's wife, a pastry cook. Certainly no magazine ever occupied such sumptuous quarters. MacAusland's life-style was unique, even among the rich and famous, many of whom were his dinner or luncheon guests. He hand-picked his editors, one of whom was the noted food and cookbook writer, Ann Seranne, who occupied the editor's chair for years.

Advertisers and readers were equally attracted to these magazines. A long list of products appeared in their pages, from microwave ovens to food processors to all kinds of gourmet cookware. It was presumed that people who loved good food also loved to travel, and so another advertising category was opened up, with ancillary goods involved in such recreation. In the first six months of 1983, these periodicals had combined advertising revenues of nearly $20 million, an increase of 25 percent over the preceding year. Their total circulation was more than 3 million. To gain such figures, their appeal had to be extended beyond food, as *Gourmet* expressed it in its subtitle "The Magazine of Good Living." Its contents included restaurant reviews and travel articles.

Bon Appetit, the other pioneer in the field, had the broadest appeal and the largest circulation, as well as being the leader in total advertising revenue. Knapp had bought the magazine in 1975 from the Pillsbury Company, its originators, and transformed it from a promotional effort to a glossy food periodical capable of pulling in $9 million in advertising revenue during the first six months of 1983.

The magazine with the most spectacular growth in this field, however, was *Food & Wine,* the beneficiary of the high-powered promotion that its parent company, American Express, put behind it and having excellent editorial content and design. Its view of food was perhaps more relaxed than the others, and it accepted advertising for prepared foods, which most of its competitors would not do, except on an extremely selective basis.

In the early eighties, *Cuisine* broadened its appeal, offering itself as more of a life-style periodical, with an accent on dining, entertaining, and traveling, and few stories on food preparation.

Cook's World carved out its own portion of the food-magazine group, appearing six times a year and focusing its efforts on the 100,000 to 200,000 people involved with cooking as a hobby, showing them how to improve their skills by avoiding shortcuts and using fresh ingredients as well as making everything from scratch. Cooking was presented as creative fun, something all these journals attempted to do in various ways.

As an adjunct of the food audience, a somewhat larger market was explored in this same period of affluence and growth, taking advantage of the growing American interest in wine—again, probably the result of European travel and the promotion efforts of California vintners. By the end of the decade, it was estimated that there were 5,000,000 or 6,000,000 wine drinkers in America, and at least thirty publications to serve them, with another seventy-five intended for the trade. Of the country's thousand or so wineries, more than 200 were publishing wine newsletters.

The home has always been a primary interest of magazine readers, as the women's periodicals particularly attest, but by the end of the eighties, changing life-styles had also changed the periodicals addressed specifically to homemaking. At the beginning of the decade the trade journal *Magazine Age* listed the leaders in this field as *Better Homes and Gardens, House Beautiful, House & Garden, Family Circle, McCall's,* and *Woman's Day,* although the latter three had a broader scope.

Since that listing was published, *House & Garden* has demonstrated how a magazine in a somewhat restricted field could be repositioned to broaden its audience, sometimes to the confusion of readers. Now called simply *HG,* it is "not what it used to be," as Susan Heller Anderson reported in the *New York Times.* Fashion watchers called it "House & Garment," she disclosed, while a leading decorator thought the initials really meant "Hot Gossip." Other readers interested in interior design, advertising, architecture, and publishing did not know what to call it.

The transformation given it by Condé Nast was sweeping. For years a traditional "shelter magazine," as the genre is known in the trade, it featured pictures of rooms without people. Now the rooms were filled with fashionably attired models, and they (the rooms) were unrecognizable from the old days. There were also photos of society ladies doing their own decorating and of a playwright sitting on his unmade bed, petting his dog. An occasional drawing room in the old-fashioned style appeared, but it might be occupied by children on rollerskates. There were also, Anderson said, "hot architects, flashy painters, dress designers, film stars, the English, and the rich, old and nouveau."

While the editor in 1988, Anna Wintour, denied that *HG* was now a fashion magazine, four of the first five covers in its new life as *HG* showed women wearing designer dresses. The whole magazine, in fact, gave some readers, at least, the impression they were looking at *Vogue.* It looked expensive, was

graphically beautiful, and was generally luxurious. The object, said Bernard
H. Leser, president of Condé Nast, was to make *House & Garden* into a living
rather than a shelter magazine and, most important, one that would be distinctly different from its competitors, chiefly *Architectural Digest* and *House
Beautiful.* That was why readers now found articles on travel, electronics,
swimming pools, food, automobiles, and real estate resting comfortably side
by side.

This was an approach that attracted new advertisers, most of them in fashion, but it meant the loss of some former ones, notably those in docorating,
who went over to the competition. After the new *HG*'s appearance in March
1988, ten to twenty of them had switched to the *Digest* by June. Nevertheless,
HG remained healthy, with 500,000 subscribers, compared with the *Digest*'s
625,000 and *House Beautiful*'s 850,000.

For those willing to leave home, Condé Nast provided a new magazine in
the late eighties, entering the crowded but growing field of travel periodicals.
Called *Traveler*, it made its debut in September 1987, presented as something
unique in its field—and so it was. For the first time, a magazine in this genre
had the audacity to view the bleak side of travel. For example, it summarized
a United Nations report on air pollution, showing that it would be a good idea
for people with respiratory problems to avoid more than fifty cities, including
Rio de Janeiro. The immediate result was cancellation of advertising by the
South American airline Varig. In another article, the food writer Mimi Sher-
aton looked upon Alitalia's cuisine and service and found them "awful." The
airline pulled its advertising for a time but returned later. The publisher, Ronald A. Galotti, stood his ground. "We don't malign the industry, we just don't
do a whitewash," he told the *New York Times.* "For years travel magazines
have been filled with nothing but pretty sunsets and brochure-oriented copy.
That's why a lot of travelers don't read travel magazines."

This commotion served to underline the fact that restless, free-spending
Americans had made the publishing of travel magazines big business, with
major media companies entering the field. Fairchild Publications' *Travel
Today* appeared in January 1988. It was followed three weeks later by *Trips*,
a product of the Banana Republic stores, which were already in the business
of selling safari clothing, books, and other accessories. Earlier, within the previous two years, the Reader's Digest Company had acquired *Travel-Holiday*,
and Murdoch Magazines had bought *European Travel & Life.* CBS Magazines resurrected *Adventure Travel*, a promising new subdivision, while the
New York Times Magazine issued its twice-annual *Sophisticated Traveler.*

In the same period, the fever spread to book publishing, where the Atlantic
Monthly Press, Vintage, and Warner Books all started lines of travel books,
and cable television opened up a Travel Channel, sponsored by TWA.

Some media experts attributed this sudden outburst of titles to the overwhelming success of *Travel & Leisure*, American Express's publication drawing on its list of affluent cardholders. Others attributed it to the baby boomers
coming on the market full strength. Sounding a cautionary note, however,
Andrew Feinberg, writing in the *New York Times*, pointed out that travel

magazines command less than 20 percent of the $270 million in annual magazine advertising placed by travel-related companies. Fortunately, these journals have other sources of advertising income, and they appeal not only to travelers but to armchair dwellers who seldom leave home, whose interests are essentially hedonistic. The proof was in *Traveler's* list of advertisers, 68 percent of whom were offering such nontravel products as cars, liquor, and apparel.

Nothing in the travel field, however, was likely to top the leader at the end of the decade, *Travel & Leisure*, with its circulation of 1.1 million, fatter than ever after eighteen years of steady growth. In 1988, its 1,400 advertising pages broke a record and represented an increase of 15 percent from the previous year. It felt some pressure, however. *Traveler's* bid for total honesty led to canceling of free trips for *Travel & Leisure's* writers. But with so much attention being paid to travel magazines, it appeared that advertisers still might consider the leader a better buy.

James Berrien, the magazine's 36-year-old publisher, did not hesitate to keep his product in tune with changing times. He expanded the "Travel and Money" department, virtually unchanged for eight years, to accommodate the differences being felt by travelers because of a lower dollar, and the magazine also took note of the fact that Americans were taking shorter vacations by revamping its six regional editions to emphasize short trips.

22

Postwar Intellectual— and Other—Currents

While it might appear pretentious in some circles to speak of newsmagazines as contributors to intellectual currents, it is nevertheless true that millions of Americans have been getting their ideas about life at home and abroad from these journals in growing numbers since they began. They have been both politically and culturally important, shaping attitudes as well as providing information influenced by their owners and editors.

At the near end of the twentieth century, these newsmagazines were still rich and powerful, but they were less confident. As Alex Jones noted in the *New York Times*, they were not at all certain what they should be in the 1990s, and some of the writers and editors producing them believed survival might be at stake. For one thing, the competition was becoming more formidable—from cable-television news, national newspapers, and the increasing number of specialty magazines.

Ironically, these self-doubts had come at a moment when the circulations of all three—*Time, Newsweek, U.S. News & World Report*—were at a peak. But the question was how to retain that audience and if possible expand it during the century's final decade. When the newsmagazines began, the information explosion of this century had not occurred; there was the Depression and the Second World War to get through. But in the forty years since 1950, Americans have become saturated with information, available to them from thousands of sources twenty-four hours a day. The question for newsmagazines was how to stand out from this welter with information entertaining enough to attract attention and yet, at the same time, not abandon the original conception that had given them their identity. At a time when celebrity and money were the primary icons, could they keep on in their role as providers of serious news and analysis?

There was some indication that the original concept had already been softened. Somehow the hard, bright tone of the news surface had been glossed over with creeping entertainment. Some readers were discovering that, with so many sources available, they could be just as well informed if they did not read newsmagazines. In his analysis, Jones observed that *Time,* the industry's leader with 4.7 million, was still keeping its focus on the traditional review of the week's news. However, he saw a growing tendency to emphasize life-styles (once dealt with in the "back of the book") in articles about important news events.

For some time, *U.S. News* had deviated from both *Time* and *Newsweek* by devoting its back-of-the-book pages to service features instead of reviews of the arts, celebrity profiles, and essays. "News You Can Use," the magazine's promotion called it. Deliberately more serious, it was facing the problem in 1989 of whether to soften its formula in order to shore up its circulation of 2.2 million. It intended to be useful, telling readers how to get their children into college as well as analyzing the week's events, steering away from breaking news whenever possible.

Newsweek, the Avis of the newsmagazines business, always number two, had a circulation of 3.2 million by 1989. From the beginning, as we saw earlier, it was a magazine with an identity problem, trying to set itself apart from *Time.* The solution arrived at for the 1990s, or so it appeared in 1989, was to be unpredictable and to depart, at considerable risk, from tradition. Some of its reports were reasonably dispassionate (and they had always been more balanced than *Time*'s), while others were unabashed advocacies, with still others representing the personal viewpoints of their writers. Whether that would place the magazine's overall credibility in jeopardy was worrying some observers, Jones noted in the *New York Times,* and he provided an example. When Corazon Aquino began her campaign to oust President Ferdinand Marcos, *Newsweek* joined in the effort, in effect. Questioned about this development, Richard M. Smith, the editor, explained: "We don't shy away when we think it's an appropriate response to a news event like the Philippine revolution." Readers, at least, were given the impression that these stories about the revolution, and others, came from reporters and not from management. The mixture now at *Newsweek* ranged from hard news analysis to sheer gossip.

Under pressure from the competition, all three newsmagazines were, in fact, giving an increasing amount of space to "soft" news. The decline of hard-news space could be seen in the statistics. National affairs had been given 30 percent of the space in 1980, in both *Time* and *Newsweek,* with *U.S. News* providing a robust 50 percent or so. By 1988, this figure had dropped to about 25 percent in the two leaders, while in *U.S. News* it had tumbled to about 28 percent, although the magazine had nearly doubled its coverage of culture and the humanities. Conditioned by television and such pseudo-newsmagazines as *People,* many readers appeared to find the "soft" material more enjoyable than the "hard" and more likely to hold their loyalty to a magazine.

Of the three, only *Newsweek* appeared to be in any financial distress. Its profits dropped to $15.3 million in 1987, with the future uncertain. At the same time, *Time*'s advertising pages were dropping by more than 8 percent, but since it could charge more because of its circulation, its revenues were $239 million.

A disturbing sign at *Newsweek* was its history of changing editorships. From 1972 to 1984, it had had six editors. There had been the colorful Osborn Elliott, the Texan William Broyles, and the veteran Richard Smith, among others. Broyles had run successful regional magazines but had virtually no background in news coverage, a deficiency that brought him into sharp conflict with some staff members. Smith, who replaced him in 1984, had the proper background, first as writer and reporter specializing in international news and then as editor of the magazine's international editions. *Newsweek* had been his only employer. As editor, he took charge of a magazine company that had known seven presidents, perhaps a reflection of its owner, Katharine Graham, head of the Washington Post Company, who had been periodically impatient with *Newsweek*'s failure to overtake *Time*.

The history of these magazines illustrates how intellectual political currents can be influenced, at least to some extent, by journals that deal with the news and have large circulations. From the beginning, *Time* had been forthrightly Republican, with Henry Luce playing a power role in party politics, while *Newsweek* had been on the side of the Democrats under its original owner, Vincent Astor, whose silent partner, Averell Harriman was a highly visible part of Franklin Roosevelt's administration. As noted earlier, Astor, who had also been a strong New Deal supporter, became disillusioned, and *Newsweek* began to sound much more right wing, especially under the editorship of Raymond Moley, who had departed from the president's early "Brain Trust."

After Philip Graham, publisher of the Washington *Post*, bought *Newsweek* from the Astor estate for $15 million in the sixties, the political emphasis in the magazine shifted to favor the Democrats, since Graham was a friend of Lyndon Johnson and had access to the Kennedy White House. He gave the journal a fresh infusion of cash, expanding it editorially both at home and abroad. For the first time, too, writers and critics were given bylines.

At that point, *Newsweek* appeared to be drawing ahead of its orthodox Republican rival. Under Osborn Elliott's direction at the time, such veterans of the magazine as Kermit Lansner and Gordon Manning explored the problems of blacks, the counterculture, and other tumultuous events of the time. When the Vietnam War began to be a national concern, *Newsweek* was ahead of *Time* in describing the way it was tearing the country apart. As the media critic Edwin Diamond put it, "*Newsweek* was the 'hot book' for Madison Avenue in the '60's."

In the seventies, it was a different story at Time Inc. After Luce's death, financial control of *Time* shifted to Temple Industries, a building materials company in Texas, which had acquired 15 percent of its stock; Henry Luce III held 7 percent. *Time* then made the same kind of turnaround *Newsweek* had done under Elliott, bringing in Roy Cave from *Sports Illustrated* to work with the veteran Henry Anatole Grunwald, editor-in-chief. The momentum shifted, and it was the effort to regain it that caused *Newsweek* to run through five editors in ten years. However, the circulation of both magazines remained substantially the same. Meanwhile, *Time* was redesigned to give it a stronger, more elegant look; bylines appeared, new essay forms were tried, and a general modernizing process took place. The emphasis was still on hard news. As the eighties progressed, however, there remained an uncertainty about the future in both camps.

Time had long since lived down some unsavory moments in its past. When Whittaker Chambers became foreign news editor in August 1944, he had in his stable a formidable array of foreign correspondents: Walter Graebner in London, Charles Wertenbaker in Paris, John Hersey in Moscow, John Osborne in Rome, and Theodore White in Chungking. Chambers did what no other editor had done on any major magazine, particularly one devoted to the news. He took the copy these correspondents sent in and edited it to conform with his own often extreme right-wing ideology, completely rewriting whenever he thought it necessary. If there were any perceived gaps, Chambers filled them himself. Luce knew that Chambers had once been a Communist, but until the Alger Hiss trial, he did not know that his membership had included being part of a spy network. That discovery led to Chambers' resignation.

How much these events may have affected *Time*'s readers, it is difficult to say, but apparently they did no real damage to circulation or advertising. Both Luce and the public were easily distracted by the appearance of Time Inc.'s new magazine, *Sports Illustrated* in 1954. W. A. Swanberg wrote of this journal in his splendid biography, *Luce and His Empire*, "It was sneeringly called Muscle, Sweatsox or Jockstrap by TLF intellectuals. The search for a new journalistic enterprise had hovered for a time on the subject of fashion (which Clare had pushed enthusiastically and which Timen [Swanberg used a Time-like word here] had feared she might appropriate), then briefly on children, and had settled firmly on sports. Luce was creaky for tennis, taking up golf, had little knowledge of other sports, but felt obligated to develop a nodding acquaintance with some of them."

It was typical of Luce that in order to make himself knowledgeable about sports, he would, as Swanberg wrote, "sit through a pouring rain to watch a soccer match between Rome and Naples at Mussolini Stadium in Rome. As ideas developed from the new staff, he even found that it was possible to make sports fit the Cold War: Russian athletes, after all, were merely employees of the state. In his dogged pursuit of knowledge, Luce would attend sporting events at Madison Square Garden and Yankee Stadium, taking with him editors who could explain to him what was happening."

For some time, *Sports Illustrated* lost money, and in 1960, William Benton asked Luce why he stayed with such a losing proposition. As Swanberg tells it, Luce said, "I think the world is going to blow up in seven years. The public is entitled to a good time in seven years." The reason for this gloomy prediction was that seers from the Rand Corporation had convinced him nuclear war was imminent. The world did not blow up, of course, and only a year after Benton's query, *Sports Illustrated* began to make money and became a profitable property.

Fortunately, Luce did not live to see the sad demise of *Life* in 1972. Only five years before, he had seemed to be at the peak of his powers, no doubt deserving of what the German weekly *Der Spiegel*, had written of him in 1961: "No one man has, over the last two decades, more incisively shaped the image of America as seen by the rest of the world, and the Americans' image of the world, than *Time* and *Life* editor Henry Robinson Luce. . . . No American without a political office—with the possible exception of Henry Ford—has had greater influence on American society. . . . Recently, at a party on board Onassis' yacht *Christina*, Winston Churchill counted him among the seven most powerful men in the United States, and President Eisenhower, while still in office, called him, 'a great American.'"

However much those estimates might be exaggerated, no one could question that Luce had built one of the most powerful communications empires in the world, and his death on February 28, 1967, sent a shock wave through the organization. Some had gloomy forebodings, and the death of *Life* seemed to bear them out, but in retrospect, it can be seen that the magazine died because of circumstances over which no one had any real control.

Summing it up in his biography, *The Great American Magazine*, Loudon Wainwright wrote that *Life* had "outlived its own strength, was dead on its feet, and its fatal weakness was apparent in almost every issue. That weakness was clear in *Life*'s pitifully skimpy contents, in its obvious uncertainty about the temper and makeup of the audience it wanted to reach in an increasingly fragmented society, in the fact that the audience it did reach generally received it with a detachment far different from the impatient order with which subscribers in better days eagerly awaited the postman and buyers snatched it from the newsstands." It had already lost $40 million in the previous four years. The last issue was dated December 29, 1972.

The departure of *Life* was a distant prelude to some further disasters in the post-Luce empire. In 1983, a year in which there were a record number of business failures, Time Inc. suffered its first flat loss with the collapse of its weekly listings guide for cable television viewers, *TV-Cable Week*. It had been alive for only six months. While its arrival and departure made little impression on the public, this gross lack of judgment on the empire's part occasioned profound excitement in the media business and four years later even produced a book about the experience titled *The Fanciest Dive: What Happened When the Media Empire of Time/Life Leaped Without Looking Into the Age of High-Tech*. Its author, Christopher M. Byron, had been an eyewitness.

As Byron told the story, since Time Inc. was already the dominant corporate power in magazine publishing and cable television, the two most rapidly growing media industries, *TV-Cable Week* was considered a natural development, solidifying its position. Some thought it might even surpass *Time* in size and become the company's largest revenue producer. Instead, failing in less than six months, it became one of the costliest and most surprising failures in magazine history, as Byron wrote.

The company blamed unworkable marketing formulas, but Byron, who had been its senior editor, pointed accusing fingers at a project without direction, whose executives were constantly quarreling, whose computers did not always work, and whose staff kept getting more top heavy by the day. It was also an example of how individual pieces of an empire can be destroyed by too many conflicting executive egos and ambitions at the top. At bottom, of course, was its lack of purpose. As Byron recalled, it was a product "conceived not for readers, or even for any known market, but developed by two young Harvard MBAs, Jeff Dunn and Sarah Braun, neither of whom knew much about the cable television industry or its customers, but both of whom were eager to move ahead rapidly in their careers by making use of the analytical techniques taught at the Harvard Business School." These two entrepreneurs, said Byron, then became pawns in a corporate power struggle.

On a broader scale, the failure could be traced "to yet another expression of financially oriented corporate management—the cult of 'growth by numbers' and the rise of the business conglomerate, the final elaboration of professional management theory, a business enterprise in which the form of decision making is not the customer or the marketplace but Wall Street, the stock market, and earnings per share." This would, indeed, be a major theme of the eighties in magazine manipulation. In the seventies, Time Inc. had expanded into fields unrelated to what the corporation was originally intended to do. The empire now began to include forestry, heavy industry, cable, and moviemaking in Hollywood. *TV-Cable Week,* in Byron's estimate, was an attempt to unite "mammon and art," a mistake others would not make.

Nevertheless, the strength of a modern conglomerate is like nothing seen before. Mistakes can be absorbed, and still the empire expands. Time Inc. proved it in the eighties. In the international aspect of its operations alone, good health prevailed. It was circulating in 191 countries, with a total figure of 1,222,000. Its *Asiaweek,* the first pan-regional newsweekly, was acquired in 1985, bringing in another 70,000. A Japanese edition of *Money* had a circulation of 450,000, and *Fortune International,* with 105,000, had the highest circulation outside the United States of any American business publication. The French edition of *Fortune* appeared early in 1988, in cooperation with Hachette-Filipacchi, circulation 50,000; and *Yazhou Zhoukan,* the world's first independent newsweekly in Chinese, was launched in December 1987 with a circulation of 38,000.

New ideas and products continued to emerge from the Time Inc. factory. *Sports Illustrated,* celebrating its thirty-fifth anniversary in 1989, announced a new magazine, *Sports Illustrated for Kids,* a monthly designed, so it was said, to fight illiteracy while making money for the corporation. These young people, it was hoped, were the parent magazine's future readers who would increase its already fat circulation base of 3,000,000. Half of the new journal's initial 500,000 circulation would be sent free, with teachers' guides, to underfunded elementary schools. Research had already defined the market—13 million children from eight to thirteen, with a total of 16 million expected by 1993. For them, the magazine would be edited at a fifth-grade reading level.

The editors intended to fill a third of the new journal with crossword puzzles and other games, while the stories would be mostly interviews with children, parents, and educators.

The Time Inc. empire seemed to be expanding in every direction as the eighties came to a close. In the summer of 1988, it bought half the stock of *Hippocrates* magazine, a six-issues-a-year consumer journal covering personal health, begun a year earlier on only about $5 million of venture capital. Time Inc. paid $9 million for its half share. *Hippocrates,* in its first year, as noted earlier, had already won a National Magazine Award for general excellence.

In the summer of 1989, Time Inc. announced that in the following February it would introduce a new magazine, *Entertainment Weekly,* its most ambitious new venture since the *TV-Cable Week* disaster. It would be a magazine with two-thirds of its contents devoted to reviews of television, movies, video, music, books, and magazines, as well as reviews of children's entertainment, and designed to be read by parents. Features would occupy the remaining third, but would be related to general entertainment rather than to the fan-magazine kind of features in *People.* This venture, the product of two years' research, expected to have 500,000 readers, about equally divided between men and women, with an average age of thirty-seven. Most would have full-time jobs and an average yearly income of $50,000. Subscriptions would be fifty dollars a year. The new weekly would not be content to treat entertainment as news, it also intended to give out grades from A to F, as *People*'s already controversial television critic, Jeff Jarvis, had been doing. Jarvis had, in fact, developed the new magazine's concept, together with its publisher, Michael J. Klingensmith, and he would be its managing editor.

Time itself was renovated in 1988. There were two new departments, "Travel" and "Interview," a question-and-answer session with celebrities. Two other sections, "Critic's Choice" and "People," were expanded. A two-page index was added, and there was a more ambitious use of headlines and photographs. Noting these changes, the *New York Times*'s Geraldine Fabrikant wrote again that all three magazines were moving away from hard news. *Time* was emphasizing stories with a universal appeal, for mass audiences. As far as media directors were concerned, they were upscale magazines competing with others in that class.

There appeared to be no limit to how much money the Time Inc. mill could turn out. In 1989, a particularly profitable effort was made to cash in even more on *Sports Illustrated*'s highly popular and profitable annual swimsuit issue. A videotape was made of the models in their designer swimsuits, and the company-owned HBO cable service carried a videotape on how the issue was made, complete with moving models. These efforts, it was estimated, would bring in $50 million, counting the issue itself and its spin-offs. Appearing for the twenty-fifth time, the special was sold in addition to the regular weekly issue. The tapes based on it produced $600,000 more in revenue, at $20 each, and another spin-off, a calendar using the same models, was expected to sell 750,000 copies at a price of $10.95 each. The issue itself carried 118 pages of advertising, bringing in still another $18.1 million, up 47 percent from the previous year. On the newsstands, it was expected to sell about 2.5 million copies at a $3.95 price.

As for readership, the swimsuit issue would muster 41 million adults, the total including 13 million women. If feminists complained that it was sexist, Time Inc. could point to the fact that its editor was a woman, who surrounded the sexy pictures with 180 pages of editorial matter, including articles about swimsuits, letters from previously outraged readers (an annual crop could be expected), and profiles of former cover models. If it was sexist, the swimsuit issue was also hugely profitable—a lesson in exploitation.

Seeking to erase the memory of *TV-Cable Week,* Time Inc. launched another new magazine in 1989, *Cooking Light,* a magazine devoted to health and food. The new biomonthly was published by Southern Progress, Inc., acquired by Time Inc., as noted earlier. If it succeeded, as appeared likely, *Cooking Light* would end the string of failures beginning with the cable weekly and including *Picture Week,* abandoned after testing, and *Discover,* sold after years of losses. Even so, the company's magazine division had produced $1.7 billion in revenue for 1988, making it the largest magazine publisher in the nation.

Time Inc.'s acquisition of the Southern Progress Company, with its multiple book and magazine outlets, was another example of how the new climate of the eighties was changing the penetration of magazine markets. It had been acquired in part to launch other magazines, like *Cooking Light,* but primarly to reach out to more female readers, since Time Inc.'s other periodicals were more male-oriented. The food magazine could supplement Southern Progress's *Southern Accents,* a home decorating magazine directed to women. Also, Time Inc. saw that Southern Progress had a tremendous regional influence, in a South where the population was expected to reach 85 million in 1989 and 100 million by the end of the century.

In search of the regional market elsewhere, Time Inc. had already launched *New York-New Jersey-Connecticut Real Estate,* a controlled circulation journal, free to the reader, joining its similarly distributed *Home Office.* The audience for these publications was carefully researched and targeted in advance. Apart from this expenditure, start-up costs were low. The real-estate magazine, for example, was directed to top executives, whose average income was $130,000, with a median age of forty-three and who had recently bought homes and could be considered as thinking of such properties as investments.

In the new age, Time Inc.'s multilayered success in America and abroad had made it a prime takeover target. In 1989, its 49-year-old president, Nicholas J. Nicholas, Jr., called it "a superb collection of media franchises in key markets." The magazines were now a separately incorporated subsidiary, one of four comprising Time Inc., the other three being HBO, one of the foremost cable programmers; the American Television and Communiction Corporation, the number two cable-television system; and Time Inc. Books, the second-largest conglomerate in the book business. Together, these four units had produced revenues of $4.5 billion in 1988. It was an achievement brought about, at least in part, by four years of layoffs and cost cutting, which did not preclude the kind of expansion described above, bringing its magazine total to twenty-three periodicals, with a total circulation of more than 30 million worldwide and an operating profit of $287 million in the periodical division alone.

Reviewing Time Inc.'s extraordinary progress in the *New York Times*, Lester Bernstein noted that it was competing in an $11 billion consumer-magazine industry populated by other giants—Condé Nast, Hearst, and Meredith, among others—whose 1,700 publications included a top tier of 160 journals eating up 85 percent of total revenues. Of Time Inc.'s six core magazines— *Time, People* (the most profitable), *Money*, the resurrected *Life, Fortune*, and *Sports Illustrated*—only *Life* was not profitable in 1989, and it was near to being so.

Aside from the basic group and its other publications, Time Inc. also owned the forty-two properties of Whittle Communications in Knoxville, Tennessee, whose products were considered nonmagazines by some but, in any case, were astute marketing devices designed for special readerships, such as its six *Special Reports* for doctors' waiting rooms, each issue containing twenty-seven minutes of reading matter to match the average wait.

In his review of the Time Inc. empire, Bernstein spoke of "the coming to power of the videonauts, and the MBA's, the diminishing role of the magazines, the growing tyranny of the bottom line." At a critical point in 1989, J. Richard Munro, chairman and chief executive officer of the company, foresaw a future in which deregulation would make it possible (within five years) for cable companies and broadcasting networks to own each other and then for networks and production companies in television and movies to unite. Costs could be spread over many worldwide distribution systems. The end result, he foresaw, would be anywhere from six to eight vertically integrated media and entertainment megacompanies owning virtually everything in the entertainment and information field. One or two might be Japanese, two others European, and two American, of which Time Inc. would be one.

The company took a step in that direction in 1989 by agreeing to a merger with Warner Communications, another monolith of movie, magazine, book, radio, and television companies. That action touched off a titanic struggle as Paramount Communications, a giant of the same stripe, sought to defeat this deal by making a hostile takeover bid. Other suitors tried to join in, and the giants went to the courts. Munro may have been correct about the future, but what the ownership of Time Inc. would be had been momentarily cast into doubt until it beat off Paramount in the Delaware courts and became Time Warner.

Certainly the potential influence, intellectual and otherwise, of these behemoths was obvious and, for some observers, frightening. The future was far from clear.

From contemplation of these mountaintops on the magazine landscape, the comparative foothills at their base offer a more comforting contrast. It is easy to exaggerate alarming trends in bigness, to envision a world of information controlled by half a dozen or so megacompanies, quite capable of becoming giant propaganda mills, with too little diversity of opinion and too much opportunity for mass persuasion. While the potential is undeniable, in practice the political influence of newsmagazines (or other journals, for that matter), while hard to measure, seems minimal. Television is a far more potent weapon, influencing even when it makes no effort to do so.

It could reasonably be argued that magazines of opinion, and those that combine opinion with literary endeavors, are sources of influence, since they reach that intellectual audience loosely called opinion makers, just as the *Nation*, in its nineteenth-century days, had an influence far exceeding its meager circulation. Today, however, this genre lends itself more to reflecting intellectual currents, both social and political, that are more generally diffused than before. In this category fall the *New Yorker*, the *Atlantic*, and *Harper's*. While the distinctions are arbitrary, these three represent the seamless web holding past and present together. Advertisers and agencies think of them as "thoughtleader magazines," as Robert Gerry has called them in *Magazine Age*, and lump them in with an organization called the "Leadership Network," under whose umbrella are the *Columbia Journalism Review*, *Commentary*, *Foreign Affairs*, *National Review*, the *New York Review of Books*, *Technology Review*, the *New Republic*, and the *Wharton Magazine*. While all these periodicals have something in common, they are quite diverse in their approaches to intellectual matters— "the egghead books," as they are sometimes called. Their sociopolitical effects have been the subject of many scholarly studies, and much more work remains to be done, but for purposes of a general history, a sampling may provide an overall view.

From the beginning, opinion magazines have been in financial trouble, and to say that they are generally unprofitable is obvious. Special interest magazines in our time have given them further trouble, and mismanagement has contributed its share in some cases. Five of the Network periodicals—*Columbia Journalism Review*, *Commentary*, *Foreign Affairs*, *Technology Review*, and the *Wharton Magazine*—have a nonprofit status. On the other hand, the *New Republic* and the *National Review* are owned by rich men, and they reflect the strong ideological opinions of their editors and owners.

The Leadership Network itself selling collective advertising is a new development in this field, begun in 1976 by Robert Bennott, with a combined rate base of 600,000; advertisers were given a 20 percent discount if they placed advertisements in all eight journals. Much of the advertising is institutional, presenting corporate images of such companies as Citibank, W. R. Grace, and Puerto Rico Industrial Development, and is usually seen in only five of the magazines. Those advertising consumer products employ eight. All these advertisers believe, mistakenly or not, that they are reaching the leaders of national thought and opinion.

It may be, however, that today more of those leaders are readers of a single magazine, the *New Yorker*, with a circulation as large as the eight others combined. This journal's career has been, and continues to be, remarkable—one of the wonders of periodical history. We have seen earlier how Harold Ross,

its founder, began it as a smart, mildly satiric comment on upper-middle-class
life mixed in with good fiction and poetry, as well as pertinent articles and,
most identifiable of all, sophisticated cartoons by the best in the business. In
carrying out this formula, if it could be called one, the *New Yorker* developed
a persona like no other magazine. Staff members considered it more a strange
kind of club than a magazine. Life in its cell-like offices became the stuff of
legend, as did Ross himself. The gathering of so many highly talented individuals was also bound to produce rivalries, disagreements, and outright
hatreds, some of which were expressed in the periodic emergence of memoirs
by present and former staff members. It is safe to say that no other single
magazine has produced so many books about itself, nearly all by those who
work or have worked on it.

When Ross died, the *New Yorker* did not suffer the usual shock, fatal to so
many others when a strong hand at the helm was removed. Its unique character was simply proved in another way when William Shawn, a man unlike
Ross in most ways, not only carried on successfully but did so for decades
without materially changing the magazine except to accomodate new writers
and cartoonists. As the media critic Ben H. Bagdikian wrote of those years,
"The *New Yorker* is almost the last repository of the style and tone of Henry
David Thoreau and Matthew Arnold, its chaste old-fashioned columns breathing the quietude of Nineteenth Century essays." There were those who
thought this verdict excessive, remembering perhaps their earlier outrage
when Tom Wolfe, in the budding *New York*, had characterized Shawn as an
undertaker, described the *New Yorker* as a kind of literary mortuary, and
spoke of the "gray rivers of type" running endlessly between the chic columns
of advertising. Those columns, however, kept the magazine comfortably in the
black for years, while its circulation remained stable at somewhere around
500,000, as Ross had intended. He did not think that just anyone could, or
should, read his journal.

A high percentage of those who did read it were affluent people, as the
luxury character of the advertising testified. But medium-income people were
reading it too, if they were sophisticated enough to appreciate the cartoons,
stories, and articles which gave the magazine so distinct an identity. Year after
year, the *New Yorker* stood first or second in the number of its advertising
pages among consumer magazines. As Bagdikian wrote, "In 1966 the magazine attained the largest number of advertising pages sold in a year by any
magazine of general circulation in the history of publishing . . . 6,000 pages
of ads."

At that point, something mysterious occurred. For reasons never satisfac-
torily explained, the magazine's advertising began to decline sharply. In only
a few years, 2,500 pages simply disappeared—nearly a 50 percent loss. Profits
shrank with them, from $3 million net in 1966 to less than a million. Share-
holders who had been getting dividends of $10.93 per share in 1955 were
receiving $3.67 in 1970.

What was the cause of this sudden collapse? Bagdikian and other media
observers believed it was the issue of July 15, 1967, which carried Jonathan

Schell's "Reporter at Large" column, a report from a small Vietnamese village. Schell had arrived in Vietnam as, ostensibly, a standard establishment product—Harvard educated, editor of the *Crimson*, his father a prominent Manhattan lawyer. He was not even a correspondent at the moment, having come over from Taiwan, where he had been visiting his brother Orville, who was doing some Chinese studies there. But to the American commanders in Saigon who were devoting a good deal of time to convincing the American public that their losing war was going well, the new arrival, with his impeccable establishment credentials, seemed an ideal vehicle to spread the word. Consequently, this young visitor, barely out of Harvard, was treated as though he were a famous journalist come to inspect the action, although he carried only a press card from the *Crimson*. He was given special privileges and taken to view those areas of combat that seemed most likely to shore up the Great Deception.

But Schell was not the sitting duck the commanders thought him to be. He was a highly intelligent observer who saw through the smoke screen and was appalled by what he saw. When he returned to the United States, profoundly disturbed by the scene in Vietnam, he went to see Shawn, a family friend, who listened to his story and asked him to write it. Schell then produced what Shawn called "a perfect piece of *New Yorker* reporting." It was not an argumentative piece, simply an account of an assault on one small village and what it meant to the attacked and the attackers. It was the beginning of a firm, quiet compaign against the war, carried out mostly in the "Notes and Comment" department that began the magazine's text pages. Much of this was written by Schell, who came on the staff.

This anti-war compaign began initially at a moment when the *New York Times* and the Washington *Post* were still supporting the conflict editorially and before the mass protests that hastened its end. This made the *New Yorker*'s outspoken opposition all the more notable, particularly since its audience was presumed to be primarily upper-class people who appeared, in general, to support the war. That audience was now being supplemented by young people everywhere in the country, who were reading from it at anti-war rallies. Their presence brought the median age of the journal's readership down from forty-eight in 1966 to thirty-four in 1974.

It was a shift that brought unexpected results. Surveying this much more youthful market, many advertisers of luxury goods concluded that the *New Yorker* had acquired the wrong kind of readers, from their standpoint. It was this perception, rather than the mass withdrawal of advertising by conservative corporations (although that did occur to some extent) that accounted for the magazine's precipitate slide in advertising pages as the war in Vietnam ground down to its bitter end. "Bad demographics" was the phrase in the trade.

In similar circumstances, any other magazine would have fired the editor, or at least made a drastic change in editorial content, but that would have been unthinkable at the *New Yorker*. For some time (until the early seventies) Shawn did not even know he was addressing a different audience. "It would be unthinkable," he said later, "for the advertising and business people to tell me that. . . . Who the readers are I really don't want to know. I don't want to know because we edit the magazine for ourselves and hope there will be people like ourselves and people like our writers who will find it interesting and worthwhile."

In the time between 1967 and 1974, Shawn did hear, or was told, that his magazine was becoming too serious, intruding too much into politics, and running pieces that were too long, but, as he said, "My reaction was that we should do nothing about it." For a time, events seemed to prove him correct. The magazine recovered from its severe setback, and with Vietnam in the past, its circulation climbed to over 500,000. More important, it was running more than 4,000 pages of advertising a year, fourth among American magazines. Profits were more than $3 million. The advertising itself was still the admiration of the trade, and there were still critics who claimed that readers paid more attention to it than they did to the text, although there was little basis for such belief.

To those on the staff, it seemed that the *New Yorker* would never change, that it would simply go on unruffled, unchanging, as it had always done. But in 1985, Shawn was seventy-eight years old, and he had been editor since 1952. He was now operating in an entirely different climate, where periodicals were being created, merged, expanded, and manipulated by high-flying corporations.

In 1985, the unthinkable happened. The Newhouse company bought the *New Yorker* for $168 million, adding it to a growing list of media properties that included Condé Nast, newspaper and cable operations, and two prestigious book publishing houses, Alfred A. Knopf and Random House. S. I. Newhouse, Jr., then fifty-seven, took charge.

His first move was to install a 35-year-old publisher, Steven Florio, who set about completely rebuilding the business side, aggressively selling advertising on a new basis, and incorporating layouts not seen before in the magazine. Two years after the purchase, it was the editorial department's turn, and something even more unthinkable occurred. Shawn was fired, to be replaced by Robert Gottlieb, fifty-five, editor-in-chief at Knopf, regarded as one of the best editors in the book business.

The announcement of his appointment and of Shawn's departure set off what amounted to a revolt in the editorial department. There were meetings and speeches among the staff, and a formal letter signed by 154 people was sent to Gottlieb, asking him to refuse the job. Writers, editors, and artists combined forces in protest. To them, the apocalypse had arrived. But the revolution petered out in the face of inevitable change. Shawn agreed to retire as of March 1, 1987, saving considerable face. He lunched with Gottlieb at (where else?) the Algonquin, and the formalities were concluded. Oddly enough, as Knopf's editor, Gottlieb had worked with several of Shawn's most admired writers, including Jonathan Schell. He declared that he approached his new task without any preconceived ideas.

In his first two years on the job, Gottlieb made a serious effort to overcome, without any radical changes, the complaints heard for so many years about the magazine; principally, that the articles were too long, the subjects often dull, and the cartoons frequently unfunny—a matter of taste, in any case. The length of the articles did not change under Gottlieb (some appeared to be even longer), and there were still dull stretches, but there were also more absorbing pieces, mostly excerpts from forthcoming books, along with minimalist fiction, much of it from new writers, and a general tone of appealing to younger audiences.

The most visible change, however, occurred with the sixty-fourth anniversay issue in February 1989. It appeared with the first color art inside the magazine in decades and an entirely redesigned "Goings on About Town" in its familiar front section. There were eight color drawings by the veteran William Steig spread over four pages, titled "Scenes from the Thousand and One Nights." Color drawings had not been seen in the *New Yorker* since 1926, when a two-color cartoon by Rea Irvin ran, and its appearance in 1989 was the result of changes in the way the magazine was printed. In common with other journals, heavier stock was being used, making color printing possible on the black-and-white pages. Color advertising had been printed on heavier stock. Although color was now possible anywhere, Gottlieb said he would use it sparingly.

In the "Goings On" section, there were major changes. The "Art" and "Photography" listings were expanded to include criticism, while brief comments pointing out highlights preceded the sections on "Theatre," "Dance," "Night Life," and "Music," as well as "Art" and "Photography." An entirely new category also appeared: "Edge of Night Life," covering the late-night clubs. Illustrations in this section had been sparse in the entire life of the magazine, but now there were sketches and photographs of people in the listings, brightening these pages considerably. For events that did not fit into the standard categories, italicized boxes were provided. The day of Shawn was truly over. By that time, he had settled down in the editorial department of Farrar, Straus & Giroux, one of the most literary publishing houses.

Clearly, the *New Yorker* had been turned deftly to a somewhat different direction in the last years of the century. It remained to be seen whether it could survive into the next one with its audience relatively intact. The man responsible for the turning, and for everything else in the Condé Nast magazine empire, was S. I. Newhouse, Jr., who at sixty-one (in 1989) could be considered the most powerful man in the publishing business, controlling what *Fortune* called the greatest concentration of wealth in private hands. He shared responsibility only with his brother Donald, a year younger, who directed the company's newspaper and cable-television interests, while he concentrated on the magazines and the book group. Writing about him in the *New York Times*, Gigi Mahon noted that as the company had expanded, it had also become "a harsher place," in which editors were often fired arbitrarily and no one's job seemed entirely safe. Since Advance Publications, the umbrella covering all the family interests, is owned exclusively by the Newhouse brothers, there are no shareholders (and therefore not even the S.E.C.)

to interfere in the administration of a communcations empire worth between
$8 to $10 billion. It included at the close of 1989, twenty-nine newspapers; the
Random House book publishing group (Alfred A. Knopf; Little, Brown; and
Times Books); the Newhouse Broadcasting Company, with one of the largest
cable systems; and the magazines *New York, Details, Parade;* the Condé Nast
publications *Vogue, Mademoiselle, Glamour, Vanity Fair, HG, Bride's, Self,
GQ, Gourmet;* Street & Smith's Sports Group; *Traveler;* and *Woman.* Abroad,
Newhouse issues several foreign versions of *Vogue,* and in Britain owns the
Tatler, World of Interiors, and 40 percent of a company publishing the
upscale magazines *Face* and *Arena.*

Outside the Newhouse empire, meanwhile, all the older magazines were
enduring the difficulties of changing times in the closing years of the twentieth
century; for the oldest of them, the *Atlantic* and *Harper's,* the problems were
perhaps even more serious than the *New Yorker's,* since they did not have
large quantities of advertising to sustain them. Change came to the *Atlantic*
in 1980 after Boston real-estate executive Mortimer B. Zuckerman bought it.
A new editor-in-chief, William Whitworth, a *New Yorker* editor, was brought
in, and two years later a new publisher and president, David Auchincloss,
appeared, brought over from *Institutional Investor.* The audience for this
nineteenth-century survivor was also changing, its 70 percent increase in circulation falling mostly in the eighteen to forty-four age category. In 1982, the
circulation rate base was increased from 325,000 to 360,000.

The new owner was not one of those businessmen currently getting into
magazine publishing simply as a venture capitalist. Canadian born and raised
in Montreal, an American citizen since 1977, Zuckerman had degrees from
McGill, Harvard Law School, and the Wharton School. While he had made a
fortune from Boston Properties, his real-estate firm, he had also become an
associate professor of city and regional planning at Harvard. He was not, as
some supposed, a financial adventurer buying respectability or hoping to
inject life into a moribund property. In fact, the *Atlantic* under its former
editor Robert Manning, who had taken over the job in 1966, had been more
aggressive and topical than it had ever appeared to be in its long history and
was much easier to read. It was still an intelligent voice, and its reputation for
literary excellence had not appreciably diminished. It could claim to rank
with the *New Yorker* in the latter respect, with stories by John Barth, Bernard
Malamud, John Updike, Joyce Carol Oates, and John Gardner.

Because it had survived so successfully, the *Atlantic* was considered in serious danger when Zuckerman bought it. There were those who said he coveted
only its valuable property in downtown Boston, overlooking the Public Gardens. Suspicion increased when he fired Manning abruptly and brought in
Whitworth. While the literary quality and tone of the magazine were
retained, it now carried more politically oriented articles, and in spite of deni-
als, they often had a distinctly conservative flavor. Nevertheless, the *Atlantic*
appeared to be livelier, and in spite of having the same financial difficulties
all of the magazines in its category perpetually suffer from, it seemed in 1990

that the magazine would survive as long as "thought" magazines themselves survived.

The later history of *Harper's,* the *Atlantic's* long-standing competitor, was much more complicated. When its editor Frederick Lewis Allen put together the centennial issue in October 1950, he could look back on his own reign, beginning in 1941, with satisfaction, confident that he had maintained the journal's traditional literary quality. John Dos Passos and Fletcher Pratt had contributed excellent reporting during the war. Bernard De Voto was still sitting in the Editor's Easy Chair. The fiction had included stories by J. D. Salinger, Katherine Anne Porter, John Cheever, Kay Boyle, and Mary McCarthy. As a social historian himself, Allen had also been aware of the necessity to keep up with the news by providing interpretation and discussion of vital issues, while at the same time opening his pages to original thinkers.

John Fischer, the editor who succeeded him and conducted the magazine's affairs from 1953 to 1967, carried on in the same manner and, as one of the most talented editors of his time, introduced provocative supplements on several topics—"The College Scene," "Writing in America," "The American Female," and "The Crisis in American Medicine." Conscious that the trend in magazines was toward nonfiction, Fischer reduced the number of short stories to one per issue on the average and said he wanted to seek "the germinal idea, the unexpected talent, the fresh mind on the growing edge of American culture." To that end, he printed the work of Peter Drucker, Arthur M. Schlesinger, Jr., Anthony Eden, and Barbara Tuchman.

With his departure in 1967, the magazine drifted, for the first time in its long history. Willie Morris, the writer, was editor until 1971, without conspicuous success, and he was followed by Robert Schnayerson, from 1971 to 1976. As the youngest editor in the history of *Harper's,* Morris had made the magazine more topical, with David Halberstam and Larry King as contributing editors. Bill Moyers also wrote for the journal, and it carried the full text of Norman Mailer's "To the Steps of the Pentagon." *Harper's* was clearly becoming much more liberal, and Morris finally had to resign, along with seven staff members, after a dispute with management over policy.

Schnayerson, who followed him, brought back conservatism. He added a section on new books, another section called "Wraparound," which was a selection of short essays and excerpts from books, and a new department, "Tools for Living," describing items for sale ranging from kimonos to kites. But Schnayerson, like Morris, had the misfortune to edit *Harper's* at a time when the spectacular rise of special-interest magazines was causing a decline in circulation in many of the old standards, including his.

With Schnayerson's departure in 1976, Lewis Lapham assumed his position, having been managing editor since 1971. Former policies were continued in much the same way. Lapham's magazine considered such problems as "The Wreck of the Auto Industry," "The Cult of Mormonism," and "Real Money" (a discussion of inflation). The "Easy Chair" continued to rock along in a somewhat uneasy way. Tom Wolfe contributed a column, "In Our Times"—

and yet, in spite of a calm surface, *Harper's* was moving inexorably toward collapse.

The crisis came in the summer of 1980, when the owner at that time, the Minneapolis Star & Tribune Company, announced it would suspend the magazine with the August 1980 issue, citing the increased costs of paper and postage and the loss of $1.5 million annually since 1977 because of rising production costs, a decline in circulation, and consequent loss of advertising revenue. With the increasing number of special interest journals, the general interest orientation of *Harper's* seemed anachronistic. As Lucy Hackman quoted him in *Serials Review*, Lapham diagnosed his own predicament accurately: "Twenty years ago an issue of *Harper's* might have contained articles or essays on topics as miscellaneous as marine biology, toy railroads, failures of American foreign policy, ecology of Yellowstone National Park, and the unhappiness of women. Each of these topics now commands a magazine of its own and the inveterate reader of periodicals can make his own catalogue of interests (in effect becoming his own editor) by subscribing to *Polo*, the *Washington Review*, *Car and Driver*, *Orbis*, *Architectural Digest*, the *Texas Monthly*, the *Wilson Quarterly*, and the *American Beagle*. The special-interest magazines have been attracting readers at the expense of the major general-interest magazines, which include *Harper's* the *Atlantic Monthly*, and the *Saturday Review*." James Alcott, his publisher, added: "if you're Head Ski Company, you advertise in the top ski magazine. There has also been a drastic change in the way people spend their time."

When it was announced that the venerable magazine was on its death bed, there was a quick response. Thousands of readers protested, and there was a flood of offers by both individuals and organizations to buy it. A more depressing result was the immediate departure of several staff members. There were editorials in newspapers all over the country expressing both displeasure and disbelief that *Harper's* would die. Even William Buckley, owner of the right-wing *National Review*, declared that *Harper's* and other "serious journals" should be subsidized if they could not get enough advertising. It would be a public cost, like public education, he said, no doubt surprising many of his own readers.

On July 9, less than a month after the announcement of suspension, the cavalry came riding over the hill at the last moment in the form of the John D. and Catherine T. MacArthur Foundation and the Atlantic Richfield Foundation. MacArthur bought the magazine for a reported $250,000, a bargain-basement price. Atlantic did not join in the purchase but agreed to pay off liabilities. *Harper's* would now be a nonprofit organization, it was said. Consequently, any profits would not be taxed but would be distributed to charity. That would also make the magazine eligible for reduced postal rates, a critical element in keeping the journal alive.

Two years later, the nonprofit Harper's Magazine Foundation was set up with a $3 million grant from MacArthur and Richfield. The magazine's new publisher was Louise Daniels Kearney, then thirty-one, who had worked ten years for the magazine, rising through the business side of the company. The new editor, also thirty-one at the time, was Michael Kinsley. These two assumed control of a magazine that had already cut back its rate base from 325,000 to 140,000, by dropping subscriptions sold through agents, and whose advertising pages had declined by nearly 32 percent in the previous six months.

Harper's had been rescued by a method unique in magazine history. Although opinion magazines had always been subsidized, at least to some extent, they had never been acquired by rich foundations, and the rescue raised a number of questions. One was motivation. As always, there were those who suspected a sinister plot, but in fact, the takeover appeared to be a matter of family interests. J. Roderick MacArthur, the founder's son, who initiated the deal, had a special interest in journalism (his uncle was the noted writer Charles MacArthur and both of his sons were journalists).

In December 1983, Lewis Lapham returned to the editorial chair at *Harper's*, even though according to the *Washington Journalism Review*, he had been "generally assigned blame for precipitating the losses of revenue and readership" that led to its announced suspension. Rick MacArthur, J. Roderick's son and a young reporter on the Chicago *Sun-Times*, who was said to have persuaded the foundation to rescue *Harper's*, had been a friend of Lapham's. They had met at Columbia University, where Lapham was making a speech. But the foundation board was not anxious to return Lapham, the *Review* reported, having more or less compelled his resignation in May 1981 (after which he left to work for the Network News Syndicate and to write a book).

Kinsley, who had been a senior editor at the *New Republic*, was approved by the board to succeed Lapham, but he soon asserted his independence, living up to his reputation as an aggressive and abrasive man. It did not help that only a few months before, he had written an article for the *New Republic* that was critical of the MacArthur fellowships. In the summer of 1982, after he had joined the magazine's board, Rick MacArthur had a confrontation with Kinsley over his acceptance of a flight to Lebanon on an Israeli organization's plane, but the board refused to fire the new editor, although one of its members resigned as a result of its inaction.

It was undeniable that Kinsley had cut the magazine's losses from $2.4 million in 1981 to $308,000 in 1982, no mean accomplishment. He had even cut his own salary from $80,000 to $55,000. Meanwhile, the quality of the magazine had remained high, especially in the talented young writers introduced. One of them, James Wolcott, stirred controversy with a critique of Susan Sontag titled "Susie Creamcheese Makes Love and War." That did not add to Kinsley's popularity with the board, which subsequently refused him a raise and a one-year contract. That led to his resignation in May 1983 and Lapham's return two months later. By that time, Rick MacArthur had become president of the magazine foundation and appeared to be in charge.

Lapham told the *Washington Journalism Review* that *Harper's* must make the "general interest its special interest. . . . *Harper's* can become the nation's op-ed page, as well as an annotated history and monthly review of the trend of events and the tendencies of the public mind." That, of course, was what the *Atlantic* and the other opinion-literary magazines were trying to do. In the new order of the 1990s, they were likely to confront fresh problems.

Facing that prospect, nearly all these magazines were making some attempt to adapt in a market that sometimes seemed to be leaving them even farther behind. Small as they might be, they were not immune to acquisition. When Arthur Carter bought the *Nation* in 1985, he added it to his *Litchfield County Times*, a newspaper in Connecticut, and the New York *Observer*, a Manhattan weekly. In the opinion of media critic Edwin Diamond, it was all "a vanity press." If so, the vanity was costly. Carter's 80 percent control of the *Nation* in 1985 cost him $2 million in assumed debts, and beyond that, more millions to be invested in the magazine over the next few years. That would be more affluence than this journal had ever known. The infusion of capital, however, resulted in making the acquisition appear to be a better deal. In the next four years, circulation rose from about 35,000 to nearly 100,000, and Carter aimed for 150,000. If the *Nation* needed more injections of cash, he had it; as a financier, he was worth about $150 million.

In calling his new property "the world's most liberal opinion journal," Carter was making no idle boast. It had also been living proof that liberalism seldom pays in terms of financial reward. The *Nation* had always tottered on the verge of extinction, saved by what Republicans would call "limousine liberals." When George Kirstein became its publisher in 1955, it was in still another state of crisis: in debt, fund-raising efforts fallen off, expenses steadily rising, stigmatized by Senator Joseph McCarthy as a communist organ. Kirstein believed he had to ask the editor, Carey McWilliams, whether he was now or ever had been a Communist party member. McWilliams protested the propriety of the question but gave the new publisher a flat no.

After studying the financial statements, Kirstein could not immediately make up his mind whether the *Nation* could or should be saved yet again. He talked it over with many people, including his brother Lincoln and his sister, Mina Curtiss, both of them longtime contributors to the *Nation*'s pages. Everyone agreed it would be a tragedy if the magazine died, but they were uncertain whether Kirstein had the qualifications to save it. But he had a plan. It was futile, he thought, to attempt existing on small contributions from a long list of supporters. It would be better to find a few people who would join him in providing substantial sums. Within a year, he had found eight people who would do so. They preferred anonymity and never held a formal meeting, nor did they make any effort to influence policy. These committed supporters included faculty members from Harvard and Columbia, a dairy farmer, a lawyer, a recent Princeton graduate, an author-scholar, two housewives with a social conscience, and a corporate executive.

In the following decade, after strenuous efforts by Kirstein and McWilliams, the magazine's deficit still existed but it was more manageable, and the quality of the journal had improved. McWilliams brought in new contributors: Fred J. Cook, Dan Wakefield, Harold Clurman, Alexander Werth. He had been editor since the spring of 1951, a Californian who had written a biography of

Ambrose Bierce and was well known as a liberal writer. Brought into the magazine by Freda Kirchwey, who had been associated with it for thirty-seven years and had been its editor and publisher from 1945 to 1955, he himself took over the magazine in that year when Kirchwey told him she would be willing to relinquish it if he could raise enough money to liquidate the indebtedness and ensure continuity. It was this endeavor that led McWilliams to Kirstein and initiated their long association.

Twenty years later, under Carter's ownership, the *Nation* appeared to prove McWilliams correct. It continued to be the voice of liberalism, an increasingly unpopular doctrine in a time of Republican presidents and right-wing supremacy. Its investigative reporting displeased the same people who had always hated it, and its politics continued to please those who had always loved it, particularly since its longtime rival, the *New Republic*, once so much like it that there had been rumors of a merger, had become a right-wing journal under new ownership.

Overlooked in this small corner of the magazine business were the *Progressive*, eighty years old in 1990, and the *New Leader*, sixty-five in 1988. The latter made a virtue of its obscurity, advertising that only one in 8,000 people in America had ever read one of its issues but asserting as its slogan, "You've been influenced by it." There was something of the early *Nation* in this claim, when E. L. Godkin had exerted an influence far beyond its 35,000 or so circulation. At 20,000, the *New Leader* was even less substantial, but it had published more than its share of historic material. The magazine had first presented Solzhenitsyn in this country, as well as Joseph Brodsky, the Nobel Prize winner in 1987; some of his first poems had appeared in its pages. When Brodsky had been tried by the Soviets in 1964 on charges of "parasitism," the magazine had published an exclusive, a transcript of the trial smuggled out of the Soviet Union. In 1956, it had also printed Milovan Djilas's "The Storm in Eastern Europe," following the Hungarian uprising. A few days later, Djilas found himself in jail, but the article was a forecast of his book, *The New Class*.

As managing editor of the *New Leader* in the early 1940s, Daniel Bell, then a Harvard professor, was in a position to assess the magazine's role. Until the late sixties, he said, it had been in the front lines of the fight against Stalinism, more so than any other magazine, with the possible exception of the *American Mercury* under Eugene Lyons's editorship. But there was a dichotomy. Taking what was then considered a politically right-wing view about most public events, it was at the same time committed to labor unionism. One of its special strengths came from the first generation of Russian émigré writers, who filled its pages and gave the West its first glimpse of the Gulag.

The *New Leader* began life on January 19, 1924, as an unofficial newspaper of the American Socialist party, with both Eugene Debs and Norman Thomas among its founders. It supported the labor movement that was just gathering steam for the struggles of the thirties and in a few years had become the voice of those liberals who were opposed to communism. It thrived in an angry climate of controversy among Stalinists, Trotskyites, Communists, and Social

Democrats, all of whom were either writing for or helping produce opinion journals, of which *Partisan Review* was the best known.

In those years, the *New Leader* was edited most of the time by Samuel M. Levitas, another Menshevik Russian émigré. He made the magazine a forum for Soviet dissidents, including Victor Nekrasov and Andre Sinyavsky, whose pen name was Abram Tertz. They went to jail for their writings. Other contributors included George Orwell, Willy Brandt, Vladimir Nabokov, Bertrand Russell, Arthur Koestler, and Sidney Hook. John P. Roche, the conservative Tufts professor, was an early contributor and among those who inhabited the offices on 15th Street in New York, where German, Polish, Yiddish, and Russian filled the air.

During the fifties, the *New Leader* lost much of its old flavor. The émigré culture was being blurred by time and change, and some of the original contributors had died. Myron Kolatch, who had been editor since 1962, when the magazine went biweekly, continued to direct the journal on its established intellectual lines, believing that it was still presenting an even-handed view of differing opinions. "The real purpose of many magazines of opinion," he told Richard Bernstein, of the *New York Times*, in 1988, "is to breast-feed the reader, enable him to find the views he already holds being expressed by others. We are the only genuine open forum."

In another corner of the literary-opinion market, the "little magazine," whose early history has been recounted, took on new life in the eighties. From their inception in the days before the Great War and through the twenties, they had progressed in the fifties with such outstanding magazines as *New World Writing*. Then, between 1967 and 1978, it was the *New American Review* that held center stage. In its pages, readers found a section of Philip Roth's *Portnoy's Complaint*, titled "The Jewish Blues." Parts of E. L. Doctorow's *Ragtime* and Kate Millett's *Sexual Politics* also appeared. The review could accommodate such diverse styles as those of John Barth, Grace Paley, and Stanley Moss, along with the work of obscure writers who continued to remain obscure. At its peak, this pocket-size paperback was selling more than 100,000 copies from drugstore and airport bookshop racks. Ted Solataroff, founder and editor of the magazine, told Caryn James in the *New York Times* that when he had begun the *Review* in the sixties, he had taken advantage of the social and political ferment sweeping the country, but a decade later, the American scene had changed again, sales of the *Review* were off, and Solataroff decided to suspend his magazine.

The literary journals today bear a closer resemblance to paperback books and are sold in bookstores and other outlets more as books than as magazines. In fact, as *Publishers Weekly* reported in March 1989, book publishers were invading the world of literary journals as they never had before. Vintage Books, a division of Random House, issues the *Quarterly;* British Penguin, owner of the American Viking Press, sponsors and sells the independent *Granta;* while Collier does the same for *Conjunctions.* Fireside, a Simon and Schuster imprint, has issued "the best of" collections from such literary publications as *Grand Street, Between C & D,* and *Ploughshares,* and other small houses have followed suit. In this category, City Lights Books of San Francisco easily won the title of the year in 1989: *The Exquisite Corpse Reader: The Stiffest of the Corpses, 1983–1988.* Publishers of both books and magazines see the new literary periodicals as an alternative to the best-seller production of the major publishing houses. *Granta,* the *Quarterly,* and *Conjunctions* are shipped to stores as books from book distributors, but the *Paris Review* and hundreds of others come from magazine distributors. Three national companies distribute literary magazines: Eastern News, Ingram, and Bernhard DeBoer.

While opinions differ, many critics believe that *Granta,* the quarterly review edited from London by William Buford, is the leader. Its direct-mail solicitation asks: "Why, in England, is *Granta* read by more people than any other literary magazine in the history of the Twentieth century? Answer: Because its editors don't like literature." However, the editors have somehow managed to print Graham Greene, John Updike, Nadine Gordimer, Doris Lessing, Saul Bellow, and Gabriel Maria Marquez, along with outstanding new works by such writers as Jay McInerny, William Boyd, Don DeLillo, Hanif Kureishi, and James Fenton. Its articles, particularly the news genre that might be called "political-travel-adventure," are of equally high quality.

Buford, its editor, is an expatriate American who was drawn into the burgeoning *Granta* in 1979 while he was a Shakespeare scholar at Cambridge. Asked by a friend to help edit one issue, he quickly became editor and changed the obscure student publication into an internationally respected magazine. Circulation climbed from 800 to more than 100,000, with about 70 percent of it in the United States. In 1989, it hoped to begin publishing five times a year and promised a Granta Books imprint in conjunction with Penguin, planning to issue about a dozen books a year. Penguin's marketing and distributing methods, as well as Buford's editorial talents, are regarded as the reason for *Granta*'s success. In describing his magazine, Buford was also describing at least the intent of other literary periodicals. *Granta,* he said, was "sort of a literary magazine for people who grew up on television. In a mass-market culture, we offer writing of intelligence and complexity, but it is also writing that deals with contemporary concerns and issues."

That could also be said of its nearest rivals. The *Paris Review,* which was present at the creation of the genre, carries on its much quoted interviews with writers and continues to reproduce the work of contemporary artists on its cover and frontispieces. George Plimpton and a small, young staff edit it from his townhouse. While it is no longer the magazine of discovery it once was, Tom Stoppard and Edmund White appeared in its pages, along with other literary figures, in the eighties.

Somewhat more radical than the others, *Conjunctions* has been banned in some southern bookstores because its cover illustrations have included Eric Field's watercolor of two nudes on a beach. It was not the nudity that offended the booksellers; the man in the picture was black and the woman white. The magazine's content barely manages to remain in the mainstream, viewing realism in revisionist terms. Its offbeat writers have included Robert Coover,

John Hawkes, and Walter Abish. Bradford Morrow, a novelist, is founder and editor of *Conjunctions,* which is published both as a paperback and as a hardcover book twice a year.

Triquarterly, originating in Northwestern University and issued three times a year, has been in the field for more than twenty years and is noted for its receptivity of new writing and minimal number of well-known authors. Since it dares more, it has tended to be uneven, but nowhere else are literary readers likely to find translations of Polish, Chinese, and Korean poetry.

The *Quarterly,* published four times a year, is edited by its founder, Gordon Lish, a Knopf editor, and is much more offbeat. All of its stories are extremely short, and critics have complained that they are too much alike. The poems are equally short, surrounded by acres of white space. Random House, Knopf's parent, is the publisher.

23

Alternative Magazines

In the sixties, "alternative magazines," as they were first called, were those considered out of the mainstream and not quite respectable because they were concerned mostly with unpopular ideas, ranging from right-wing dissent to far-left radicalism. Since then, however, they have greatly broadened their range, and the term is now considered to mean simply "out of the mainstream" and not necessarily political or ideological.

If we accept this spectrum (as it is historically convenient to do here), we have a category that extends from established conservative magazines like the *National Review* to the so-called "punkzines," devoted to trashing everything. There are more than a thousand independent, small-circulation magazines, journals, and newsletters extant with only one thing in common—their devotion to a minority political, social, or cultural interest.

Those who want to know what the radical press is saying without reading it can buy the *Reader's Digest* of this category, a magazine called the *Utne Reader*, which digests what it believes is the best in the field. ("Utne," rhyming with "chutney," is a Norwegian word meaning "far out," says its publisher, who is named Utne.) Published every other month, the *Reader* covers the feminist press, regional publications, the "punkzines," the journals of the American Buddhist movement, baseball fanzines, rock music magazines, peace and disarmament advocates, those devoted to mind research, the new right, the radical right, animal rights—and more.

A single issue of the *Reader* encompasses an astonishing range of interests and public personalities. Here, for example, is Jerry Brown, the former governor of California, interviewing Daniel Ortega. On another page, two writers, Ken Kesey and Alice Walker, meditate on death. There are articles on liberation theology and alternatives to television. Tom Wolfe writes about the worship of art. There is advice on whether or not to have children, essays on marathoning and herpes (not in conjunction), Jan Morris on Boy George, Ray Bradbury on managing from within, and observations on the frontier mentality and on what the next step in feminism may be.

The titles of the magazines excerpted are unfamiliar to most Americans, but all have their devoted readership, and again their titles suggest a broad range of interests: *American Demographics, American Spectator, Audubon, Boston Observer, Christianity and Crisis*, the *Churchman, Country Journal, Cultural Survival Quarterly, Dissent, Earth First!, East-West Journal, Food Monitor*, the *Guardian, Jewish Currents, Journal of Humanistic Psychology*, the *L.A. Weekly, New Age Journal, Nutrition Action, Skeptical Inquirer, Vegetarian Times*, the *Whole Earth Review*, and *Yellow Silk*. But the *Utne Reader* is not simply a digest of far-out publications to justify its title. It also includes opinion from *Harper's*, the *Nation*, the *New Republic*, the *New York Review of Books*, and the *Smithsonian*, as well as other relatively establishment publications. The newspaper *USA Today*, surveying this smorgasbord, called it, "A magazine junkie's heaven. It slogs through the newsstand muck to save you the trouble, and picks out the delectable and provocative for your attention." Even the *Wall Street Journal* was prepared to admit that it "sifts the good from the goofy."

The *Reader*'s habitat is in keeping with its mission. It is published from a hole-in-the-wall office in Minneapolis. A three-person staff includes the executive editor, Eric Utne; Jay Walljasper, an editor; and a female "general factotum." Friends and relatives help when it is necessary, including Utne's wife and, as the magazine's promotional material describes them, "a revisionist history teacher, a socially responsible philanthropist, a neighborhood organizer, an Amish quilt merchandiser, and a retired hog farmer." No other editorial staff could boast such varied talents.

At the upper end of the alternative magazine spectrum is the *National Review*, the nation's leading conservative journal and the voice of its owner, William Buckley, whose money and ideas have made it viable. Willi Schlamm, a German journalist, is credited by the *Review* itself with originating the magazine; later he became the editor, publisher, and chief correspondent of another controversial journal in West Germany, *Die Zeitbuhne;* Schlamm died in 1980.

From the beginning, the *Review* has been the voice of the conservative right wing, staffed by a succession of people with solid credentials in promoting their causes. Its first managing editor was Suzanne La Follette, whose name evokes liberalism but whose earlier career encompassed both Lawrence E. Spivak's *American Mercury* and the old *Freeman*. In mid-1957, Buckley took on William A. Rusher, who had been assistant counsel to the Senate Internal Security Subcommittee. Rusher eventually became publisher. In the late fifties, James McFadden took over as head of the circulation department, entrusted with the difficult task of attracting from 25,000 to 30,000 new subscriptions every year (or 25 percent of its total circulation) to stay even. An annual renewal rate of 70 to 75 percent testified to his abilities. For a long time, Buckley refused to permit use of his own name in soliciting subscriptions but changed his mind in the late seventies.

James O'Brien came from the art department of Hearst's late New York *Mirror* in 1962 and in time transformed the appearance of the *Review*, bringing it out of dull drabness into a contemporary design that won him several awards. A stable of cartoonists acquired over the years was led by the veteran C. D. Batchelor, whose right-wing jabs at everything remotely liberal had decorated the pages of the Chicago *Tribune* for years.

There were major staff changes in 1976. George Will, soon to become a prominent New Right columnist and television personality, became book editor and head of the Washington office. John Coyne left the staff to write speeches for Spiro Agnew, Richard Nixon, and Gerald Ford, but returned to be the Capitol correspondent and conductor of the magazine's "Cato" column. In 1977, conservative writer James Burnham, now seventy-two, went into semiretirement but continued to contribute another column, "The Protracted Conflict," and occasional editorials. He had been with the magazine since its first issue. Jeffrey Hart, a syndicated conservative columnist, took his place as editor of the editorial section.

Rusher retired as publisher at the end of 1988, replaced by Wick Allison, a 40-year-old Texan who, improbably for the *Review*, had been publisher, editor, and minority owner of *Art and Antiques* magazine. With an MBA from Southern Methodist University, he had started the Southwest Media Corporation, whose backers included H. L. Hunt's son, Ray. This corporation published four magazines.

The *Review*'s new editor in 1988 was John O'Sullivan, formerly the editor of *Policy Review*, published by the right-wing Heritage Foundation (Buckley retained his title as editor-in-chief.) O'Sullivan, an Englishman, had proper conservative credentials, having worked for the Tory government and the *London Daily Telegraph*, besides serving as editor of the editorial page of Rupert Murdoch's *Times of London* and his New York *Post*.

With such talent, and backed by the Buckley fortune, it was not surprising that the *Review* in 1990 was by far the most successful of the opinion magazines among the alternative journals. Its gross advertising revenues in 1987 were more than $1 million; participation in the Leadership Network had been a major help. But the *Review* was still a comparatively small magazine in 1990, with only 118,000 circulation. However, that represented a gain of 30 percent over the previous three years. This dramatic rise was the result of a subscription solicitation television commercial employing the talents of Charlton Heston, Tom Selleck, and Ronald Reagan.

The *National Review* was not seriously challenged by any other right-wing publication until September 1985, when the Washington *Times*, a newspaper owned by the Reverend Sun Myung Moon (whose church denies any responsibility for policy) began a news magazine as right wing as the paper itself called *Insight*. When it appeared, the *Washington Journalism Review* observed; "The marching orders, apparently, are to turn out a conservative national newsweekly that is an alternative voice to *Time* and *Newsweek*— much as the Washington *Times* is billed as an alternative voice to the Washington *Post*." Robert Dillingham, former advertising director of *U.S. News & World Report*, was publisher. Key members of Moon's Unification Church were prepared to underwrite substantial losses for at least four years, but in that time *Insight* was able to move from its original part-controlled, part-paid circulation basis to being an all-subscription magazine with nearly a million claimed subscribers. By that time, too, it appeared to be established as what Dillingham had said it would be, a spearhead of the anticommunist crusade. "They have a tremendous fear," the *Review* reported in 1986, "they see a softening of the media toward communism." There were crossovers from newspaper to magazine; some writers wrote for both, and *Insight* borrowed some of the *Times*'s more successful features, including its attractive graphic design. The stories themselves appeared to rely heavily on unidentified sources, possibly indicating more ideological speculation than fact. While *Insight* might never have the popular appeal of *Time* and *Newsweek,* in 1990 it appeared to be potentially profitable and to have carved out its own corner of the conservative market.

Other right-wing magazines (and a host of newsletters as well) tend to be cut from much the same cloth, covering similar territory. The Liberty Tree, a distributor for most of them, presumably permits them to describe themselves for a direct-mail catalog covering the field. A sampling indicates the character of these journals.

The *American Spectator,* an equivalent of the British journal, carries on the same kind of right-wing commentary on political, economic, and social issues. It was begun in 1967 as an underground magazine to counter the New Left.

Consumer's Research (not to be confused with the liberal *Consumer's Report*) is "a consumer magazine with a difference, showing the hazards of government regulation, including the dangers of overinsulation, unsafe cars by federal mandate, food and energy shortages by design, the bilking of small savers, the airbag boondoggle, how government causes airline congestion, and much more."

The *Free Market* is "the lively and accessible monthly of the Ludwig von Mises Institute," a right-wing think tank set up in the name of the ultra-conservative Austrian economist, carrying on (with the aid of a covey of right-wing writers on economics) von Mises's personal crusade for free markets and the total separation of the economy from the state.

The *Freeman,* published monthly by the Foundation for Economic Education, in Irvington, New York, a nonprofit educational institution, is intended to promote "improved understanding of liberty in relation to private property rights, the free market, and limited government."

Human Events, a news magazine with a heavily right-wing viewpoint, is, it says, "objective; it aims for accurate presentation of all the facts. But it is not impartial. It looks at events through eyes that favor limited constitutional government, local self-government, private enterprise and individual freedom. These were the principles that inspired the Founding Fathers."

The *Independent Report,* published by the Independent Institute, "traces social and economic problems to government's dominant role in society, and fosters new and effective alternatives based on individual choice and free market processes." The Institute also issues books, tapes, and "media programs."

Liberty "offers analyses of culture, politics and the arts, plus fiction and humor, written mostly from a libertarian perspective." *Liberty* was among the first to present the politics of Robert Bork, before his Supreme Court nomination failed, and it also carried such enticing reports as how to live tax-free in the Bahamas.

Private Solutions, a monthly publication of the National Center for Privatization, "provides coverage and commentary on government inefficiencies and current federal, state and local efforts to reduce spending and taxes by privatizing government programs and services."

The *New American* "is written from a conservative anti-totalitarian perspective and rejects the accidental view of history. Our Founding Fathers devised the best form of government—a constitutional Republic. Let's keep in that way by limiting taxes and the power of politicians."

Nomos features "studies in spontaneous order." It is a bimonthly "for and about those interested in freedom. Governments have made drastic and violent inroads into the individual's personal and economic freedoms. Articles explore these areas of intrusion and emphasize the individual's capacity to live fully."

Policy Review, the Heritage Foundation's quarterly, is "widely recognized as the intellectual cutting edge of the conservative coalition. . . . An indispensable tool for understanding conservative thought, trends, and people."

Reason "reports on past, present, and future free market alternatives to government's oppressive and idiotic bureaucracy. . . . The best investigative journalism, humor, and popular culture."

The *Southern Partisan* is the "unreconstructed voice of the South," in the tradition of John C. Calhoun and Robert E. Lee, covering history, politics, and literature.

The Teaching Home, as its subtitle says, "A Christian Magazine for Home Educators," offers "essential news, information, and encouragement to families considering or already educating their children at home."

Conservative Digest is "feisty and fearless. For the past 14 years it has served as the conscience of the conservative movement."

Modern Age celebrated its thirtieth anniversary in 1989 as "the primary principal quarterly of intellectual conservatism," with articles by such authors as Henry Regnery and Russell Kirk.

Plain Truth offers "exclusive insight into world trends. Fifteen million people read it for answers and solutions. Through Bible prophecy, it discusses what's next and reveals the prospects for a better world tomorrow."

The *Public Interest,* founded in 1965, is "the nation's most influential forum for discussion of public policy on issues that affect our lives and livelihood." Certainly it has the largest galaxy of right-wing stars: George Gilder, Nathan Glazer, Irving Kristol, and Michael Novak, among others.

With such self-advertisements, no further analysis of the conservative magazine press would seem to be necessary.

On the left, the situation is quite different. While the right-wing magazines have grown in number, strength, and power since the Second World War, well financed as many of them are by rich conservatives, the Left clings to the precarious perch it has always occupied, without well-heeled "public interest" institutions to support it. The Left's journals are obviously more colorful, but they are perpetually on the verge of collapse, and they have had to fight off repeated attempts by the federal government to sabotage them. That situation was first disclosed in 1981 by Angus Mackenzie, writing in the *Columbia Journalism Review*. Using documents obtained under the Freedom of Information Act, Mackenzie unearthed a pattern of activity by governmental agencies (primarily the CIA, the FBI, and the army) against the underground press in the late sixties and early seventies, clearly violating the First Amendment.

Mackenzie estimated that about 150 of the approximately 500 underground publications opposing American involvement in Vietnam were affected by this crusade. Government agencies infiltrated those publications and reported on them, paradoxically arousing the suspicion that they were CIA fronts. Staff members and contributors were harassed by government-ordered audits, and advertisers were pressured not to use the journals. In some cases, the government resorted to violence against property and persons, Mackenzie reported.

In this unprecedented attack on First Amendment rights, the CIA became the first federal agency to operate against underground publications in the United States. When *Ramparts* magazine exposed the CIA's involvement, the IRS moved in to audit the backers of the periodical. The CIA was assisted from time to time by the army's Counterintelligence Analysis Branch. As Mackenzie reported, it "kept track of underground periodicals and maintained a microfilm crossfile of writers and editors affiliated with them." FBI and army infiltrators supplied much of its information. With all this surveillance, no evidence of backing by foreign powers was ever found.

The CIA's campaign was well named Operation Chaos, but the title of its program to close down dissident journals was even more apropos—Project Resistance. Most of its efforts were directed at persuading advertisers not to use the journals, and they succeeded in shutting down some of them, including the Washington, D.C., *Free Times* and a six-paper chain in Wisconsin. It was the FBI, according to Mackenzie, that appeared to be in charge of harassment, accompanied by occasional violence. "In San Diego," he reported, "the paramilitary Secret Army Organization, led by FBI informant Howard Godfrey, assaulted the offices and staff of the *Street Journal* on December 25, 1969. . . . FBI documents released under the FOIA show for the first time that the Secret Army Organization's operations extended as far east as Wisconsin, where the organization threatened to kidnap Mike Fellner, editor of the radical Madison paper *Takeover*."

After the miasma of the war in Vietnam had lifted, this kind of government guerilla activity appeared to diminish, if not disappear, and the left-wing magazines returned to more familiar ground—trying to stay alive financially. *Ramparts* was a case in point. Founded by Edward Keating in 1962 to advance reform in the Catholic Church, its editor was a former reporter and public relations man, Warren Hinckle II. Keating was a rich real-estate operator who had converted to Catholicism and had his own agenda, but Hinckle greatly expanded the magazine's coverage, spent Keating's money freely, and finally attracted enough other backers to force the founder out. Analyzing the magazine's history in the *Washington Monthly,* Adam Hochschild, a former staff member, summarized Hinckle's abilities: "To be a good editor you need not only a skillful pen but also the tact of a diplomat and the patience of a wetnurse. Hinckle had the pen but not the rest." He and his managing editor, Robert Scheer, pursued their vision for the journal with boundless energy and a great deal of money, but their management operations were disastrous. Manuscripts were rewritten without asking permission from the authors, payment to contributors was slow, and the actual production of the magazine was continuing chaos. The result was so much staff opposition that both Hinckle and Scheer had to resign in the late sixties.

In the early seventies, collective ownership by the staff was established, but that failed to prevent the circulation from falling to about 70,000 by 1974. Nevertheless, *Ramparts* was printing lively and provocative articles, including, in Hochschild's words, "an extraordinary account of the electronic spying by the National Security Administration, penetrating radical critiques of American foundations and the politics of population control, and some long overdue self-criticism of New Left strategies." *Ramparts* survived for a time because it was willing to publish articles and exposés not to be found in other journals or newspapers. Perhaps the most memorable of these was its disclosure that the CIA had supported several organizations the public had thought were private, including the National Student Association. Such stories boosted the circulation to more than 200,000, but the magazine was losing from $300,000 to $500,000 a year, with an eventual debt totaling more than $2 million.

Hochschild believed that *Rampart's* muckraking exposés opened the way for the wave of investigative reporting that began in the seventies, but if so, it was not enough to save the magazine's life, and it died in March 1976. Its subscription list was turned over to a New York magazine called *Seven Days.* Reviewing its career in *Quill,* the organ of the Society of Professional Journalists, Tom Bourne wrote that Hinckle had led "perhaps the most important muckraking revival since *McClure's.*" After he left the magazine, Hinckle became a free-lance writer and columnist for the San Francisco *Chronicle,* while Scheer went farther down the Coast to be a staff writer for the Los Angeles *Times,* where he produced some unorthodox page-one interviews with major American politicians.

Something of the flavor of *Ramparts* can be gleaned by turning the file pages to the magazine's more memorable moments, such as the December 1967 issue, with its cover showing four of the magazine's editors burning their draft cards. In the April 1966 edition, the cover depicted Madame Dinh Nhu, looking much like a college cheerleader; inside was a scathing examination of Michigan State University, where the CIA was recruiting professors to help sell the Vietnam War. *Ramparts* was also a leader in the civil-rights movement. Eminent writers appeared in its columns: Howard Griffin, author of *Black Like Me;* Eldridge Cleaver; Jack Newfield, later of the *Village Voice;* and Jessica Mitford.

Bourne saw *Mother Jones* as the logical successor to *Ramparts,* pointing to its investigative stories on the crimes of corporations but regarding it as too self-congratulatory and ideological. No magazine, however, could be quite like *Ramparts* in full cry, and it moved many newspaper and magazine reporters to share in its sense of mission.

Mother Jones has proved to be more durable than *Ramparts.* It was one of four new politically oriented magazines spawned in the mid-seventies by the Vietnam War. *Seven Days* was first to appear, publishing tryout issues as early as 1975 but taking three years to work up to full speed; David Dellinger, an anti-war activist, was its editor. Taking over some of *Ramparts's* abandoned subscribers, it announced itself not as the voice of a movement but as a newsmagazine—an alternative to *Time* and *Newsweek.* Disdaining corporate advertising and profits, if any, it was published by the Institute for New Communications, a nonprofit organization. Reviewing *Seven Days* in the *Nation,* James Boylan wrote that it "lacks the depth of personnel to do much more than others. It does provide alternatives when covering stories that the major news organs fail to cover." Measured against what the major magazines could accomplish with the same story, Boylan said, "it often does not succeed."

Two of the other new magazines were distinctly transitory. One, *Politicks & Other Human Interests,* published biweekly beginning in 1977, was edited by Thomas B. Morgan, who had once tried to buy the *Nation.* He assembled an impressive array of writers—Barry Commoner, Edwin Diamond, Walter Karp, and Ronald Steel, among others—but Morgan became disillusioned with his own product. He had hoped to arouse the public from its political apathy, but nothing *Politicks* printed seemed likely to overcome voter inertia.

Inquiry, appearing in November 1977, sponsored by the Cato Foundation, advertised itself as "America's most outspoken magazine," but its tone, in Boylan's words, was one of "concerned indignation." It appeared to have no firm ideology but was simply opposed to abuses of governmental power and seemed to have no real direction.

All three of these new magazines appeared to be in the same category as two other political journals already extant, the bimonthly *Working Papers for a New Society* and the socialist tabloid from Chicago *In These Times.* Summarizing their role in American culture, Boylan concluded that there was "a tendency among them to sublimate the anger that has been the classic emotion of the Left," and in doing so, they seemed once more to lack direction.

Mother Jones, however, was a different matter. Although it was published ten times a year by a nonprofit organization, it had the appearance and tone of a commercial magazine from the beginning. While its focus was on the poor and working classes, its specialty was investigating the evils of corporations, in the best muckraking tradition. "Investigative consumerism," its method was called. By the spring of 1986, it had a new publisher and a new editor, who were busy at once trying to attract new advertisers. But Michael

Moore, the editor, assured everyone that *Mother Jones* would continue to be investigative. In any case, there was no direction to go but up, since at that point the magazine's debt was growing and circulation was dropping.

Founded in 1975 by Adam Hochschild, Richard Parker, and Paul Jacobs, all former *Ramparts* editors, *Mother Jones* had begun with the intention of making left-liberal ideas more palatable for larger audiences, and for a time, it had exceeded the founders' expectations. Its circulation climbed above 100,000, and it won three National Magazine Awards. Five years later, however, the magazine's progress had stalled; circulation dropped from a peak of 233,000 in 1980 to 150,000. By that time, Adam Hochschild was the only one of the founders remaining on the magazine. Even at its best, Hochschild said, *Mother Jones* still lost money, and the deficit was now $3 million. Hochschild's father, president of a large international mining company, had saved the journal from bankruptcy.

It was the great overhaul of 1986 that provided new hope. The format was redesigned to give the magazine a more elegant tone, stories were shorter, and even a new name was considered ("Mother Jones" was a turn-of-the-century labor organizer, Mary Harris Jones). News service features were added, and political reporting was aimed at the average reader. Plainly, *Mother Jones* was gearing itself for a new and different decade, at the risk of at least bending its original principles.

By 1989, the transformation was nearly complete. In their San Francisco offices, the editors were wrestling with a problem that had not unduly concerned them in earlier days—how not to offend readers. With the actress Susan Sarandon's face on the cover of the February 1989 issue, the magazine once billed as designed for "the rest of us" had gone post-revolutionary, as Randall Rothenberg reported in the *New York Times.* The profile of Sarandon inside was testimony to the magazine's new subtitle, "People, Politics, and Other Passions." Establishment ideas were given a slightly different twist, as in a travel column directing readers to "a more authentic travel experience in the developing world," and another column instructed readers in "socially responsible consuming." The collective intent was to move *Mother Jones* out of the sixties at last. The profile of Sarandon was proof, however, that a basic intent remained but had been given a different spin. Readers were told about the actress's political and social activities, not about her new picture or her personal life. *Mother Jones* intended to track the social involvement of musicians and actors, while still carrying at least one investigative article in every issue.

As the magazine entered the nineties, its financial health had improved. There were now 200,000 extremely loyal readers, who (as surveys indicated) spent at least two hours with every issue, and more than half of whom had been subscribers for at least three years. Moreover, the renewal rate was high, and according to the *Times,* the conversion rate of first-time subscribers to long-term ones had increased more than 40 percent in the previous two years. Advertising pages had grown by 11 percent in 1988, and there would have been an even greater growth if both staff and readers had not firmly opposed making any concessions to potential advertisers by adding more service features and celebrity articles. *Mother Jones's* readers were now approaching middle-age affluence, with a $47,000 average household income and a median age of thirty-nine.

Of all the alternative left-wing magazines, *Rolling Stone* has been the most successful. Jann Wenner launched it on the proverbial shoestring in 1967, when he was only twenty-one. Aimed at reflecting current pop music and culture, it was brash and youthful, full of rock-and-roll's vitality. It was also on the scruffy side, as Gauri Bhatia noted in *Folio* magazine, printing the four-letter words at which other journals drew the line and flaunting opinions seemingly for the sheer sake of their controversial nature.

Fifteen years later, it had toned down, in keeping with the different climate of the eighties. By that time the baby-boom generation had grown up, and the record industry was experiencing a flat period. Both circulation and advertising had fallen off sharply. Still, the magazine's progress had been remarkable. The first issue, containing only seven pages of advertising, sold just 5,000 copies at twenty-five cents. By 1982, there were nearly 100 pages in the magazine, about half of them advertising, and 775,000 copies were selling twice a month at a cover price of a dollar and a half.

Wenner no longer had to depend on the magazine alone for income. He had spun out a small empire, including a book division, issuing twenty titles a year; a radio division; and a licensing division, busy syndicating material from *Rolling Stone*. His company also had a new magazine, *Record*, aimed at rock fans. Wenner had experienced two failures—an ecology magazine and a British edition of *Rolling Stone*—but the entire company was recording more than $20 million in sales every year.

Early in 1982, Wenner repositioned his magazine to make the necessary adjustments caused by demographic and economic changes. The results were extremely effective, beginning with a complete redesigning to create a streamlined format. Paper quality was upgraded at the same time. There was also a shift in editorial emphasis, with less space for music and more for politics, movies, and the arts. As *Folio* reported, Wenner asserted he was simply altering his magazine to "reflect the changing tastes of readers who have grown up."

How much of a shift there had been in left-wing ideology since the gaudy days of the sixties was evident in the later careers of David Horowitz, a Marxist theoretician of the New Left, and Peter Collier, with whom he had been co-editor of *Ramparts* in its early days. In July 1989, the *New York Times Sunday Magazine* disclosed that these onetime radicals had followed the same route taken by the disappointed socialists of the twenties and thirties who joined the ultra-conservative Radical Right. Now fifty years old, Horowitz and Collier had launched what they called the Second Thoughts Project, "a kind of peripatetic think tank," as the *Times* described it, "that promotes the revisionist views of some twenty former radicals who drafted key New Left texts and led anti-war protests. The project has developed into one of the most spirited anti-left, pro-American campaigns since the lapsed Communist Whittaker Chambers turned on his former comrades in the 1940s." Horowitz and Collier still thought of themselves as "guerilla intellectuals," according to the *Times,* but no one was likely to be fooled.

Other left-wing magazines are inclined to be much more exotic than those previously described. For example, *Yellow Silk* describes itself as a "Journal of the Erotic Arts," with stories and articles by Ntoke Shango, Jean Genet, Pierre Louys, Marge Piercy, Eric Gill, and Octavio Paz, along with other less-noted writers.

One of the better-known alternative magazines, the *Washington Monthly,* could not be described fairly as either right or left but rather as professionally iconoclastic, an alternative to both sides. Celebrating its twentieth anniversary in 1989, Charles Peters, its founder and mainspring, recalled that he lost $45,000 during the first five years. He began with the barest minimum: castoff furniture and $150,000 from five investors, including Jay Rockefeller, who had been with Peters in the Peace Corps and was now a senator from West Virginia.

The *Monthly* is a reflection of Peters's active mind. He is a lawyer and former state legislator in West Virginia who started his monthly on the premise that nothing in the government worked well. He intended to explore, as he said, "almost as an anthropologist does, the bureaucratic factors in decision making." The result, as of 1990, was a magazine with 33,000 subscriptions, paying its own way, partly because the owner pays himself only a $24,000 annual salary and his eight staff members $10,000 each. "People come and work for two years, then go on to better things," he says. Some of the more noted who came and went include Taylor Branch, author of *Parting the Waters;* James Fallows, the *Atlantic*'s Washington correspondent; Michael Kinsley, editor of the *New Republic;* Suzannah Lassard, a *New Yorker* staff writer; and Peter Braestrup, editor (until 1989) of the *Wilson Quarterly,* a magazine begun in 1977 by the Woodrow Wilson International Center for Scholars, which acts as a transmission belt between scholars and nonscholars—"the upper-middle-brow's *Reader's Digest,*" as it has been called.

Another group of magazines are only somewhat political, leaning much more to literature. They include, *Witness, Margin, Paintbrush, Grand Street, Crosscurrents, Parabola,* the *Uncommon Reader,* and the *Portable Lower East Side.*

The *Childbirth Alternative Quarterly* advocates childbirth reform, returning to traditional methods, including midwives, and opposes "unnecessary obstetrics." Advertising itself as the most widely read alternative parenting publication, *Mothering* is a quarterly also advocating midwifery and home births, as well as education at home. The *Animals' Agenda* is devoted to animal rights, while *Clothed with the Sun* is a nudist magazine, which calls itself "not just another pretty face among magazines."

Looking ahead are the *Futurist,* a New Age bimonthly featuring easy ways to explore the future, and *Artifex,* offering an "integrated view of the meaning of psychic research, from a Jungian perspective."

Contemporaries is a magazine of first-rank literary quality, "dedicated to the reporter's art and the American beat." It "reports from the cutting edge of New American Journalism," the arts, satire, literature, and politics, with such writers as Frances FitzGerald, Bobbie Ann Mason, Joe Klein, Ellen Willis, and J. Anthony Lukas.

Gnosis, as it says, "bridges the gaps between the spiritual, the occult, and depth psychology."

Liberty analyzes culture, politics, and the arts, adding fiction and humor, "written mostly from a radical libertarian or anarchist perspective."

One of the most intriguing titles among the alternative magazines is the *Journal of Polymorphous Perversity*, which says it is a humorous journal of pscyhology and psychiatry. The *Wall Street Journal* called it "a social scientist's answer to *Mad Magazine*." Among its most recent articles have been "The Etiology and Treatment of Childhood" and "Psychotherapy of the Dead."

A music magazine, *Option*, concerns itself with "music you're not hearing on commercial radio," including underground rock, jazz, reggae, world music, and avant-garde compositions.

One of the better-known alternatives is Paul Krassner's the *Realist*, a magazine of social-political satire recently reincarnated after a previous life as an irreverent counterculture journal from 1958 to 1974. *People* magazine called Krassner the "father of the underground press," but he demanded a blood test.

Semiotext describes itself as the best underground journal in English, containing in its pages "anarchists, unidentified flying leftists, neo-pagans, zeroworkers, foreign agents, monarchists, tax resisters, vampires, mimeo poets, prisoners, unrepentant faggots, witches, hardcore youth, the lunatic fringe of survivalism, xerox pirates, cultists, mad bombers, ban-the-bombers, paraphysicians, and poetic terrorists."

In revolt against technology, *Processed World* says it offers "sharp commentary from the underside of the information age. Explores the fluorescent dungeon with esthetic vengeance."

Operating at the edge of the law, *Sinsemilla Tips* advertises itself as "the trade journal of the domestic marijuana industry," picking up "marijuana news where the straight media leave off." Subscribers receive it in a sealed brown envelope. "We do not sell, rent, or give away our mailing list," *Tips* prudently asserts.

Scientia "emphasizes theories and discoveries other science media miss or ignore. Alternative evolutionary theories, sociobiology and morality, folk medicine, health benefits of drugs, higher consciousness, artificial intelligence critics."

The *Socialist Review* asserts succinctly that it is the "informative, insightful, innovative, and iconoclastic journal of the American Left."

Nearly all of the alternative press, except for the well-established journals noted earlier, are subject to change without notice, dependent as most of these magazines are on beneficence of some kind. The right-wing alternatives are much better financed than the left in a conservative era and at a time when it takes more and more money to keep a magazine in operation. The alternative press is important, however, because of its tremendous range, reflecting a nation more varied in its politics, opinions, and tastes than any other.

24

Pulps and Science Fiction

"Day Dreams for the Masses," the critic Marcus Duffield titled his 1933 *Vanity Fair* article about pulp magazines, an accurate description of the ten-cent journals, which, at their peak, reached more than 10,000,000 readers a year. At an affordable price, they provided entertainment of the kind that would make television a staple in American households. Their content was pure escapism, but at the same time, they did make magazine readers out of many who otherwise would never have opened the pages of a periodical.

Although they flourished in the twenties and thirties, before the arrival of television killed many of them, the pulps had their roots in the nineteenth century, when Frank Munsey, who may have been first in the field, changed his *Argosy* into an all-fiction journal and in 1896 began to use cheap pulpwood paper in an effort to cut costs. At about the same time, such dime-novel publishers as Beadle and Street & Smith were beginning to print their books in a pulpwood magazine format, making them eligible for low second-class mailing rates.

Munsey's *Argosy* soon had to complete with other pulp magazines—*Popular, Adventure, Short Stories,* and *Blue Book,* among others. Competition and sheer necessity combined to keep prices at a stable ten cents. Titles were killed off mercilessly overnight if the bottom line failed to show a profit; only the hardiest survived. In 1908, for example, Munsey merged his *Live Wire* with the *Scrap Book,* and four years later, that magazine absorbed the *Cavalier.* In 1914, however, the *Scrap Book* itself became a part of *All-Story Weekly.* Once a title proved itself successful, publishers began to look around for a new pulp to buy or else created another one. The economics of printing and distribution dictated such empire building. Harold Hersey, a veteran editor of scores of pulps, wrote, "The fight to launch a new pulp and keep an old one going is a never-ending one."

These early pulps contained general fiction, usually adventure, but later they became specialized, dividing into the subcategories. A critical element in the success or failure of each magazine was its cover, which not only had to sell the contents but also stand out as much as possible on the newsstands. They were usually done in three colors because anything other than the primary shades tended to smear when there were large runs of pulp paper. These journals contained little advertising, relying on quantity purchases and low costs. Budgets for art work were rigorous, and payment to authors was about two to four cents a word. The goal was to keep expenses so low that even if high sales failed to materialize, the magazine could still turn a profit.

Advertisers who did use the pulps were often on the edge of respectability. Correspondence schools offered a smorgasbord of careers, unhappy readers could buy booklets on "The Philosophy of Love," and a variety of products was available on the installment plan.

Pulp editors had to stay close to their readers, trying to anticipate their preferences even when the readers themselves were not sure what they wanted. It was not the pulp editor's role to elevate, lead, or somehow "improve" his constituency. Describing the editorial process, Harold Hersey observed that "what seems to be a hasty pudding affair is really an intricately worked-out problem repeated with every issue, consisting of the title and subtitle which must never be lost to view, the painting and the way it is handled, the selection of authors and titles—the result a homogeneous entity that, unless it sells the magazine, is worse than useless. The reader is led into buying without knowing why."

Pulps survived the Great War successfully, and in the twenties they began to specialize. Subtypes breaking away from the general adventure formula appeared, most prominently mystery, romance, and Westerns, all of which proliferated especially after Street & Smith created prototypes in *Detective Story, Love Story,* and *Western Story.*

With the appearance in 1921 of *Love Story,* the romance pulp category was launched. It was the brainchild of Anita Fairgrieve, who had been instructed by her employer to produce an idea for a new pulp. After examining hundreds of old dime novels, she created the love pulp and naturally named it *Love Story.* Starting as a quarterly, this magazine was soon so popular that it went semiweekly, then weekly. That was a typical pattern for the pulps. A publisher would test the waters with a monthly, then move to more frequent publication by degrees if his entry was successful.

Love Story quickly produced a swarm of competitors, including *True Love Stories, Pocket Love, Romantic Range,* and *Real Love.* By 1938, there were eighteen pulps dealing with this ever-popular subject whose total collective circulation was more than 3,000,000 purchasers per month. Their stories featured immaculate heroines who guarded their virginity scrupulously. The appeal, obviously, did not lie in descriptions of sexual adventure but in standard narratives about the agonies of loving, being loved, or being ignored or otherwise traduced by the love object. Later in the twenties, a different category of love pulps appeared in which there was more explicit sex (of the mildest variety by today's standards), in keeping with the changing moral standards of that decade. *Breezy Stories* and *Young's* were two of the most successful in this category, but the writing in them was on a lower level than in the love stories printed by the major women's journals.

For all its reputation, the decade of the twenties was not as morally unfettered a period as people today tend to believe. Flappers might roll their hose, wear the first mini-skirts, smoke and drink, and even appear to be careless about their virginity, but it was also a time when Homer Croy's popular novel *West of the Water Tower* was attacked and removed from bookstores and libraries in some places because of a single sentence: "He turned out the light and the bedsprings creaked." Censorship and attempted censorship were rife.

Westerns were much safer for a publisher. They had always been the bestsellers among the dime novels, and they did equally well in the pulps. Magazines devoted wholly to Western stories began to appear in the early twenties and found an immediate audience. John Dinan, a historian of this genre, observes that "the significance of the Western pulp magazine clearly lies in the number of publications produced, and the obvious size of the public appetite for this material." He links the popularity of the pulp Western to the nostalgic desire for the ideals of a mythic West where individualism reigned, there were no restrictions, and men and women had clearly defined roles.

Between 1920 and 1950, several hundred Western titles existed. Among the earliest were *Cowboy Stories, Ace High, West, Frontier, Ranch Romances, Lariat, Triple X Western*, and *North West Stories*. The first all-Western pulp was *Western Story Magazine*, and it continued to be the favorite for a long time. It had its origins in Street & Smith's "Buffalo Bill" dime-novel series, which was translated into magazine format in 1919. *Western Story* printed the works of writers whose names became household words among fans of the genre—Frederick Faust (his pen name was "Max Brand"), William Colt McDonald, Fred Glidden, Hugh Cave, Walt Coburn, and W. C. Tuttle.

As always, the success of *Western Stories* spawned a flood of imitators, among them *Wild West Weekly*, an instant favorite with its tales of Billy West and the Circle J Ranch. By this time, Street & Smith was getting competition from such other publishers as Fawcett, Fiction House, Clayton, and Doubleday, to whom Westerns were only a lucrative sideline.

Targeted to males, most pulp Westerns were written by men. Even Western romances, a further refinement of this category, pioneered by Harold Hersey, were written primarily by males. Christine Bold, who has studied the rise and fall of the pulp-Western writer, found that his decline in status "changed the dime author from a hired storyteller conforming to broad outlines, to a minor member of a collaboration who followed orders in the very writing of the fiction, and finally to a worker who was openly shown to hold second place to the reader in the editor's regard."

Pulp magazines reached their zenith in the early thirties, partly because of the numerous men made jobless by the Depression, many of whom appeared able to produce a dime for a magazine offering escape and a way of passing idle time. With the Depression ebbing away in the late thirties, however, after the recession of 1937, new competitors began to appear. Besides the increased rivalry from radio and movies, comic books began cutting the profit margins of pulps. Attempting to counter this trend, *Wild West Weekly* began including a comic strip series in 1935. Other pulps made efforts to improve their appearance, displaying new designs and trimmed pages. There was one enemy they could not fight, however. Fiction was declining in all the consumer magazines that carried it, and since the pulps were all fiction, they were hardest hit.

Then came the paper shortages of the Second World War to plague the survivors, followed by the rise of television in the fifties and a new paperback revolution. Production costs continued to rise, and they could not be offset by advertising revenues large enough to make a difference. Many of the pulps died in this period, but a few managed to survive by changing themselves into slick-paper publications or converting to such specializations as men's magazines, science fiction, or comics.

Of these subdivisions, science fiction was not the most profitable, but it was by far the most interesting. Tales of interplanetary travel and the supernatural had been printed frequently in Munsey's *Argosy* and the other pulps, but it was not until 1926 that the first science-fiction pulp appeared. It was *Amazing Stories*, the product of Hugo Gernsback, a talented scientist and inventor. As early as 1911, he had published a science-fiction story of his own in *Modern Electrics*.

Introducing his new pulp, Gernsback declared: "*Amazing Stories* is a new kind of fiction magazine: it is entirely new—entirely different—something that has never been done before in this country." It would print "scientifiction," he said, "the Jules Verne, H. G. Wells, and Edgar Allan Poe type of story—a charming romance intermingled with scientific fact and prophetic vision."

In its early issues, *Amazing Stories* reprinted the work of all these classic writers, but it also introduced new ones, including Edward E. Smith, whose "The Skylark of Space," published in three parts in 1928, was both popular and influential in developing the genre. *Amazing Stories* soon became so successful that Gernsback brought out *Amazing Stories Quarterly*, a fifty-cent pulp magazine, which often published complete novels in one issue.

As an editor, Gernsback made certain demands on his writers. Stories had to be at least loosely based on scientific fact, they had to move at a rapid pace, and they had to be constructed so that they could be cut into sections for serialization. Unfortunately, Gernsback encountered legal problems and had to sell his magazine to Teck Publications, but soon after, he started a new pulp, *Science Wonder Stories*. With that as a base, he went on to build a magazine empire that included *Radio News, Your Body,* and *Tid Bits*. To ensure the scientific accuracy of his stories, he paid various eminent scientists to review the material, including Clyde Fisher, of the American Museum of Natural History; William Luten, of the Harvard Observatory; and T. O'Connor Sloane, an inventor and science writer who became an associate editor.

Reviewing the history of science fiction, one of its most distinguished writers, Frederick Pohl, recalls that there was still a minimal amount of it being published in America in book form during the early thirties. There were

Edgar Rice Burroughs's novels about life on Mars and about that other world
At the Earth's Core, plus a few children's stories. There was no science fiction
on radio until Buck Rogers appeared near the end of the decade. Only occasionally were there motion pictures—"Just Imagine," or "Transatlantic Tun-
nel," imported from England. In the magazine market, Gernsback's pulps and
Weird Stories were the staples, along with *Astounding Stories of Super-
Science.*

Science-fiction pulps had only marginal circulations. Pohl estimates that,
collectively, they published about 2 million words a year, for which writers
were paid at an average rate of half a cent a word. These writers were mostly
young men in their early twenties, and they were largely male. The audience
included men equally young who were fascinated by science and pointing
toward it as a career. That was why these pulps sold best at newsstands near
college campuses. However, the largest number of readers were boys whose
average age might have been fifteen, although there were readers under ten.

No audience quite like it has existed in magazine history, according to Pohl.
The science-fiction pulps created a community of writers and readers who felt
extremely close to each other and to the magazines as well. They shared the
same interests and outlook and tended to isolate themselves from the world as
much as possible. They yearned to talk about the ideas in the stories and about
their own ideas as well, so they began to form science-fiction fan clubs. *Wonder Stories* even launched a correspondence club called the Science Fiction
League.

After Gernsback, probably the most outstanding editor was John W. Campbell, Jr., who began writing science fiction while he was an undergraduate at
M.I.T. and rose quickly to become editor of *Astounding Stories,* a Street &
Smith pulp. Pohl observes that Scientology was provided with most of its early
impetus through this magazine, later renamed *Analog.* Campbell edited his
magazine for thirty years, during which time he introduced more than half
of the most noted science-fiction writers, including Arthur C. Clarke and Isaac
Asimov, all of whom made their first appearance, or their first significant one,
in the pages of *Astounding Stories.*

Pulps continued to be popular through the Second World War, but a slump
came in the pulp market when the conflict ended. The dramatic rise of paperbacks was one factor, but mostly it was the change in social climate that follows every great war. The Westerns were first to slump and then virtually to
disappear, followed by the war and sports pulps. Several of the large publishers either suspended quietly or branched out into specializations.

But science fiction did not leave quietly with the others. Its fans not only
read but participated, and to keep their interests alive, they began to reissue
the great classic novels in the field that had gone out of print. Many had been
serialized in the science-fiction pulps. Fans contrived to find printers and binders to reissue them and saw that they got into bookstores. Trade publishers,
observing excellent sales figures, began to issue science fiction of their own.
The amateur fan publishers were soon out of business, but they had started
such a strong revival of the genre that by the eighties one novel in every four

was either science fiction or fantasy, some (Clarke's *2001* and Frank Herbert's *Dune*) becoming best-sellers. Today science fiction has a literary vogue. It is even taught in colleges, and the Modern Language Association has made it the subject of seminars.

There were more than twenty-five science-fiction magazines existing in the early fifties, including such new ventures as *Galaxy Science Fiction* and the *Magazine of Fantasy and Science Fiction,* the latter edited by Anthony Boucher. At their postwar peak, however, there were never any more than forty science-fiction journals, although the revival that began in the sixties produced a new audience that appeared to be still increasing in the eighties. Those reading such fiction today tend to be a much more heterogeneous group than the audiences of the thirties, forties, and fifties. They are still young and better educated than the population as a whole, but they are no longer stereotypical males working in engineering or at scientific jobs. Today's audience is far more diverse, and the appeal of the fiction itself has expanded proportionately.

The reasons for science fiction's continued, even increasing, popularity are obvious. It offers escape from a world infinitely more complex than when its readers' lives began. It frequently offers hope of a new and better world at a time when it would be hard to find that promise anywhere. Then, too, it teases and stimulates the mind in a society increasingly technological. It is no longer considered "pulp" in a derogatory sense; often it competes favorably with other high-quality fiction. Like all good literature, Emily Alward has noted, "SF features unusual events which no reader is likely to have experienced, but which he or she believes, according to the best wisdom of the time, *could* happen. Major values of our culture are expressed in this fiction, most notably our infatuation with science and technology and our orientation toward the future."

25

The Triumph
of the Business Press

Business publications have existed in the United States since colonial days, and they have been growing steadily in number since, but only in our own time have they become the dominant part of the magazine industry, a status not understood by the general public. Consumer publications are so visible, so much a part of people's reading lives, that it is easy to overlook the numerical superiority of business magazines, not to mention their power and influence, reaching into every segment of American life.

One problem is how to define this incredibly broad mass of magazines, numbering at least 12,000 (and probably several thousand more) out of the approximately 22,000 magazines extant in America; there are no absolute statistics in this field. People tend to think of the business press as "trade papers," or "technical journals," not realizing that these are only parts of the whole. They are not the house organs produced by manufacturers; that is a separate genre. General business magazines, like *Business Week* and *Forbes*, do not precisely belong in this category either, because they verge into the area of general interest and perform a different function.

In search of a definition, the Association of Business Publishers, the major organization in its field, representing 746 member publications (624 domestic, 122 international) as of 1989, produced in the seventies a comprehensive description of what the business press is: "Independent, specialized periodicals of a business, technical, scientific, professional, or marketing nature, published in either a magazine or newspaper format and issued in regularly specified frequencies to serve special fields of private or public enterprise and not directed to the public at large."

This is a workable definition for the ABP's members and for the nearly 3,000 publications, with a total distribution of more than 70 million, falling within the organization's scope. If these figures seem to be a puzzling distance from the 12,000 or more cited earlier, it is because the latter figure would include thousands existing outside the ABP definition—house organs, those not issued "in regularly specified frequencies," newsletters, and other hybrid forms of enterprise serving every part of the American economic and business structure.

Just as a relatively small number of consumer magazines, no more than 2,000, represent the most important part of that side of the total industry, so do ABP's members constitute the cream of the business press, although there are other important and influential magazines in the field. The ABP serves its members in a variety of ways. Its Neal Awards, given annually, are often called "the Pulitzer Prizes of business journalism." In 1988, it began a Publishing Management Institute, in conjunction with Northwestern University, as a plan to provide intensive training for business publishing professionals. Faculty was recruited from the University's Kellogg Graduate School of Management, and the Medill School of Journalism. The ABP also gives an annual McAllister Fellowship to a top executive of a member company. Fall conferences and winter roundtables are sponsored by the ABP, and it has established the Business Press Educational Foundation, which promotes internships and helps schools start courses in the business press. That press began with the "price-current" papers, hardly any more than pamphlets, which advised colonial businessmen of changes in commodity prices, the movements of ships, and the fluctuation of money exchange rates. As the country began to develop and business expanded in the years before the Civil War, much more diversified information was required, and it was then that the specialization taking place in general or consumer magazines was duplicated in the business and industrial publications. In a sense, this was a free market of business information, a forum for the open discussion of business and professional problems.

There were more than 160 of these antebellum journals, of which *Railway Age* is generally considered the progenitor; it was published before an operating railroad existed. Growth of these magazines was especially evident in such fields as mining, petroleum, the shoe and leather industry, banking, pharmaceuticals, drygoods, hardware, and tobacco.

The industrialization of America after the Civil War immensely extended the range and variety of the business press. The trend was toward ever more specialization, and it happened at an even more rapid rate than in the consumer field. For example, while one magazine was once enough to serve the infant electrical industry, hundreds now rose to cover communications, light and power, distribution, and wholesale and retail selling. When electronics were introduced, specialized publications multiplied again.

A historian of the business press's rise, Harney Harkaway, observes that "the work of the great business publication editors and publishers in the years before World War I set a pattern. It prepared the business press for an era of even greater expansion following World War II. . . . The business or professional man, the engineer or merchant, recognized that without reliable business or technical information he was in danger of standing still or moving sideways while the mainstream passed him by."

Harkaway finds several conditions which account for the eventual dominance of the business press. A primary factor is the American thirst for printed
information, which began with the establishment of the Cambridge Press in
1639. A second factor was the end of the apprentice system and the explosive
growth of new inventions that revolutionized manufacturing and distribution.
At the same time, there was a corresponding increase in the dissemination of
ideas about scientific management, as well as a constantly increasing demand
for news of new developments. The business press's unique role is also the
result of a lack of formal education among the early artisans, craftsmen,
mechanics, and other workers who found in the trade publications a substitute
for the classroom. Editors were teachers of business then, and today they still
provide a postgraduate education for their readers, even those with MBAs.

One of the remarkable things about the development of the business press
is the fact that, beginning in the late nineteenth century, there has been such
a steady growth in the dependence of people in business, industry, and the
professions on the magazines that serve them. There is nowhere else to go for
the information these journals provide. Their reliability was guaranteed by the
separation of editorial services from advertising, which prevailed in all the
successful publications. For their part, the advertisers in time came to appreciate that the credibility of the magazine enhanced their own. All this, of
course, did not happen overnight.

To those exploring the business press for the first time, it is often startling
to see the editorial courage these magazines display. At first glance, one would
think that periodicals directly dependent on a particular business or industry
(or profession, for that matter) for both readers and advertising would become
uncritical defenders of whatever might be wrong, or at least would simply
overlook controversial issues. Although it has not been universally true, either
yesterday or today, the business press as a whole has been fearless in attacking
ills within its own constituencies. It has been well ahead of newspapers and
other magazines in exposing whatever kind of problems exist. Nursing periodicals do not hesitate to view doctors and hospitals critically when required.
Trucking magazines are quick to attack practices that hurt the industry as well
as the public. The same thing is true in virtually every other field. Apologists
and uncritical defenders exist, but they are a small minority.

Because of this editorial courage, the business press has been a positive factor in American life. Construction publications, for instance, have pointed to
industry failures that were endangering lives. Mining journals led the way in
the fight to make companies adopt safety measures. "Automation" was a word
coined by metalworking journals, and "electronics" originated as the title of
a magazine. George Westinghouse found the idea that led to his invention of
the air brake in an industrial magazine. Similar examples abound.

The business press penetrates its audiences more deeply than the consumer
magazines. Surveys show that 97 percent of doctors, scientists, technicians, and
businessmen in general get the information they need from specialized periodicals. These are the primary, and in some cases the only, sources. As Robert
S. Stevenson, former president of Allis-Chalmers, put it, "The business press

serves us by providing us with authoritative and knowledgeable information relating to a specific business—a type of information which is just not available from any other source."

The range of these magazines is impressive. There are nearly 200 categories of business publications, and these are subdivided into hundreds of others. It is safe to say that there are none in industry, the professions, trade, or in any vocation or geographical location who do not have a publication directed to them.

In the eighties, the business press came fully into its own. In 1982, according to a report in *Folio,* these magazines constituted a third of the nation's leading 400 revenue-producing periodicals. The difficulty of assessing this category statistically was evident in the report, which cited about 4,000 publications doing an annual business of $2.6 billion. Whatever its exact size, the business press has not only become an amazing success story in itself, but it has influenced the other media as well. As the quality of these periodicals has been steadily raised by astute publishers and editors, the superior reporting of business and industrial news has compelled many newspapers to enlarge their business sections, not so much to compete as to cover what was plainly a large and important news area that they had failed to develop adequately in their own coverage. Business news appears in special segments, too, in both local and national newscasts with much greater frequency.

By 1985, *Newsweek* was declaring a "Boom in the Business Press," illustrating the scope with two rather far-out examples. One was *New York CityBusiness,* a biweekly tabloid in that gray area between newspaper and magazine, which had just, as *Newsweek* put it, "reported on the economics of local prostitution rings as if it were analyzing widget companies. Its competitor, a brand-new weekly called *Crain's New York Business,* came back in one of its prototype issues with a story on making it as a *mohel*—the men who run the ritual-circumcision industry. Nowadays, *everything* seems to be business."

In keeping with that impression, two magazines for entrepreneurs had emerged, *Inc.* and *Venture.* The former was begun by Bernard Goldhirsh, a Boston-based publisher, who had raised *Sail* magazine from its start as a mimeographed newsletter to a journal so successful that it was sold to the Meredith Corporation. Goldhirsh called his new venture "The Magazine for Growing Companies" when he launched it in 1979. (He followed it quickly with two others, *Technology Illustrated,* for a nontechnical audience, and *High Technology,* for more knowledgeable readers. The former expired in 1983.) *Inc.,* found an audience immediately, and it was a large one—small businessmen, who had been relatively neglected. Describing *Inc.* in the *Wilson Library Bulletin,* Gail Pool wrote, "Recognizing that the interests and concerns of small business differ from those of large corporations, *Inc.* provides information directly relevant to the smaller enterprise. National and international trends are not ignored, but they are viewed from a more localized perspective. Texas, legislation, and the shifting stock market do not have the same impact on a small manufacturer as they do on General Electric. . . . *Inc.* has found its place as both reporter and interpreter of news and issues for owners of small businesses." Pool also noted that "by its selection of articles and by its sophisticated presentation, it clarifies the broader implications of issues without sacrificing practical elements." In October 1983, for example, it printed an article titled "Management Is Walking Away," relating the methods of a manager who gave substantial responsibilities to his employees. Other stories discussed various managerial styles, the working environment, and "How To Manage Your Company—Ten Easy Steps."

With such excellent editorial material and substantial advertising, *Inc.* had the small-business field virtually to itself, but in 1979 a rival appeared when Joseph D. Guarraputo founded *Venture: The Magazine for Entrepreneurs.* Its chief investor, Arthur Lipper III, took over the magazine later, and Guarraputo departed. The magazine was not successful immediately, but in a single year, turned itself around. Its November 1983 issue was only 108 pages, with little advertising, but a year later the November issue was 176 pages, advertising from major companies was heavy, and editorial matter was much improved.

A major development in the business press during the eighties was the steady growth and importance of local and regional magazines. Here, again, is a category illustrating how difficult it is to deal with periodicals statistically, since the local journals, especially, are often in the twilight area between magazines and newspapers.

Beginning roughly in the seventies, local business publications began to establish themselves in their area news markets, opening a specialization almost untouched before. Their audience was drawn from the affluent upper-echelon of management—of immediate value to advertisers.

The local and regional business-magazine phenomenon was analyzed in 1984 by three researchers—Bruce G. Vanden Bergh, Alan D. Fletcher, and Mary A. Adrian—reporting their results in the *Journalism Quarterly.* They concluded that "the total business publication industry has been one of the fastest growing sectors of the media market, paralleling the growth of metropolitan magazines. . . . Although some local business publications have existed since the turn of the century, approximately two-thirds of them were established within the last ten years. The greatest growth period occurred between 1976 and 1980, when nearly half of the publications were founded." There were enough of them in 1979 to create the Association of Area Business Publications, with more than sixty members by the mid-eighties.

Largest of the publishers, according to the survey, was the Cordovan Corporation of Houston, with eleven business publications across the country by 1984 and more planned. Others included the Business Press, Inc., of Indianapolis, with four journals; and Crain Communications of Chicago, with two. Based on circulation, the largest magazine in this category was the *Pennsylvania Economy,* with *Oregon Voter Digest* the smallest.

There is a wide variety of both content and style in these journals, some resembling full-size newspapers and others slick, four-color magazines, with still others somewhere in between. The researchers characterized the style of the magazines as "breezy and charismatic," while those more like newspapers adhered closely to standard news forms.

In 1985, *Newsweek* pinpointed some of the more notable local and regional business magazines. *California Business*, for example, struggled for several years before it became successful. In Philadelphia, three local business publications were fighting it out for circulation and advertising. In New York, *CityBusiness* was struggling against the publications of a much larger rival, Crain Communications, already owners of *Advertising Age* as well as its highly successful *Chicago Business*, with a circulation at that point of 47,000 and considered the best of its breed. In Pittsburgh, the *Business Journal* and the *Business Times* began competing in 1981.

How business press organizations can grow is illustrated by the Cordovan Corporation, owned by the E. W. Scripps Company. Cordovan was begun by a former newspaper reporter named Bob Gray, who in 1959 began turning out, from his garage, a journal called *Horseman Magazine*. In 1971, he began to expand and launched the *Houston Business Journal*, followed by three other similar ventures before he merged with the Scripps organization in 1980. He made that move so that he could command enough capital to go into other markets, which he did immediately, starting a new journal every three months until he had ten business weeklies with a combined circulation of 118,000 and a total of nearly 3,000 advertising pages. Only three were immediately profitable, however.

A major reason business journals on the local and regional levels can be started so easily is their comparatively inexpensive start-up costs. Crain began *Chicago Business* with an investment of only $3 million, but other entrepreneurs have gotten into the field with no more than $1 million or $2 million. About $500,000 was needed in the early eighties to buy the necessary computer equipment, and the only other major expense, then and now, was staff, usually hired in small numbers. Perhaps because of comparative costs, the newspaper-style business journals appear to be outnumbering the authentic magazines.

But the boom has continued, spurred on by the declining competition among general circulation papers, offering new outlets for reporting and analysis directed to specific local and regional audiences. The key to their future, Rip Keller observed in *Magazine Age*, "may be in how attractive they can make their rate cards to national advertisers. These magazines are picking up on a trend toward regional identity."

Local and regional magazines are not without their problems. In common with the consumer press, advertising support is always a critical factor, and that fluctuates with economic conditions. There is also the necessity to reach new people and expand territory. Most of these journals have controlled circulations, but many are moving to paid, and others are already operating with a mixture. Still another problem is the competition with local newspapers for advertising dollars, augmented by further inroads from national business magazines and city periodicals.

Finally, the local and regional journals have, in a somewhat different form, a problem every magazine has to contend with: the necessity to maintain a high editorial level. With them, the difficulty is not in getting enough material but in maintaining the high standards expected of them by both readers and advertisers. Exceptional staff people are often hired away by larger publications, and there is the constant difficulty, noted by business magazines generally, of finding people with the necessary language skills, a problem they attribute to low educational standards.

In the nonregional business press, these and other problems are being surmounted with increasing success. The fact is that business magazines are extremely profitable, and they command high figures when they change hands. In 1984, Rupert Murdoch bought a dozen of Ziff-Davis's trade publications for $350 million, including such highly profitable titles as *Meetings & Conventions* and *Aerospace Daily*.

As in all the media, conglomerates sweep up large numbers of magazines in the business field. The largest sweeper has been a multifaceted book publisher, Harcourt Brace Jovanovich, which in the late eighties had more than a hundred business publications on its list, with titles obscure to the public, such as *Plastic Focus, Pit & Quarry,* and *Food Sanitation.* A reader not involved in any of these fields would find such magazines excruciatingly dull and even incomprehensible in some cases, but their specialized audiences read them carefully, with complete understanding, and the editors and writers speak to them with a directness not always present in the consumer press. It is this rapport that makes them so attractive to advertisers, who are willing to pay as much as $2,500 for a full-page color advertisement in such a magazine as *Milling & Baking News.* "That's not much money in absolute terms," David Owen noted in an *Atlantic* survey of what he called the Fifth Estate, "but it works out to nearly half a million per subscriber, or about what it would cost to read the ad aloud to each one over the phone. A similar advertisement in *Time,* in contrast, has a cost per paying reader of less than three cents."

Yet, in spite of the general aura of success, business magazines can fall on perilous times, or expire, as easily as any other kind. A case in point is Arthur Lipper's *Venture.* According to Albert Scardino, reporting on its career in the *New York Times,* Lipper permitted his magazine to slip out of step with readers and advertisers, no doubt because he was a Wall Street operator with no magazine experience, who did not take advice and went through five editors in as many years. As a consequence, *Venture* never had a profitable year, and in less than half-a-dozen years slid from being the "hottest magazine in America" into deep trouble. A crisis came about because of the turmoil that struck the computer industry, *Venture*'s chief source of advertising, in 1985.

All business magazines were affected by the stock-market crash three years later, but *Venture*'s competitors rode out the storm. *Entrepreneur,* in new hands, was redesigned and became successful all over again by focusing on new small companies. *Success* did much the same thing, but *Venture* was struggling and Lipper looked for someone to buy it, after laying off much of its staff and skipping two issues. He found a most unlikely buyer—Drake Publications, Inc., of New York (owners of *Playgirl, High Society, Cheri,* and several adult pictorial magazines), who were diversifying not only with *Venture* but a new music journal. Drake had what Lipper needed most at that point, expertise in newsstand distribution as well as capital.

The business field is so attractively profitable, however, that all kinds of players get into it. For example, Warren, Gorham & Lamont, a company acquired in 1980 by the International Thomson Organization, Ltd., had never created a magazine, even though it published more than 300 titles for professional readers, including books, annuals, newsletters, and computer software, all based on circulation alone. In 1985, it bought two magazines, *The Practical Accountant* and *Computers in Auditing,* and three years later created one of its own, *Financial Manager,* "the magazine for financial and accounting professionals." Advertisements were solicited for the first time in the company's history. At the beginning, the new journal could offer a subscription list of 15,000, out of 400,000 accountants, each willing to pay sixty-four dollars for six issues, considered in the trade a most promising start.

New industries and new trends generate new magazines, often in large quantities. The most recent (largest in number and most unstable) have been journals for the computer industry, particularly for personal computers. In the seventies and eighties, the rapid develpoment of these machines for individuals resulted in a tidal wave of magazines that were subjected, as the market itself changed, to consolidations, acquisitions, failures, and expansions too numerous and complicated to record here. The rush had begun modestly in the late seventies with a few independent journals intended for hobbyists, but by 1988, after a decade of turmoil, two large corporations were dominating the field: International Data Group in Framingham, Massachusetts; and the Ziff-Davis Publishing Company in New York.

Beginning in 1967, IDG began introducing new computer magazines and newspapers at a rate of one every two months, on the average, and by 1989 had produced 120 publications, with only six failures. Its major successes included *Computerworld, PC World,* and *MacWorld.* There was even *PC World USSR,* and *China Computerworld.* In fact, IDG in 1989 was publishing in forty countries, in twenty-eight languages, with total circulation of all of its magazines at 4.4 million and no end in sight.

In 1989, Ziff-Davis was publishing nine computer magazines, including *PC Magazine, MacUser,* and *PC Computing.* Both companies were moneymakers in substantial figures—IDG with revenues expected to reach $500 million by 1990 and passing $1 billion by 1992. Ziff-Davis registered a more modest $300 million in 1987, but it was climbing too. (Because these companies are privately held, all figures are estimated.)

Analyzing the giants in the *New York Times,* Lawrence M. Fisher noted that these magazines, which could be classified as both trade and consumer publications, were different in style and personality but shared "a format devoted to product reviews and how-to material." He observed a "marked absence of visionary technology-based stories that fill pioneering publications like *Byte,* a McGraw-Hill publication, and *Personal Computing.*" The editor-in-chief and publisher of *PC Magazine,* Bill Machrone, told Fisher: "We write for one type of person only, the business professional who must decide what to buy," a kind of reader called in the trade a "brand specifier." With a circulation of 500,000, *PC* was in a close race with *PC World*, San Francisco based, which had 475,000. There was a flaw in this success story, according to Fisher: "Close relationships between magazines and the manufacturers call into question their role as consumer guides. The computer press has never really acknowledged that it's a trade press, beholden to the industry it covers. Others say the conflict is more apparent than real."

If there was any sign of a trend in personal computer magazines as the eighties ended, it might be found in Ziff-Davis's *PC Computing*, which was taking a different approach, more in the spirit of a consumer periodical, with fiction, human interest articles, a slick-paper appearance, and well-known writers. Whether this would truly establish it after an uncertain early stage remained to be seen.

One of the endearing qualities about computer magazines, as far as non-users are concerned, is their titles. Magazines for users of Apple computers, for instance, can buy *Apple Orchard, inCider, Nibble,* and *Peelings.* Tandy owners, while not possessing so obvious an image, may read *80 Micro, Rainbow,* and *Hot CoCo.* For the IBM users there are, more prosaically, *PC Week, PC Magazine, Magazine World,* and *Soft Talk IBM.*

The largest magazine in this field is *PC,* which IBM introduced with the advent of its personal computer in 1981 and which Ziff-Davis acquired in the following year. Philip Dougherty, reporting in the *New York Times* in 1988, called it the largest American magazine in either the trade or consumer category, on the basis of the number of advertising pages carried annually, with 5,500 pages in 1987, somewhat ahead of *Vogue.* With a circulation of 550,000 subscribers at that time, it was gaining at the rate of 10,000 a month and was expected to be at 700,000 by 1990. This was all the more remarkable considering that subscribers must pay $44.95 for twenty-two annual issues. What they get for their money is advice on the best equipment available and how to obtain the maximum performance from it. A red-and-white *PC* logo identifies products in the advertising that are the "editors' choice," like the *Good Housekeeping* Seal of Approval. Every reader becomes a potential buyer, since the journal contains only product reports.

Advertising itself is served by its own periodicals, and we have quoted from them frequently in these pages. *Advertising Age,* by far the oldest, remained unchallenged until *Adweek* appeared in the eighties, dividing the audience between them. At the end of the decade, two others were launched. One, *Target Market News,* is a monthly journal covering black-consumer marketing. Ken Smikle, its publisher, had been arts editor of the *Amsterdam News,* a black newspaper in Harlem, and also music editor of *Record World.* Smikle had been struck by the scant attention devoted by the trade press to black media and black-oriented advertising. He tested the waters first by writing several articles on the subject for *Black Enterprise* magazine before starting his own publication. Based in Chicago, it began with a controlled circulation of 10,000 and was directed largely to executives of black communications companies as well as to businesses serving or marketing to the black community. This consumer group, as Smikle well knew, spends more than $200 billion annually.

An entirely different kind of magazine was David Breznau's *Body Copy*. As a professional working on the creative side of advertising, Breznau had become convinced that the major trade publications were simply ignoring people in his side of the advertising business. He described *Body Copy* as "a magazine that gets involved with the people behind ads and the reasons why we're seeing the ads we're seeing." Beginning as a bimonthly, this journal proved to be both slick and colorful, focusing on personalities. Its interviews covered media planners, account executives, and clients. Breznau was not quite alone in his specialized market, however. *Advertising Age* issues a supplement called *Creativity*, and *Adweek* publishes *Winners*, devoted entirely to the creative process. Breznau, like others before him, hoped to offer more depth, but whether he could successfully challenge the two leaders when so many others had failed remained to be seen at the end of the decade. *Body Copy*'s first issue was sent to a controlled list of industry people and vendors, mostly in Los Angeles and its environs. Twenty advertisers took the initial plunge.

How entrepreneurs work in the business magazine field can be seen in the later career of Bernard Goldhirsh. After launching *Sail* and *Inc.*, Goldhirsh then turned to *Dun's Business Month*, which he bought in 1986. Targeted to leading business executives, it was a struggling publication, losing about $1.5 million a year, even though at ninety-six years it was the oldest business magazine in the United States. Its controlled circulation went to 300,000 executives every month, but even this affluent market was not enough for advertisers; in 1986, the magazine carried only 500 pages, half of its 1980 figure.

Goldhirsh dropped "Dun's" from the title, increased the distribution to 425,000, and invested in giving the magazine a sharper focus. Critics thought the new circulation was too large for such an approach, according to Geraldine Fabrikant, in the *New York Times*. It was already half the size of *Business Week*. By contrast, its rival in the same field, *Chief Executive*, had only 32,000 circulation. But Goldhirsh persisted, shunning the kind of investment articles appearing in *Forbes* and attempting to minimize the resemblance to *Business Week* and *Fortune*. Again, the outcome was not predictable.

Another phenomenon of the eighties was the increasing number of company-sponsored magazines, many of which carried material unrelated to the sponsor. "Custom magazines," Gail Pool called them in the *Wilson Library Bulletin*, predicting that conflicts of interest were probably inevitable. These journals were seen as still another extension of specialization. By controlling both audience and contents, custom magazines combined a journal's credibility with the targeting capability of direct mail. Wisely, however, most avoided the hard sell. Instead, they were largely devoted to providing product information with the hope of attracting new customers, while keeping old ones informed. Some appeared to be devoted mostly to spreading goodwill.

There was, of course, specialization within this specialty. For example,

Sourcebook was published by the 13-30 Corporation, but it was sponsored by the United States Army, directing its content to the youth market. While there were no stories or articles about the army, all the advertising was directed to it.

Automobile companies tend to publish travel magazines in this field; *Ford Times* was a pioneer. Today, *Datsun Discovery* and BMW's *Destinations* are new and well-edited entries. The former is sent by Datsun dealers to custom-ers, while the latter goes to a selected list of possible customers. As Pool noted, "Both keep promotion and advertising low-key, fostering instead the sense that the company is giving the consumer an interesting magazine as a gift."

Custom magazines have other characteristics. Sometimes they are sent to college students, a market difficult to reach; here product advertising is the motive. Unlike independent magazines, they are not required to earn money directly, but the sponsors hope they will do so indirectly, through sales. Adver-tising, however, is not a major factor. For some, it would be too difficult and costly to sell; others recognize that their advertising appeal is minimal. A few see product advertising as the only reason for their journal's existence, although they would never dilute its impact by soliciting advertising from any other company, however unrelated its product might be. Airline in-flight magazines are a substantial part of the custom-magazine category.

A few of these company sponsors have made attempts to move into the independent field. Mattel's *Barbie Magazine*, for example, was a natural to make such a move, and so was *Health Picture*, published by Miles Labora-tories. Such ventures raise the conflict of interest questions noted earlier.

The major battle in business magazines, aside from takeovers and acquisi-tions, has been the continuing fight for readers and advertisers on the part of the Big Three—*Business Week, Fortune,* and *Forbes*. All of them are aimed at substantially the same market—upper-level managers with high purchasing power—and so attract the same kind of advertiser. In 1980, their combined circulation was 2.2 million and their combined advertising revenues more than $230 million. These figures increased substantially during the ensuing decade.

Of the three, *Business Week* is the leader in both circulation and revenue and publishes weekly, while its rivals are biweekly. It is the star in McGraw-Hill's extensive stable of business magazines, and its advertising revenues sur-pass those of any other magazine, trade or consumer. Begun in 1929, just two weeks before the Great Crash, it has risen to first place since then against what it considers its chief rival, the *Wall Street Journal*, since both are in the news business and differ only in format. Its formidable staff gets material, like any newsmagazine, from bureaus here and abroad.

We have examined *Fortune* as part of the Luce empire. Those who remem-ber it in its earlier days recall the impressive graphics in its large pages, before all magazines came down to smaller standard sizes in the effort to cut costs and find space on newsstands when the number of magazines exploded. It was an early model of layout and design, important writers wrote for it, and it appeared to be relatively fearless in displaying the bad as well as the good side of business. In the seventies, it suffered a sharp decline, but it was redesigned to give it more "class" and continued to print in-depth examinations of major companies, some of which articles were much more critical after Luce's death; the founder could find little wrong with American business. *Fortune*'s annual listing of the 500 leading corporations, "the Fortune 500," as it is known, became an institution, widely quoted. The magazine still spends whatever is needed to provide the first-rate research materials its writers employ to cover the corporate scene.

Of the three, *Forbes* is the oldest, begun in 1917 by B. C. Forbes, who started as an immigrant printer's devil and became a doughty professional Scotsman (with authentic credentials), who loved to dress in kilts on social occasions and who possessed a keen, penetrating business mind. After his death, the magazine was directed by his son, Malcolm, who became an international celebrity, was seen with the rich and famous, and was noted both for his hobby, balloon flights, and his friendship with Elizabeth Taylor. Malcolm Forbes, who died in 1990, was also the magazine's hands-on publisher, establishing *Forbes* as a "capitalist tool," as its promotion called it. James Michael became its editor in 1960, and between the two, *Forbes* grew more rapidly than either of its competitors.

While *Fortune* sometimes exposed sharp practices and washed business linen ordinarily unwashed, *Forbes* established itself as a magazine that was both feisty and irreverent, like its publisher, and quite willing to report with excruciating honesty on the corporate world. This kind of enterprise won the publisher a designation as "publisher of the year" by the Magazine Publishers Association in 1984. His journal had already surpassed *Fortune* in advertising pages nine years earlier and by this time was read by more business executives than either of its rivals, not to mention Ronald Reagan's White House, which had four subscriptions (compared with two for *Fortune* and one for *Business Week*).

Forbes's utterly frank approach to business news made it highly successful but unloved in many quarters, particularly since the publisher's overwhelming public presence led to the publication of *Sayings of Chairman Malcolm*, which has nearly approached the circulation of Chairman Mao's. Like Hearst, Forbes was a rich man who knew how to enjoy his wealth, as only a man who owned sixty-eight motorcycles alone could understand. But perhaps beyond price are James Michael and the other editors of *Forbes* who have made the magazine such a distinctive success. As the publisher once said, "Our function in the magazine is to deal with and evaluate management." In doing so, friendships were strewn along the way and enemies made, by both publisher and magazine, but nothing stood in the path of success and the good life. The success inspired both *Fortune* and *Business Week* to sprightlier formats and writing. *Forbes* continued to advance, not only with its own "Forbes 500," to rival *Fortune*'s, but with a listing of the nation's 400 richest individuals, attracting an audience of non-business readers. According to Chairman Malcolm, his rivals had become his imitators.

26

Designing and Marketing the Modern Magazine

In marketing the modern magazine, specialization and quality of readership are the hallmarks. A good many publications have given up fighting for greater mass readership and are no longer trying to prove they can reach as many individuals as television does. Publishers remember what happened to *Look* and *Life* in the fifties when they engaged in a suicidal race to lure new subscribers. Now the trend is toward establishing elite audiences who will be loyal to the magazine, readers with the willingness and means to consume advertised goods. This trend is reflected in the titles of many new periodicals, obviously aimed at specialized audiences.

At the same time, there is a similar trend in the advertising business, as clients become more and more involved with targeted marketing and look for media able to offer clearly defined groups. Cable television and direct marketing are getting an increasing amount of advertising dollars, and that has affected the magazine publishers. In this particular feeding frenzy, magazine publishing has reached out to so many specialized markets that a periodical now exists for virtually every human interest.

"Targeting" began as a buzzword, but is now approaching the status of a science. Rather than trying to deliver as much of a particular group as possible, publishers today want to reach special audiences that are defined by specific demographic and geographic characteristics. According to publishing consultants David Foster Associates, the number of magazines issuing special geographic or demographic editions had risen to more than 200 by 1990. Cutting rate bases once was seen as a sign of decline, but analysts are viewing it increasingly as a smart move. The cut in circulation bases by both *TV Guide* and *Time* confirmed the trend. To succeed, new magazines may have to prove they can penetrate the market they have targeted, rather than rely on ever increasing circulation.

Combinations of paid and controlled circulation have long been commonplace in the business press, but trying to combine these two methods in various ways is comparatively new for consumer publications. For example, three magazines—*Insight*, the *Walking Magazine*, and *Inc.*—started as controlled circulation publications, then moved over to paid circulation types. Three other new publications—*Prevention, European Travel and Life*, and *American Health*—are entirely paid, but distribute controlled circulation copies to attractive markets they want to penetrate.

A somewhat different case is *Business Month*, previously titled *Dun's Business Month*. With a nonpaying circulation, it was running a distant fourth in the business-magazine field (behind *Business Week, Fortune,* and *Forbes*) when it began to trim its circulation to include only those corporate chiefs who requested the magazine. Narrowing its sights still further, *Business Month* specified that, to make the request, an executive must be working for a company with 500 or more employees or one grossing at least $50 million in sales. Advertisers were told that even though the magazine's circulation was lower than that of the leaders, its readership was composed of real executives and decision makers.

The argument over controlled circulation continues unabated. Norman Raben, president of Raben Publishing Company, issuer of the *Walking-Magazine,* has said that "while controlled circulation, carefully targeted, is an appropriate mechanism no matter how much publicity or success the device garners, it still confuses Madison Avenue, particularly junior-level media people." Some publishers use it to reach desirable reader markets, while others employ it when the advertising base is narrow and circulation is targeted at a very specifically defined group.

Although their eyes are often focused on advertisers, publishers are more and more concerned about the nature of their readers. Obviously, if these people are affluent and likely to buy advertised goods, it would be difficult to ask much more. But when reader demographics shift from types favored by advertisers, it can be a cause for concern.

Having the proper demographic readership can be important not only in obtaining advertising pages but also in selling a magazine's subscription list to other publishers, mail-order catalog houses, and direct mail companies. *Folio,* the magazine-industry trade journal, carries regularly a "List Watch," detailing newly available subscription lists (with their salient consumer characteristics). Subscription-list management agencies have also sprung up; for a fee, they will promote and handle a publisher's list.

One of the most difficult problems that magazines dependent on advertising revenue have always had to face is how to measure audiences. In the first decades of this century, simple circulation numbers were enough to satisfy advertisers. But as time went on, space buyers increasingly wanted to know exactly who and what those numbers represented. As we have seen earlier, such publishers as Curtis, Crowell, and Butterick pioneered in making readership studies, during the teens and twenties. They were joined by Time Inc. during the thirties, and by the forties, under pressure from radio and the imminence of television, magazines were not only measuring readership; they had moved to a broader concept of the audience. Now they measured how many people had seen or read the magazine, not simply how many had bought it. This kind of study continued as magazine publishers sought to compete with the wide reach of television. Several kinds of measurements were developed, including how long people kept a magazine in the house, how well they remembered advertising in it, and whether they recognized material in the publication.

In this era of specialized audiences, such measurements have continued with greater intensity and sophistication, and that has led to controversy in the industry. Advertisers and their agencies were suspicious of research figures produced by publishers. They came to believe that independent agencies could provide them with the objective, hard data about circulations and audiences they needed for making decisions. But the two firms measuring audiences for the industry over the past decade had already created a problem. Simmons Market Research Bureau (SMRB) and Mediamark Research, Inc. (MRI) had provided numbers that differed by as much as 30 percent for some magazines, as the result of using two different audience-measurement techniques. That difference only illuminated the fact that estimated audience figures can vary by millions, and the very portability of magazines makes readership difficult to measure. Nevertheless, publishers whose magazines were given wildly different ratings from the two services, including Time Inc., had reason to complain, and they did. Even those who were getting comparable figures found reasons for uneasiness.

Stepping into this controversy, the Advertising Research Foundation (ARF) established a committee whose five-year goal was to develop a "gold standard" for measuring consumer-magazine audiences, both paid and pass-along. To do so, in 1985 it devised a technique designed to get at the truth of magazine readership. Rather than relying solely on interviewing for measurement, as both SMRB and MRI were doing, the ARF committee's plan called for observation of individuals in such public places as doctors' and dentists' offices, beauty parlors, and barber shops. Researchers then set up interviews with people who had read the titles being measured.

Interviewees were shown full copies of the magazine, rather than the stripped-down issues used by the two research firms, and subjects were asked about their reading of the previous day, not of the preceding week or month, an extended measurement that both the agencies were using. The final phase of this study was due to be completed in 1989–1990, using spouses to "spy" on the reading behavior of their mates. While these techniques were too expensive to be performed regularly by a syndicated research company, ARF hoped that its audience numbers would come closer, consistently, to validating the necessary figures.

Market research has also become a valuable factor in the creation of new magazines. Time Inc. developed *People* after its readership studies disclosed that the "People" department in *Time* attracted a high number of readers. *Consumer Reports* launched its *Consumer Reports Travel Letter* after research showed an exceptionally strong interest in travel among readers of the parent magazine. Developing steadily over several decades, market research is now more widespread and pervasive than ever before.

As for magazines themselves, constantly advancing techniques in print technology have caused tremendous changes in appearance and content. Computerization of editorial and production processes has improved, and this has transformed periodicals. Still, important decisions about printing processes continue to face publishers, who must choose between letterpress, gravure, and offset lithography. Letterpress, in which the print is captured from a raised surface, is as old as Gutenberg. In the gravure process, images are etched into a cylinder, then reproduced from these depressions. With offset lithography, images are printed from a flat surface, which allows for camera-ready work, much increasing the possibilities for graphic design.

All three of these processes are used in the magazine business, and sometimes more than one is employed in the same journal, as is the case when a cover is printed from offset and the body of the periodical is done with gravure or letterpress. More and more, however, offset and gravure processes are coming to dominate, while use of traditional letterpress continues to decline. Gravure has proved to be best for print runs that are long and uninterrupted. Otherwise, web offset (using web presses with rolls of paper rather than sheet-fed presses) or letterpress are better choices. For example, when *Life* changed from gravure printing to web offset, its closing date was pushed back a week, permitting the magazine to offer advertisers split runs at a lower rate. (This was no casual change. *Life*'s printer for fifty-two years had been R. R. Donnelley; now it would be World Color Press, Inc.)

Today the new technologies of ink-jet printing and selective binding are changing the face of the magazine industry. These techniques (which have the side benefit of making sorting for the postal system easier) were developed in the late seventies and early eighties by Donnelley, the Chicago firm *Life* abandoned. Further technological advances in computers and printing will soon permit publishers to match profiles of their readers with the specific needs of advertisers, who will be able to target selectively, within a publication, those readers with the most potential as customers. List matching, ink-jet printing, and selective binding combine to make this possible. By late 1989, Time Inc. was already using ink-jet and selective binding in its publications.

New typesetting technology has also speeded up the process of magazine production, as well as increasing the complexity of design decisions and enhancing the status of magazine designers. In the past, designers and art directors had more narrowly defined jobs, limited mostly to choosing art for the magazine. *Fortune* was one of the first periodicals to exploit the further possibilities of design, under the direction of T. M. Cleland.

Fashion magazines, always concerned with the trendy, also gave their art directors a major design role. Alexey Brodovich, Russian-born art director of *Harper's Bazaar* in the thirties, once described poetically what he wanted to give his magazine. It was "a musical feeling," he wrote in *Print*, "a rhythm resulting from the interaction of space and time." Brodovich wanted the

Bazaar to read like a sheet of music. He and Carmel Snow, the editor, bemused inadvertent onlookers by dancing around layout pages spread before them on the floor, trying to pick up the rhythm (much as Hearst once danced among his empire's papers spread over the floor—he was not looking for rhythm, however, but mistakes made by his editors).

In the past fifteen years or so, with advances in technology providing greater flexibility, designers have become increasingly important in the magazine business. They work more easily now with the staff as it takes over functions previously performed by typesetters, using Compugraphic or similar equipment in the phototypesetting process that is now the most common method. This new process has proved easy to use, and as more and more publishers have installed it, cost of the machinery has decreased.

Inevitably an entirely new category of magazines has emerged to provide information (and advertising) about these new processes, enabling design staffs to keep up with the latest developments. Among them are *Graphic Arts Monthly, Computer Graphics World, Technical Communication, Magazine Design & Production, Typeworld,* and *U & Ic,* as well as a section in *Folio* called "Tech Trends."

Desktop publishing has been another innovation in the magazine world. This is made possible by personal computer systems which automate typesetting and layout, enabling the production of magazines of high quality at a relatively low price. It is an advance comparable with the low-cost techniques introduced around the turn of the century that permitted such publishers as Munsey, McClure, and Curtis to compete effectively with the older, high-quality, elite journals. Periodicals using desktop publishing programs to create their own product include *Contemporary Magazine, PC Publishing,* and *Capital.*

There are also a few of these desktop programs designed for art directors, including PageMaker, Xpress, and Venture. Still others allow artists to make line drawings on screen. Laser printers turn out high-quality copy for proofing and for the actual product. Desktop color programs for layout and line-art production are available, and computer experts are working on improving digital color photo processing. In fact, desktop publishing has not only greatly influenced magazine publishing and other fields but also promises to bring even greater changes in the future.

While the periodical industry will always be risky, as it is by nature, the flexibility provided by new technologies and consequent lower costs are likely to make it healthier than it was in the past. It has been an across-the-board revolution: writing, editing, drawing, typesetting, printing, and delivery.

However, it is still the editorial content and purpose of magazines that are the primary factors in whether or not they become successful, or stay so. A speculator looking for winners in the early nineties could be reasonably certain that specialized industry publications, offering summaries and analyses in particular fields, are best bets. But there are other growth categories, including magazines focusing on automotive interests, city and regional journals, travel magazines, the growing field of women's periodicals, the computer world, which is growing even faster, and an often overlooked category, crafts and games.

The paramount importance of covers continues. Publishers seek headlines, graphics, and logos likely to catch the eye on crowded newsstands. Circulation directors are often involved in the decision process, deciding which covers will most attract buyers. As we have seen so often in the pages of this history, the name of a magazine often has a decisive effect on its future. One consultant, John Klingel, advises that a title should be descriptive, one not easily confused with similar publications (health-magazine titles are particularly confusing), and, of course, one not in use or having been used by some other magazine.

In the intensely competitive magazine world of today, redesign has become a vital tool in keeping up with changing times and audiences. The formula for success is often elusive. There are conspicuous successes and failures. Anthea Disney, for example, was given barely a year by Condé Nast to give *Self* magazine a new appearance before the company decided to look elsewhere. *Ms.* was completely redesigned by its new Australian owners, without conspicuous success, and was ultimately sold to Dale W. Lang and Citicorp Venture Capital. In 1989, *Mother Jones* was redesigned under a new editor, Doug Foster, but it took a year of weekly staff meetings. The goal was to make a serious magazine more fun to read by adding a dynamic logo and employing a looser format. These were strategies calculated to increase readership, and they did.

In 1982, Condé Nast's *House and Garden* (now *HG*) underwent a major redesign and reconceptualization, including raising the cover price from one-and-a-half to four dollars, intended to reduce the readership to an affluent elite. *Apartment Life* was transformed to *Metropolitan Home*. As the decade advanced, there were revampings of *Rolling Stone, Psychology Today*, and *Redbook*, to name a few.

Redesigning is a delicate operation. The publisher and his staff have to weigh the risk of possibly alienating old, faithful readers against the possible gain of attracting new ones. Sometimes the approach is gradual and cautious, others do it in a single, bold new issue. As in other areas of magazine marketing, specialized agencies have sprung up to advise on the redesign process in the hope of avoiding serious mistakes.

But, as we have noted so often, it is specialization that has come to be the dominant factor in the magazine business, and one of its most striking aspects is the concept of "personalization." The idea first appeared in the mid-seventies, when, as Jerry Calabrese wrote in *Folio*, "the idea was justifiably heralded as a truly meaningful marketing vision—an impending breakthrough of such consequence that it would affect the ways in which publishers sold magazines as surely as offset printing changed the ways in which magazines were manufactured." Personalization has been compared with cable TV because it offers niches to advertisers and specialty goods to purchasers. While the reality has not yet lived up to the original vision, several magazines have implemented the personalization strategy. In *American Baby, Farm Journal,* and *Games*, for example, an advertiser can choose the reader segments most attractive for a product and arrange for the advertisement to run only in copies reaching those individuals.

Regional editions of magazines are another example of personalization, as are special editions directed to selected groups. An example is the *New Republic*'s 1986 "policymakers" edition, which advertisers were told would be sent to the magazine's 7,000 subscribers in the Washington, D.C., area. The journal's space salesmen claimed that this specially defined edition would reach Capitol Hill's policymakers, an audience they hoped corporate advertisers would find attractive.

Farm Journal, a magazine read by farmers all over America, pioneered a more sophisticated version of this strategy. Its subscribers receive customized editions, based on geographic region and farming specialty. A hog farmer with 50 acres in Iowa gets an edition with material different from the one that goes to the corn farmer with 150 acres in Kansas. Subscribers provide the *Journal* with demographic information that is then plugged into a database. Donnelley's Selectronic printing system enables these customized editions to be printed for various market segments. *Farm Journal* has mailed as many as 8,896 personalized editions for a single issue.

An even more radical marketing innovation is elective editorial content, in which a supplement can be bound into the regular edition of a magazine. It is selected and paid for by some but not all of the magazine's subscribers. For example, if the reader of a home magazine was an avid recipe collector and wanted more than those offered in any given issue, that person might be willing to pay an extra amount to get a supplement filled with additional recipes. This method was used by *Needlecraft for Today*, a craft magazine printing patterns for knitting, crocheting, home decorating, and a variety of craft products. In its July/August 1989 issue, it offered a supplement on knitting and crocheting, called (not surprisingly) "Knitting and Crochet Classics." Readers who wanted the supplement paid a surcharge for it; the insert contained no advertising, although it would not be difficult to market this kind of supplement to makers of related products. In another instance, *Games* magazine brought out a supplement called "Pencilwise," an expansion of one of the journal's regular features.

The production cost of developing an existing feature is small, as is the additional postage, since there is no new "per piece" charge as there would be for an entirely separate magazine. It is the binding of the additional material into an existing magazine that makes this strategy unique and a departure from the past, when publishers often issued related pamphlets, brochures, and books for their subscribers separately.

Such tactics have led to a further segmentation and refinement of the reader market. Jerry Calabrese, president of PSC Publications, which issues both *Games* and *Needlecraft for Today*, believes personalization is "a theory that a single magazine can, within its basic editorial franchise, service a number of discrete and varied reader interests by offering a number of elective editorial options to those readers." That kind of thinking could change the definition of a magazine, or at least the identity of different titles. Publishers employing this strategy obviously hope it will be profitable.

Specialization within a title also affects circulation figures, and reporting services may have to incorporate differentiated readership figures into their calculations. The end result is that a product line will exist within a given title, from basic to top of the line—in other words, coach to first class. R. R. Donnelley, the printers who developed this process, call it Selectronics. The company points out that the new technology not only will direct both advertising and editorial matter to selected audiences but also will help circulation directors with their renewal systems. Renewal notices can now be printed on magazine wrappers, or on a letter included with the magazine on the correct date for each subscriber. The system will also help production managers with mailing and sorting problems.

Still another innovation affecting the definition of a magazine is the video magalog, combining a magazine and a mail-order video catalog in one—recalling some of the old mail-order catalogs of the past. Among the experimenters with video magazines have been Time Inc., Hearst, *Family Circle*, and *Consumer Reports*. They have prepared shows using a magazine format, or devised programs based on magazine titles for presentation on cable channels.

On another front, the increasing price of consumer magazines, the result largely of paper and postage costs, has become a publishing problem. Many publishers in the late eighties decided to raise the percentage of the expenses paid by readers, thereby lessening that paid by advertisers. As a result, industry experts were predicting in 1989 that circulation revenues for consumer magazines would grow at a rate almost twice that of advertising revenues, which would decline because of less spending for advertising by several important industries, including computer, cigarette, and liquor producers. In fact, revenues from circulation dollars, as a percentage of total revenues, have been edging up for several years, reaching 52 percent in 1987, and at the end of the decade, they appeared to be still rising.

In promoting magazines, publishers use a variety of methods, including extensive advertising that heavily emphasizes the contents. Direct mail, print, and outdoor displays (bus-stop shelters, for example) are primary tools. But other promotional strategies, not focused on content alone, are also pursued vigorously, including premiums, coupons, and sweepstakes offers.

Special editions of magazines, produced on an annual basis, have drawn an audience consisting not only of regular subscribers but also of those who buy the annual one-shot for its own sake. The most successful of these ventures has been *Sports Illustrated*'s annual swimsuit edition. Other major successes have been *Seventeen*'s back-to-school issue, and *TV Guide*'s fall preview number. Varying percentages of these single-issue buyers become regular readers.

In magazine publishing, it is vital to be sure that the journals reach subscribers. For this, subscription fulfillment services are a help, keeping track of renewals, invoices, address changes, and providing reader service programs, as well as solving other subscription problems. Nevertheless, delivery perfection has not been reached and no clear solution has emerged. Increases in rates and poor delivery records by the Postal Service have compelled some publishers to look for alternatives. Recently, private delivery systems have offered one solution. When postal rates soared in the seventies, many publishers turned to private carriers. After the rates leveled off, partly because of the competition, some publishers returned to the regular system, but others stayed with the alternative distributors. Time Inc. and Meredith, for example, were two of the major companies continuing to use private carriers for part of their deliveries. By 1980, private services were carrying an estimated 59 million items to homes, but the Postal Service still carries most of the burden, even though costs are comparable in the two systems.

Reporting on private delivery services in 1988, *Folio* observed that "benefits go beyond cost. Catalogs and local ad matter can be polybagged with the magazines, as can subscription, renewal, and moving notices." Of the private entrepreneurs, United Delivery Systems is the largest, focusing on the Middle West. Such a system lends itself best to larger publishers. For the smaller ones, it is an option only if their readership is primarily regional. *Seven Days*, targeting a New York City audience, is an example.

Obviously, maintaining timely delivery to subscription customers remains a high priority for publishers, but single-copy sales present a different problem. There the object is to gain shelf space and strategic positioning. Convenience stores and supermarkets are some of the most important outlets. In an effort to help publishers obtain good placement and adequate display space, the Magazine Publishers Association has sponsored studies showing that magazines are among the most profitable items in these stores. The studies used a measurement system specially devised by the association called Direct Product Profit. While retailers in general have not been entirely convinced by the information it produced, some publishers believe the results have made it easier to maintain their current space. Gaining retailer acceptance, in general, is a key marketing factor for magazines trying to sell single copies.

Since the fifties, publishers have been offering retail discount allowances to outlets, usually those part of a large chain. A percentage is paid the merchant to display the magazine prominently—typically near a checkout counter. This percentage is in addition to the normal profit the retailer gets. In the seventies, the extra payment became standard practice for the major magazines, and by 1978, an estimated $78 million was being spent to ensure prime display space.

A new trend may be emerging, however. In September 1989, *Reader's Digest* announced that it was abandoning its retail-display allowance program. Instead, it would put that money toward straight discounts off the cover price for retailers, thus following the established practice of those selling non-magazine products in handling retailer discounts. The *Digest* made this move in an effort to make itself more competitive with other items featured in retail stores. While no other publisher had yet followed suit by the end of the eighties, industry experts think many will do so in time. Some objections to abandoning the old system have been raised. One is that the publisher will lose some contact with the retailer, since wholesalers pass on the discounted cover price.

Waiting rooms have long been known by magazine marketers as prime placement spots, and in the eighties they were being exploited even more. Research estimated that as many as twelve readers may see an individual copy of a publication while they wait for a doctor or dentist. *Vogue* and *Life* (in its reincarnation) have routinely claimed large readership in such public places. Advertisers, however, have less faith in these figures. They argue that the demographics of waiting-room readers are unknown, and they are also skeptical about how much a reader can concentrate on a magazine and its advertising while worrying about the outcome of a visit. One media director has observed: "To pass time reading is one thing. To seek out a reading opportunity is something a whole lot more attractive to advertisers." Nevertheless, there is certain to be more research into the nature of public-place magazine readers.

Whittle Communications led the industry in penetrating the waiting-room audience in the late eighties. The company's requesting a guarantee of semi-exclusivity was a new approach; it signed up doctors in family and obstetrics/gynecology practice to place its new publication *Special Reports* in their waiting rooms, the doctors agreeing to display no more than two other titles. Advertisers responded with some enthusiasm to this concept of exclusivity, and other publishers were quick to follow, including those issuing *Family Media, Medical Economics,* and *Healthteam Communications.*

Once begun, such ideas are likely to develop rapidly. A newsstand consulting agency, Publishing Management Services, soon launched Waiting Room Subscription Service, distributing existing magazine titles, encased in plastic binders, to waiting rooms—free of charge. The backs of the binders were sold to advertisers, providing the revenue for this service. Exclusion of other titles was not required in this operation.

For many publishers, targeting advertisers must be the primary objective. Ink-jet printing and selective binding have helped, as we have seen. These processes permit advertisers to personalize advertising and that, in turn, expands the possibilities offered by magazines as an advertising medium. Commenting on the implications of this technology in 1989, David Braun, media and promotion director for General Foods, remarked, "If a mass-audience magazine like *Time* has the ability to merge and purge its subscriber list with a mail-order list, it becomes a more relevant media choice for advertisers who previously would have dismissed the magazine as irrelevant."

Another technique to attract advertisers is to offer them merchandising services, giving them added value without cutting rates. Some publishers provide such programs free of charge, but others expect the advertiser to pay, either wholly or in part. Most magazines providing merchandising services do so to help maintain current customers, not to attract new advertising accounts.

(Exceptions may be made for an unusually promising prospect.) Merchandising efforts need be limited only by the publisher's imagination. Good programs promote both the advertiser's product and the publisher's magazine. The route is not without its dangers, however. The object is not to become the promotion department for an advertising client. The kind of merchandising services provided depends on the periodical. Point of purchase displays, contests and sweepstakes, special issue tie-ins, in-store events, fashion shows— these are only a few of the services consumer magazines can offer. Some journals hold out promises of free television commercials to those who buy a certain amount of advertising space.

There is no limit to the ingenuity displayed by the promoters of merchandising services. *San Francisco Foods* sponsors a Golden Shears Award Program each year, awarding prizes to fashion designers whose clothes are then featured in the "Style" supplement of the magazine. *Better Homes and Gardens* offered advertisers a chance to have their recipes and products shown in model kitchens set up by the journal in malls across the country; advertisers were charged only for the food. *Family Circle* sponsored "The Crisco Great American Pie Celebration," bringing the attention of consumers both to the magazine and to their advertiser, Crisco. *Builder* erects an "Idea house" at the National Association of Home Builders Show every year, filling it with products advertised in its magazines. Business magazines often provide less elaborate programs for advertisers, offering events at trade shows, reader services, reprints of articles or extra copies of magazines, and market research studies.

Merchandising support by publishers for advertisers is not entirely a new idea. The research departments of magazines provided research reports as a service in the early twentieth century. Even earlier, the Curtis Company was using end-of-aisle displays for products advertised in the *Ladies' Home Journal*, with a poster board advising the public that these items were "as advertised" in the magazine.

Today, however, such services can become a crucial selling point in the competition for advertisers. As Les Ziefman, *Rolling Stone*'s associate publisher in 1988, observed: "For magazines like *Rolling Stone* that won't go off the rate card, value-added service is critical. It tells advertisers that while no one else is getting a better rate, they're still getting added value for their investment. We think that's a far more constructive approach than negotiating discounts." However, negotiating rates continues to be a fact of life in the magazine world.

Looking toward the future, a new strategy is envisioned for a select group of media owners—the "multimedia buy." Such giants as Hearst and Time/Warner could offer advertisers a package including space in several of their properties (magazines, cable, books, videos, newspapers). Such offerings are still relatively rare, but they will become more common in the future, squeezing publishers who own only periodicals.

One strategy for such publishers would be to become part of a magazine advertising package. These owners could form a network with other publishers of single titles (or groups of titles that are not heterogeneous from the advertising viewpoint). One example of such an arrangement already in operation is the Leadership Network, composed of nine "thoughtleader" titles, offering discount buys to national advertisers. The Network does sales and marketing for all the titles, thus spreading the expense of selling space to national advertisers, who can buy several or all of the titles.

Advertising revenues have been rising steadily since the Network began in 1978. The idea is portable, too, as city and regional magazines have demonstrated. On the West Coast, the California City Magazine Network brings together *San Francisco Focus, Los Angeles*, and *San Diego* for the purpose of selling to national advertisers. The magazines, of course, are also sold individually to local advertisers.

For the publisher of several strong titles, corporate packages of magazines can help in the startup of new titles. Condé Nast's LTD Plus, for instance, offers a package that includes *Vogue, GQ, HG* and *Gourmet,* and this piggybacking helped launch such new titles as *Vanity Fair* and *Traveler.* The Hearst Corporation has four packages to offer advertisers: Gold Buy, Woman Power, Man Power, and the Hearst Home Buy. Individual publishers can offer corporate discounts, based on total insertions in all the publisher's magazines. While this basic strategy is as old as the turn of the century, when Curtis, Butterick, and others were offering space in several of their publications at a discount, it is still a powerful persuader. Today single advertiser editions of titles are also appearing, in which one advertiser sponsors an issue of an entire magazine.

As technology advances and magazine marketers continue to devise new strategies, the production, design, and sale of magazines will unquestionably continue to change and be even more innovative. For those who might ask whether creating and marketing a magazine, or a group of them, is worth the tremendous effort and the intense competition, one has only to look at the astronomical figures involved in the wheeling and dealing of the present-day giants in the field. To begin with, companies today are worth billions, not millions. Individual profits from sales of periodicals run upward from $100 million. Moreover, profits from sales have increased the already high number of new publications launched every year; at least 500 were expected in 1989 alone.

When Peter Diamandis sold his Diamandis Communications, the group he bought from CBS, to Hachette, the French conglomerate, he got the largest part of a four-way split of $100 million. In this case, he remained with the organization to operate it and to introduce new titles. Diamandis had paid $650 million in 1987 when he bought these periodicals from CBS, and after first selling some of them for $200 million, he sold those remaining to Hachette for $712 million. He had owned the whole group for only eight months.

Individuals with only one magazine can do as well, comparatively. Gerry M. Ritterman had owned his *Soap Opera Digest* for only three years before selling it to Murdoch Magazines, retaining more than $50 million for himself out of the $70 million price. Only a day later, he offered to buy another publishing company with annual revenues of approximately $40 million.

Industry analysts are skeptical about the future, but some publishers, Diamandis for one, believe values will continue to rise in the nineties. The tremendous prices appear to be encouraging new floods of magazines, and where it all will end, as Wolcott Gibbs observed of Time Inc., "knows God."

27

New Horizons

As the magazine industry, celebrating in 1991 the 250th anniversary of its beginnings, moved into the last decade of the twentieth century, the vibrancy, innovation, and constant change that have always characterized it were more evident than ever. Trends toward consolidation continued, but the emphasis was on even further specialization.

On newsstands, it had become more of a struggle to survive than ever before, and yet, in spite of all odds, new magazines rolled off the presses every year. Samir A. Husni, a professor of journalism at the University of Mississippi, who specializes in the study of new magazines, reported that 491 publications were introduced in 1988, of which few could be expected to last another year. Only two out of ten last longer than four years. A major reason is that the average reader spends only two to five seconds per magazine at a newsstand.

Still they come. When 477 new consumer magazines were launched in 1987, they represented a 28 percent increase over 1986, and growth shows no sign of slackening. There are so many entrepreneurs wanting a piece of the action that an association of magazine executives called the Successful Magazine Publishing Group has set up an information hot line for prospective publishers who need advice. As Wayne Curtis wrote in the *New York Times*, on January 22, 1989, "The competition promises to turn from fierce to savage." It could hardly be otherwise, with more than 3,000 consumer publications alone fighting for space on the newsstands. While reliable statistics on the total number of magazines are difficult to come by, the 1989 edition of the National Directory of Magazines listed 17,000, counting the trade press, academic journals, and those printed in Canada. By other estimates, that figure still falls about 5,000 short.

The explosion of magazines which began in earnest during the eighties can be attributed to the overall boom in the production and dissemination of information, seemingly the key word of the century. On-line data bases and cable television are proliferating as rapidly. The demand for information is so great

371

today that readers were willing to pay $3.25 for the average new magazine in 1987, which was $1.25 more than for those already existing. That has made it possible for some to rely a little less on advertising.

Hardcover books were once the primary source of most information, and they, too, continue to grow in volume, but they have to share the market with magazines, which provide the same broad variety. There is a crossover, too, as magazines have invaded the book business with annuals that advise consumers on buying everything from computers to sports gear. Inevitable specialization has also provided a proliferation of newsletters and data bases for subjects that may not be covered by the other media. Newsletter enthusiasts once believed they would rule the media world, but experience has shown that readers are too conditioned to magazine formats and are not likely to foresake them.

Specialization has now gone so far that some experts believe, with John Fitz-maurice, president of the Periodicals Institute, that "the day of the big block-buster magazine is over." The titles of new publications tell the story: *Southern Bride, Car Stereo Review,* and *Hispanic Entrepreneur* are examples of the infinite possibilities in subdividing markets. Sports, music, automobile, and puzzle journals were the most popular categories among the new 1987 magazines; they accounted for 26 percent of new titles, as Curtis reported in his *Times* survey. Again, the titles are revealing: *Sports History, Croquet Today, Celebrity Wrestling, Power Metal, Thrash Metal,* and *Metal Madness.* (The "metal" titles, of course, refer to rock music, not industrial products.)

The hot markets of the early nineties, according to Professor Husni, will be both ends of the age scale, elderly Americans and teenagers. *Modern Maturity,* the giant leader, has already been joined by *Second Wind, Longevity,* and *Access Chicago.* A new journal for teenagers seems to be born every month. Husni warns that too much specialization will reduce the pool of potential advertisers, and so produce a major shakeout.

In 1988, the magazine industry thought it saw a weakness in television and launched a $30 million advertising campaign that was not anti-television but pro-magazine. Created by the Magazine Publishers Association, it was the largest campaign the industry had ever put behind its own product. As a prototype, the advertising created Maggie, an attractive superwoman intended to be the personification of all periodicals. She could be seen, for example, in a vacant, rubble-strewn convention hall with one line of copy: "Maggie fills in gaps the conventions leave behind." Other advertisements focused on various advertising categories that were particularly lucrative. For example, Maggie was shown riding a bicycle on a boardwalk, the rising sun beyond her. "Maggie shows me how to look great—and stay in shape," the copy read, directed to advertisers of health and beauty aids. In still another, intended for food purveyors, Maggie was shown serving a meal on a patio, with a line reading, "Hot food news (or cold)—Maggie knows how to dish it up." Fashion, travel, automotive, and business-financial categories were also covered in this campaign. To emphasize that she was only a synonym and symbol, Maggie's face was never clearly shown. The moral: Magazine advertising can do things everybody believes TV has always done—that is, create awareness and generate the switching of attitudes.

In spite of these efforts, however, the drop in advertising that began in 1987 continued to afflict magazines two years later, and circulation directors were experiencing a reprise of industry history, when subscriptions and newsstand sales were primary sources of revenue. According to the *New York Times*, trade organizations and industry analysts were reporting in mid-1989 that these revenues were running ahead of advertising. In the mid-seventies, advertising sales accounted for nearly two-thirds of magazine income; circulation, one-third. By 1987, however, circulation in consumer magazines was recording $6 billion in revenues, compared with $5.5 billion in advertising.

Responding to this persisting trend as the last decade of the century began, publishers were being compelled to spend more on production and editorial content to keep reader demand as high as possible. There have been, and apparently will continue to be, continued improvements in the quality of all the components that make up a magazine. This trend has also given circulation directors an unprecedented role in making such editorial decisions as the content of covers.

Several factors have fueled the trend toward more dependence on circulation, including pressures resulting from the growth of direct-mail advertising and consolidations in the retail industry. Fortunately, audiences have not resisted higher cover prices, even though they have more titles than ever from which to choose. The figures are astonishing. Paid circulation in the entire industry was 357,000,000 copies an issue in 1989, 35 percent higher than it was in 1979, and there are now more than 3,000 consumer magazines (with the statistical qualifications noted earlier).

Subscription prices have been rising since the early seventies, as publishers sought to offset increasing postal rates and paper prices. John M. Thornton, circulation director of *Forbes*, believes that both *Look* and *Life* failed in their original formats because subscription prices were so low that when advertising diminished, nothing was left to support these periodicals. Steven T. Florio may have had that lesson in mind when he became publisher of the *New Yorker* in 1985. He found a magazine whose policy was to reject any substantial discounting of its subscription rate, even though both circulation and advertising were declining. While he went about increasing promotion efforts sharply, Florio began to offer cut-rate subscriptions, but the cover price remained high enough so that he could show a profit on every issue. This policy paid off; circulation rose dramatically, even as advertising sales declined further.

No one can predict with confidence what the economy will do, but in 1990, most publishers expected that the long slump in advertising would reverse itself in time, and consequently there would be less reliance on circulation. Only a handful of magazines, in fact, could exist simply on circulation.

With magazines aggressively seeking new outlets to increase circulation, they were finding themselves in a much closer partnership with bookstores. Customer demand in the eighties had compelled booksellers to carry an ever-increasing range of journal titles, popular as well as obscure, Catherine Whitney reported in *Publishers Weekly*. At the supermarket checkout racks, fewer than five magazines were getting about 30 percent of the sales. Presumably, bookstores could give the others a better opportunity, offering them to people who were already certified readers. Booksellers welcomed the idea because they believed that selling magazines would attract younger, upscale shoppers. Magazines, it was thought, had something in common with quick-reading, romantic novels, since both appealed to those traveling in the fast lane. They were "stress busters," or "light reads."

As early as 1979, the two major bookstore chains, Waldenbooks and B. Dalton, were carrying magazines in all their outlets, and other chains were compelled to follow. Independent stores realized that they would have to join in if they meant to stay in a competitive position.

Assessing the result in 1988, Whitney reported that in some ways magazines had fulfilled their promise. More traffic had been generated, and the bookstore mix was better. On the other hand, booksellers have had to adjust to an unfamiliar distribution system, more complicated than the one used in book publishing. As for sales, there were few surprises. Magazines that were national best-sellers on the newsstands repeated their success in the stores. A closer examination of the figures, however, shows that regional magazines have become an important, growing category in the bookstores, with specialty journals close behind.

Bookstore sales reflect general trends in the magazine business. Women's magazines, more than fifty of them available, are the staples month after month, as are the newsmagazines. In bookstores, a *Publishers Weekly* survey reported, the best-selling categories in order were women's general magazines; news; art/literary; home, garden; how-to; computer; business-finance; social/political; men's general; entertainment/celebrity; and regional. Outside the mass market, bookstores have done particularly well with magazines devoted to art and literature. Best sellers include the *Atlantic, Connoisseur, Mother Jones,* the *New Yorker, Town & Country,* and *Vanity Fair.*

Among all the categories, booksellers especially favor the home and how-to journals, perhaps because they supplement and encourage the sale of book titles in the same category. These home-oriented magazines had been slipping in the late seventies and early eighties, but they revived handsomely in the late eighties, especially the how-to group. A strong element in this revival was the new baby boom at the end of the decade, spawning an entirely new category of local magazines for parents, appearing in more than seventy American cities. Their titles were inventive and various: *Child's Play,* in Springfield, Massachusetts; *Pierce County Parent,* in Tacoma; *Mother's Network,* in St. Petersburg; *All Kids Considered,* in Detroit. They told readers how to find pediatricians and tennis instructors, reviewed movies from the perspective of the children themselves, and covered the latest in children's fashions. Even though such national publications as *Parents, Parenting,* and *Child* were also experiencing growth, the local journals did equally well during the eighties. Most are free. Their circulation ranges from 5,000 to 90,000, depending on the community's size, and they appear once a month or at longer intervals. The outstanding success among these new magazines has been *L.A. Parent,* which had revenues of $2 million in 1988. Magazines in this category that are well managed can expect profits running as high as 20 percent before taxes, on revenues of $500,000 or more.

Expansion in the field was quick to come. Wingate Enterprises, Ltd., publishers of *L.A. Parent,* had two other magazines for parents in Southern California and planned still another start in Phoenix. By 1989, there were enough of these small magazines, thirty-seven all told, to form an association seeking national advertising and forming a clearing house for ideas. The magazines in this group had a total circulation of more than 1.5 million, only 300,000 behind the paid circulation of *Parents.*

Environmentalism offered another market for magazine entrepreneurs at the end of the century. Most of the existing journals had depended on such organizations as the Sierra Club and Greenpeace for support, but in 1989 the new quarterly *Buzzworm* appeared as an independent journal. Published in Boulder, Colorado, and taking its name from the old frontier term for a rattlesnake, it attempted to raise the level of the standard environmental-information journal to the appearance of a slick magazine. Selling 35,000 on the newsstand, at a cover price of three dollars, it had an initial circulation of 50,000. The founder, Joseph E. Daniel, began with a small investment from a Colorado entrepreneur and about $100,000 from a private stock offering. After selling $150,000 more in stock and raising an additional $200,000 in short-term loans, *Buzzworm* was still not making money in 1989. However, this periodical and others in the field are reaching into an expanding market that numbers about 20 million people, according to *American Demographics* magazine, who contribute to environmental causes and belong to conservation groups. Articles for this audience range from practical pieces on civil disobedience to the use of chimpanzees in AIDS research.

Ironically, at least one other environmental-magazine editor took a negative environmental view of *Buzzworm.* Garth Smith, editor of *Earth Island,* published in San Francisco by the Earth Island Institute, pointed out that the materials in slick paper magazines cannot be recycled and often contain heavy metals. *Buzzworm* responded that slick paper could, indeed, be recycled, and in any case, its use was necessary to compete with other special interest magazines.

As the decade of the nineties began, environmentalism had become a "hot" topic in the magazine business, with entrepreneurs scenting a relatively untapped market of readers and advertisers, ready for mass-market magazines dealing with a subject already producing numerous books and newspaper articles. A forerunner of things to come, perhaps, was *Garbage,* a bimonthly calling itself "the practical journal for the environment," which appeared in August 1989. Its editor and publisher, Patricia Poore, said she had created it to answer basic environmental questions. *Garbage* demonstrated what starting a magazine on a shoestring means these days. Its October issue carried only thirteen pages of advertising out of a sixty-eight-page total, with a cover price of $2.95 and a guaranteed circulation of 50,000.

At the end of the year, another entry in the field appeared, a bimonthly called simply *E*, intended to combine the best characteristics of a general periodical and one appealing to environmental activists. A nonprofit journal, it was financed by bank loans, foundation grants, and advertising, with a guaranteed circulation of 50,000.

In the periodical business, as in physics, for every (or nearly every) action there is an equal and opposite reaction. That was demonstrated in mid-1989 at a time when the weight of advertising was shifting to trendy magazines. Suddenly there appeared a journal heading in exactly the opposite direction, whose backers were convinced that a market existed for an old-fashioned, folksy periodical, like the old *Post*, or *Collier's*. They invested $3 million to launch *Wigwag*, a monthly whose first issue was only eighty pages but whose cover price was three dollars, with a guaranteed paid circulation of 60,000. The journal's odd name was derived from one of several definitions of that word: "to signal someone home."

Alexander Kaplen, the 29-year-old editor and founder, had been a staff member at the *New Yorker* and assembled seven other refugees from that magazine on his staff of nine. While *Wigwag* was modeled somewhat after the *New Yorker*, it was described as "livelier, funnier and sweeter." The editorial mix was much the same: "Letters from Home," (reports from correspondents' home towns), profiles, investigative articles, humor, reviews, political commentary, and fiction—even a children's bedtime story. Among its early contributors were Ralph Ellison, Tina Howe, Peter Matthieson, and Garry Wills.

In the search for new markets, all sorts of oddities were developing. One, emanating from Washington, was the *World & I*, which advertised itself as "the ultimate reading experience," offering in every month's issue "more than 700 pages of perceptive analyses, reliable reporting and expert opinion," besides a "wealth of color photographs and fine reproduction." In this self-described "encyclopedia" magazine could be found coverage of the arts, modern thought, scientific developments, and reviews of books. For this abundance, premium prices were asked—ten dollars per monthly issue at Walden bookstores, selected B. Dalton and Crown stores, as well as college and other independent outlets.

Still the search goes on. With magazines available for men, women, brides, parents, and children, somehow the newly married had been overlooked until 1989, when *Newly Wed* appeared, to be published three times a year by the J/C Publishing Company in Westport, Connecticut. Its announced circulation of 1.25 million was achieved by its distribution through Gift-Fax, Inc., an organization sending sample products to brides-to-be about two weeks before their weddings.

The push toward new markets had become more international by 1990, when several American publishers whose concerns had been primarily domestic began to look with new interest at Europe, from where conglomerates had been steadily invading the American market for some time. Hearst, for example, had licensed several of its titles abroad—*Harper's Bazaar, Esquire,* and

Cosmopolitan—but late in 1989 announced it was establishing a new operating unit, Hearst Magazines International. Licensing had required no investment, but now the corporation was ready to spend money to develop international advertising packages. At about the same time, Meredith announced it was entering a joint venture with Harmsworth Magazines, Ltd., a new subsidiary created by Associated Newspapers P.L.C. in an equal partnership to produce, first, *Metropolitan Home U.K.* as a companion to its American counterpart. This was a notable departure for Meredith, which had not been involved in any European operations, and indicated that perhaps America had learned overseas ventures could be as profitable as the Europeans had proved with their acquisitions in this country.

If there are further markets to be explored, as of course there are, one of the chief explorers is likely to be Christopher Whittle, who began his publishing career in 1968 in partnership with Phillip Moffitt, as the 13-30 Corporation, now a part of Time Inc. Based in Knoxville, they began with four annuals, *Nutshell, Graduate, 18 Almanac,* and *New Marriage,* all of them built around a single idea, "a point in time." Their market was readers between thirteen and thirty.

By 1989, as N. R. Kleinfield reported in the *New York Times,* Whittle Communications had brought forty-two projects into the world. Their titles disclosed the diversity of their marketing: *Go!, Girls Only, Dental Health Adviser, Pet Care Report,* and a variety of wall posters. There were also the *Special Report* magazines, designed for doctors' waiting rooms, and Whittle's controversial plan called Channel One to bring television news into the classroom, subsidized by commercials. The company began to publish books in the autumn of 1989, was preparing to start several new magazines, and planned television ventures for the Hispanic market.

Whittle himself is as unusual as his publications. A multimillionaire at forty-one, he had lived in a two-room log cabin in Knoxville until he moved into an eight-room apartment in 1989. But he also owns a five-room apartment in the Dakota, the historic apartment building on Central Park West in New York, and a dairy farm in Vermont. Kleinfield describes him as a "slight, mophaired, bow-tied man with an intoxicating smile and a Southern drawl. He's as polite as the corner grocer." His peers in the business consider him a brilliant salesman.

Looking into the future, Whittle believes that the media business is disassembling itself, and he intends to go on inventing things and making structural changes. Noting that the average number of readers per copy of most magazines is from three to five, Whittle points to his fifty readers per copy for *Special Reports* and to more than a thousand for some of the company's "wall media," as they are called. Whittle believes there is a fifth communications medium somewhere, which will be a "giant step beyond newspapers, magazines, radio, and television."

To explore such possibilities, and to be editor in chief of all his magazines, Whittle hired William Rukeyser, former editor of *Fortune* and *Money,* in 1988. Like others venturing into the brave new world of the next decade and next century, Whittle is willing to take risks, and in fact he has had between twelve and fifteen projects that have failed during his career in the business. But in 1988, Whittle also had sales of $106 million and pretax profits of $14 million, with sales of $315 million predicted by 1991. Whittle himself still owns 11 percent of the company after its acquisition by Time Inc.

Others are looking to the future of magazines, trying to predict what may happen and how publishers can take advantage of changing times. The difficulty of accurate forecasting was evident in 1989 not only by the advent of *Wigwag* but also by a report from Southern California, the most unlikely part of the country, that serious magazines were beginning to compete with the merely trendy publications. There were several new, or recently revived, journals, some of them national, dealing with public policy, politics, international affairs, and art, growing fast and embracing the entire Pacific Ocean community of nations. Their view was toward the twenty-first century.

One of these is *New Perspectives Quarterly*, more conveniently called *NPQ*, described by its editor as a post-liberal quarterly dealing with global issues. It began as a thin newsletter, published by the Center for the Study of Democratic Institutions, in Santa Barbara, Robert Hutchins's original creation. The magazine was financed under unique circumstances. Stanley K. Sheinbaum, a rich contributor to liberal causes, sold a Willem de Kooning portrait for $1.6 million to establish *NPQ*.

On the opposite side of the political fence is *Reason*, a libertarian monthly in Santa Monica, also with a national circulation. It, too, began as a mimeographed sheet, at Boston University in 1968, moving to California with its publisher, Robert W. Poole, Jr., a former aerospace engineer. With a circulation of 28,000, it is a publication of the Reason Foundation and calls itself "the only national think magazine outside the Northeast corridor." What it thinks is that many government functions should be turned over to the private sector. It is sent to Congress and other "power centers."

Two other new Southern California magazines are local in circulation. *L.A. Weekly*, begun in 1979, is now a thick tabloid, leftist in its politics but also covering fashion, style, and the arts, with the most comprehensive listing of music, theater, and art in the city. Advertisers are likely to be psychics and purveyors of New Age materials, but the magazine itself intends to concentrate more on Los Angeles as a "crucible for social experiment."

In an ambitious attempt to cover the culture of the entire Pacific, a former New York advertising man, Robert D. Crothers, has started a bimonthly magazine called *Artcoast*, whose editor is the well-known New York art critic Kay Larson. It began with a 20,000 circulation.

Perhaps the most successful of the new West Coast magazines are those published by California Magazines. The company's flagship, *California*, has printed provocative articles that have won it several National Magazine Awards. Its *Angeles Magazine*, in Los Angeles, has not only won awards but claims to have more millionaires per thousand subscribers than any other major publication in the world. As 1989 ended, this success was being translated to San Francisco with *SF*, targeted to that city and tailored to its interests as *Angeles* has done in L.A.

Reaching out in another direction, Boston businessman Robert Friedmann and a Russian émigré impressario, Natan Slezinger, were planning, in 1989, two bilingual publications that would receive the cooperation of two Soviet organizations—yet another product of *glasnost*. One was to be a quarterly trade magazine and import-export guide called *Moscow Business Journal*, published in both countries. The other was a consumer quarterly called *Gosconcert Magazine*, which would track Western and Soviet performers touring in each other's countries. It would be published in the Soviet Union initially, possibly in both countries later. Advertising, a relatively new institution in the Soviet Union, was to be the chief source of revenue for both magazines, with the *Journal* carrying both Western and Soviet advertisements, while *Gosconcert* would, at first, carry only Western products. Gosconcert, the Soviet organization that arranges performances, had already ordered a million copies of the magazine, while the *Journal*, printed in Moscow, was starting with an initial circulation of 100,000 in that country and probably 50,000 in the United States.

Whatever ventures are made in the future, they will be conditioned by new technology. While only futurists would care to predict how profound the changes may be in the industry, they are not reluctant to do so, and the outlook is—futuristic. Some changes are already underway. In the competition for advertisers, the computerized process called selective binding is creating custom editions for a few pioneering publishers.

Selective binding, as we have seen, enables publishers to take precision aim at targets. *American Baby*, for example, collects information from its million subscribers about when babies are expected, so that they can be sent either a prenatal or a postnatal edition. Another edition has inserts sent only to mothers with babies three months old, a period when advertisers believe brand loyalty is established. The benefits to advertisers in this kind of targeting are obvious.

There may be some drawbacks, however. The circulation director of *InfoWorld*, a weekly tabloid distributed to 186,000 computer users, warns that advertisers might shift to only one part of a press run. *InfoWorld* itself targets two of its own editions, one for those who own Apple MacIntosh computers, and one for readers who want to know about local area networks. In the future, if smaller magazines and new arrivals offer selective binding, advertisers may begin to demand the same service from magazines already established. In any case, as the result of this invention alone, publishing will certainly not remain the same.

Another innovation is the practice of delivering magazines on videocassettes. There were only a few of them in the United States in 1989, most of them devoted to sports, but the possibilities are plain to publishers. Already there is a quarterly video magazine called *Set Sail*, published by Passage Home Communications, running for eighty minutes and including segments on racing in the Caribbean, sailing to Hawaii and the Grenadines, and on how to select a sailing school. There are 5,000 subscribers, and those who buy it retail pay $10.95. Advertised in sailing magazines, it is distributed primarily through boat shops.

This medium is easily applied to any audience. *Travelquest,* for example, sells eight issues of its video magazine per year for $99.95, for which viewers can vicariously visit exotic destinations; included are five minutes of soft-sell commercials. Distributed initially, at least, by travel agents, the circulation of these videotapes could in time match the circulation of travel magazines in printed form.

But again, there are limitations. Expanding circulation is difficult because it is hard to drop prices, since it costs more to make the tapes than similar articles would cost in print journals, and duplicating videos is also expensive. The size of the market depends on the number of those in the audience who have VCRs. Nearly half of American households do not, although they may be as common as television itself in another decade or so. Then there is the question of reading habits. Watching television is one thing, but watching a magazine on television may not be something readers want to do at this juncture. The video publishers believe they can overcome this objection through hand-held VCRs, already available, but that remains to be seen. Whether commuters, for instance, would rather take a portable player with them on the train or bus instead of the more convenient magazines is doubtful, many experts believe. On the other hand, Japan has already seen a rise in video magazines as a result of Sony's portable player.

Publishers of video magazines believe that when their circulations rise to 20,000 or more, advertisers will take notice, but if so, there will have to be a change in the cost of producing video advertising. Basically, too, there is the problem of attracting enough subscribers to attract enough advertising. A step toward the future was taken by the Audit Bureau of Circulations in 1988 when it agreed to audit its first video magazine, *Travelquest.*

One of the qualified experts who has looked into the future is Steve Kurtzer, vice president and media supervisor at BBD&O. Writing in *Marketing & Media Decisions* in 1985, he could see no radical changes in the next decade but concluded the growing need for information would be met in new ways. He foresaw issues of a magazine supported by a single advertiser, a natural extension of insert supplements. There would also be, he said, several new categories by 1998 that had not existed before: magazines for people working at home, life-style journals for aging baby boomers, science and technology journals covering new areas, among others. General circulation magazines would experience some reduction in readership, he said, but they would survive because they served a purpose.

Assessing things to come in 1989, Albert Scardino wrote in the *New York Times* that "in the not too distant future, each magazine will be a one-of-a-kind collector's item. Target marketing has already given us advertising customized for each zip code. Now the process has grown so refined that magazine advertisers can vary the message household by household." He recalled that *Newsweek, Time, American Baby, Harper's,* and *Games* had already used new printing and binding technology permitting advertisers to personalize messages on business reply cards, with the subscriber's name printed on it, bound into the magazine next to an advertisement.

The next step will be printing directly on the magazine page. Readers will find a personal message directed to them on a full-page advertisement, perhaps reminding them that a product they might have been using needs replacing—at a bargain price. All that is needed to make this development reality is the invention of a method to do it at reasonable cost. When that happens, advertisers can utilize the information available to them through credit companies about the buying habits of customers, enabling them to deliver their message at the right moment.

There are further possibilities. It may be possible to have one version of a single magazine for the elderly, another for the young, still others for urban and rural audiences. The shape of the future can already be seen in the catalogues of mail-order houses that are packaged to look like magazines, with articles reinforcing the images. Whittle's specialty magazines, too, have blurred if not erased the line between editorial and advertising information.

One result of targeting is that entire neighborhoods will be eliminated as publishers and advertisers zero in on the most affluent markets, and that raises serious social questions. Are the bands that hold society together about to come apart, at least in this area? Some publishers, according to Scardino, "worry that distributing different information to each subscriber may dissolve the common bonds that hold a democracy together." Shared information enables democracies to function effectively.

But change moves on, irresistibly. We are already accustomed to magazine advertizing that employs pop-ups, scented strips, and musical microchips, but the near future holds the promise of advertisements that move and light up, as well as microchips that will make them speak to us. Thus, advertising will become entertainment. The only obstacle holding back a flood of these devices is the inability of magazines to handle more than one per issue. Major advertisers do not mind the cost when they measure it against the returns in publicity and sales. When the first musical microchip appeared in *New York* and the *New Yorker* during the 1987 Christmas season, it cost a million dollars to produce, but publicity was measured at $4 million and its sponsor, the Carillon Company, importers of Absolut Vodka, reported the largest percentage increase in sales in its history.

Other ingenious devices followed. In the spring of 1988, Camel cigarettes celebrated a seventy-fifth birthday with a three-dimensional advertisement in which the viewer could watch the familiar dromedary jump off the page to the tune of "Happy Birthday." Later, a champagne advertisement showed a woman on the first page of a four-page spread opening a bottle of champagne. When the page was turned, an origami cork, folded tightly and attached to a rubber band, popped out. "Smelly" advertising has already given us perfume, chocolate, bacon, and the odor of Rolls-Royce leather. Diaper companies have given us pull-out facsimiles and tapes. McDonald's punched five holes in an advertising page so a hand could be inserted to grip a pictured sandwich. Toyota Corolla advertising has used 3-D glasses in several magazines, while cosmetic companies have offered eye-shadow and blusher samples in their advertising.

The possibilities are endless. Automobile companies, for example, will one day be running advertising in which the reader presses a part of the page to light headlights on a pictured car, while another touch on the same page will bring readers a fifteen-second talk on why they should buy it. Another touch brings the sound of car doors closing and the engine starting up. Touch again and the wheels turn. Still another touch uses a thermal element transferring heat to change the car's color. All these miracles are technologically available; only present costs stand in the way. Microchips, powered by three-volt batteries capable of 4,000 repetitions, constitute the technology that make possible the effects described above, and they are also able to mimic the sounds of thunder, someone coughing, a telephone ringing, a jet roaring overhead, a bark or a meow—almost anything. On the horizon are movable parts in advertising, their advance spurred on by surveys showing that people are six times more likely to read such a specially inserted advertisement and that pop-ups increase positive reader response by 10 percent.

In the opinion of some experts, mass circulation magazines and television networks are already doomed. At a convention of advertising agencies in 1989, one of the largest in America, Lintas USA, described a communications system so decentralized that it would make the mass media irrelevant, as well as erase any distinction between media advertising and other marketing disciplines. Personalized magazines and narrowly segmented television channels will bring this about, it was said. The nation may have more than 90 million television sets, up from 59 million in 1970, but the audience has changed, according to Nielson research. The family no longer gathers around to watch a favorite show in the numbers that once characterized the medium. Both program loyalty and attention spans have declined, and the audiences are being fragmented, just as they are in magazines. Cable, syndicated programming, and VCRs have all contributed to this phenomenon. Lintas concludes that an inexorable transition in the way people get their entertainment and information is taking place, a verdict in which other industry experts concur.

Looking back on 250 years of magazines in America, as these pages have done, makes clear how powerful a social and cultural influence they have been, not only editorially but also through their advertising pages. Looking into the future confirms their probably even greater influence. It could be said that magazines have shaped a consumer society and done much to impose many kinds of conformity. On the other hand, the sheer numbers and incredible variety of magazines have guaranteed that all the interests of the most diverse population in the world will be served. It may be that magazines are the most democratic institution we have yet created in America.

Bibliography

This bibliography is broken down into four sections corresponding to the divisions of the text. It also includes a listing of some of the most useful general works on the subject of magazines. Throughout we have tried to incorporate the newest scholarship on magazines as well as such mainstays of magazine history as Frank Luther Mott. Because so few general works have been published on the period from World War II to the present, we have relied heavily on newspaper articles, scholarly research, and the trade press. The length of the bibliography for the fourth section reflects this, as well as our feeling that perhaps our greatest contribution to journalism scholarship lies in this part of the book.

General

Mott, Frank Luther. *A History of American Magazines*. Cambridge: Harvard Univ. Press, 1938. Four volumes.

Paine, Fred K., and Nancy E. Paine. *Magazines: A Bibliography for Their Analysis, with Annotations and Study Guide*. Metuchen, N.J.: Scarecrow Press, 1987.

Peterson, Theodore. *Magazines in the Twentieth Century*. 2nd ed. Urbana: Univ. of Illinois Press, 1964.

Taft, William. *American Magazines for the 1980s*. New York: Communication Arts Books, Hastings House Publishers, 1982.

Tebbel, John. *The American Magazine: A Compact History*. New York: Hawthorn Books, 1967.

Wolseley, Roland. *The Magazine World*. New York: Prentice-Hall, 1951.

Wood, James Playsted. *Magazines in the United States*. 2nd ed. New York: Ronald Press Co., 1956.

Part I: The Creation of Magazine Audiences

Bullock, Penelope. *The Afro-American Periodical Press, 1838–1909*. Baton Rouge: Louisiana State Univ. Press 1981.

Endres, Katherine L. "The Women's Press in the Civil War: A Portrait of Patriotism, Propaganda and Prodding." *Civil War History* (Mar. 1984), 31–53.

Finley, Ruth. *The Lady of* Godey's. Philadelphia: J.B. Lippincott Co. 1931.

Garnsey, Caroline J. "Ladies Magazines to 1850: The Beginning of the Industry." *New York Public Library Bulletin V* 58, no. 2 (Feb. 1954).

"*Godey's,* Past and Present." *Godey's Magazine* CXXV, 363 (Oct. 1892).

Martin, Lawrence. "The Genesis of *Godey's Lady's Book,*" *New England Quarterly* vol. 1, no. 2 (Jan. 1928).

Murphy, James E., and Sharon M. Murphy. *Let My People Know: American Indian Journalism, 1828–1978.* Norman: Univ. of Oklahoma Press, 1981.

Nord, David Paul. "A Republican Literature: A Study of Magazine Reading and Readers in Late Eighteenth-Century New York" (ms).

Orians, G. Harrison. "Censure of Fiction in American Romances and Magazines, 1789–1810." *PMLA* (1937).

Rogers, Sherbrooke. *Sarah Josepha Hale: A New England Pioneer, 1788–1879.* Grantham, N.H.: Tompson and Rutter, 1985.

Stearns, Bertha Monica. "Early New England Magazines for Ladies." *New England Quarterly vol. II, no.3 (1929): 420–51.*

Stearns, Bertha Monica. "Early Philadelphia Magazines for Ladies." *Pennsylvania Magazine of History and Biography* vol. LXIV, no.4 (Oct. 1940): 479–91.

Stearns, Bertha Monica. "Early Western Magazines for Ladies." Mississippi Valley Historical Review vol. XVIII, no.5 (Dec. 1931): 319–50.

Stearns, Bertha Monica. "New England Magazines for Ladies, 1830–1860." *New England Quarterly* vol. III, no.4 (1930): 627–56.

Stearns, Bertha Monica. "Philadelphia Magazines for Ladies: 1830–1860." *Pennsylvania Magazine of History and Biography* vol. LXIX, no.3 (July 1945).

Stearns, Bertha Monica. "Southern Magazines for Ladies (1819–1860)." *South Atlantic Quarterly* vol. XXXI, no.1 (Jan. 1932): 70–87.

Tassin, Algernon. *The Magazine in America.* New York: Dodd, Mead, 1916.

Zboray, Ronald J. "The Letter and the Fiction Reading Public in Antebellum America." *American Quarterly* (Mar. 1988).

Part II: The First Great Change

Best, James. "The Brandywine School and Magazine Illustration: *Harper's, Scribner's* and *Century,* 1908–1910." *Journal of American Culture* (Spring 1980), 128–44.

Blum, Stella, ed. *Victorian Fashions & Costumes from* Harper's Bazar, *1867–1898.* New York: Dover Publications, 1974.

Britt, George. *Forty Years, Forty Millions*—The Career of Frank A. Munsey. New York: Farrar & Rinehart, 1935.

Chalmers, David. *The Muckrake Years.* New York: D. Van Nostrand Co., 1974; reprinted 1980 by Robert E. Krieger Publishing Co., Huntington, N.Y.

Chase, Edna Woolman, and Ilka Chase. *Always in* Vogue. Garden City: Doubleday, 1954.

Conklin, Groff, ed. *The* New Republic *Anthology: 1915—1935.* New York: Dodge Publishing Co., 1936.

Cote, Joseph A. "Clarence Hamilton Poe: The *Farmer's Voice,* 1889–1964." *Agricultural History* (Jan. 1979).

Daniel, Walter D. *Black Journals of the United States.* Westport, Conn.: Greenwood, 1982.

Dennis, Everette E., and Christopher Allen. "*Puck,* The Comic Weekly." *Journalism History* (Spring 1979), 2–7.

Faderman, Lillian. "Lesbian Magazine Fiction in the Early Twentieth Century." *Journal of Popular Culture* (Spring 1978), 800–817.

Fishbein, Leslie. *Rebels in Bohemia: The Radicals of the* Masses, *1911–1917*. Chapel Hill: Univ. of North Carolina Press, 1982.

Greene, Theodore P. *The Changing Models of Success in American Magazines*. New York: Oxford Univ. Press, 1970.

Harlan, Louis R. "Booker T. Washington and the *Voice of the Negro*, 1904–1907." *Journal of Southern History* XLV (Feb. 1979): 42–62.

Harper, J. Henry. *The House of Harper*. New York: Harper & Brothers, 1912.

Heller, Steven. "The Art of Sensationalism." *Print* (Mar.–Apr. 1984), 58–64.

Hower, Ralph. *History of an Advertising Agency: N. W. Ayer & Son at Work, 1869–1939*. Cambridge: Harvard Univ. Press, 1939.

John, Arthur. *The Best Years of the* Century, *1879–1909*. Urbana: Univ. of Illinois Press, 1981.

Johnson, Abby Arthur, and Ronald M. Johnson. *Propaganda and Aesthetics: The Literary Politics of Afro-American Magazines in the Twentieth Century*. Amherst: The Univ. of Massachusetts Press, 1979.

Johnson, Charles S. "The Rise of the Negro Magazine." *Journal of Negro History* vol. XIII, no. 1 (Jan. 1928): 7–21.

Joyce, Donald Franklin. "Magazines of Afro-American Thought on the Mass Market: Can They Survive?" *American Libraries* VII (Dec. 1976): 678–83.

Kern, Donna Rose Casella. "Sentimental Short Fiction: Women Writers in *Leslie's Popular Monthly*." *Journal of American Culture* (Spring 1980), 113–27.

Kornweibel, Theodore, Jr. *No Crystal Stair: Black Life and the* Messenger, *1917–1928*. Westport: Greenwood Press 1975.

Lichtenstein, Nelson. "Authorial Professionalism and the Literary Marketplace, 1885–1900." *American Studies* (Spring 1978), 35–53.

Liess, William, Stephen Kline, and Sut Jhally. *Social Communication in Advertising*. Toronto: Methuen, 1986.

Makosky, Donald. "The Portrayal of Women in Wide-Circulation Magazine Short Stories, 1905–1955." Dissertation, Univ. of Pennsylvania, 1966.

Masel-Walters, Lynne. "Their Rights and Nothing More: A History of the *Revolution*, 1868–70." *Journalism Quarterly* 53 (Summer 1976).

Masel-Walters, Lynne. "To Hustle with the Rowdies: The Organization and Functions of the American Woman Suffrage Press." *Journal of American Culture* (Spring 1980), 167–83.

McDonald, Susan Waugh. "Edward Gardner Lewis: Entrepreneur, Publisher, American of the Gilded Age." *Bulletin of the St. Louis Historical Society* (Apr. 1979), 154–63.

Mott, Frank Luther. *The Magazine Revolution and Popular Ideas in the Nineties* (pamphlet). Reprinted from the Proceedings of the American Antiquarian Society, Apr. 1954.

Ohmann, Richard. "Where Did Mass Culture Come From? The Case of Magazines." *Berkshire Review* 16 (1981).

Pollay, Richard W. "Twentieth-Century Magazine Advertising: Determinants of Informativeness." *Written Communication* (Jan. 1984), 56–77.

Pope, Daniel. *The Making of Modern Advertising*. New York: Basic Books, 1983.

Presbrey, Frank. *History and Development of Advertising*. Garden City: Doubleday, 1929.

Reynolds, Robert D., Jr. "The 1906 Campaign to Sway Muckraking Periodicals." *Journalism Quarterly* (Autumn 1979), 513–20.

Rowell, Frank. *Forty Years an Advertising Agent.* New York: Printer's Ink, 1906.

Schell, Ernest. "Edward Bok and the *Ladies' Home Journal.*" *American History Illustrated* (Feb. 1982), 16–23.

Searles, Patricia, and Janet Mickish. "'A Thoroughbred Girl'": Images of Female Gender Role in Turn-of-the-Century Mass Media." *Women's Studies* no. 3 (1984): 261–81.

Seebohm, Carolyn. *The Man Who Was* Vogue—*The Life and Times of Condé Nast.* New York: Viking, 1982.

Snow, Carmel, with Mary Louise Aswell. *The World of Carmel Snow.* New York: McGraw Hill, 1962.

Socolofsky, Homer H. "The Capper Farm Press in Western Agricultural Journalism." *Journal of the West* (Apr. 1980), 22–29.

Steinberg, Salme. *Reformer in the Marketplace: Edward W. Bok and the* Ladies' Home Journal. Baton Rouge: Louisiana State Univ. Press, 1979.

Suggs, Henry Lewis, ed. *The Black Press in the South.* Westport: Greenwood Press, 1983.

Towne, Charles Hanson. *Adventures in Editing.* New York: D. Appleton & Co., 1926.

Waller-Zuckerman, Mary Ellen. "Marketing the Women's Journals, 1873–1900." In *Essays in Business and Economic History,* William J. Hausman, ed., 2nd series, vol. 18 (Fall 1989).

Waller-Zuckerman, Mary Ellen. "Old Homes in a City of Perpetual Change: The Women's Magazine Industry, 1890–1917." *Business History Review* (Fall 1989).

Waller-Zuckerman, Mary Ellen. "Selling Mrs. Consumer: The Role of Women's Magazines in the Development of Advertising." Proceedings of the Third Annual History of Marketing Conference, Apr. 1987.

Whalen, Matthew D., and Mary F. Tobin. "Periodicals and the Popularization of Science in American Culture, 1860–1910." *Journal of American Culture* (Spring 1980), 195–203.

Wilson, Christopher. "The Rhetoric of Consumption: Mass-Market Magazines and the Demise of the Gentle Reader, 1880–1920." In *The Culture of Consumption,* T.J. Jackson Lears and Richard Wightman Fox, eds. New York: Pantheon Books, 1983.

Wolseley, Ronald. *The Black Press, U.S.A.* Ames: Iowa State Univ. Press, 1971.

Wood, James P. *The Story of Advertising.* New York: Ronald Press, 1958.

Zurier, Rebecca. *Art for the* Masses: *A Radical Magazine and Its Graphics, 1911–1917.* 1938. From a review of the book by David M. Oshinsky. *New York Times Book Review,* Sept. 4, 1988.

Part III: Developing New Audiences

Allen, Frederick Lewis. "The Function of a Magazine in America" (pamphlet). *University of Missouri Journalism Series,* no.101 (Aug. 1945), 3–12.

Baughman, James L. *Henry R. Luce and the Rise of the American News Media.* Boston: Twayne Publishing, 1987.

Block, Bernard. "Romance and High Purpose: The *National Geographic.*" *Wilson Library Bulletin* (Jan. 1984), 333–37.

Burnshaw, Stanley. Letter to John Tebbel, Mar. 30, 1989.

Callahan, Sean. "*People.*" *American Photographer* 12 (June 1984): 39–53.

Cooney, John. *The Annenbergs: The Salvaging of a Tainted Dynasty.* New York: Simon and Schuster, 1982.

Elson, Robert T. *Time Inc.: The Intimate History of a Publishing Enterprise, 1923–1941.* New York: Atheneum, 1968.

Goldberg, Vicki. *Margaret Bourke-White: A Biography.* Reading, Mass.: Addison-Wesley Publishing Co., Inc., 1987.

Halasz, Piri. "Art Criticism and Art History in New York, the 1940s versus the 1980s." Part Two: "The Magazines." *Arts Magazine* (Mar. 1983), 64–73.

Heller, Steven. "Photojournalism's Golden Age," *Print* (Sept.–Oct. 1984), 68–79.

Hoffman, Frederick J., Charles Allen, and Carolyn F. Ulrich. *The Little Magazines: A History and a Bibliography.* Princeton: Princeton Univ. Press, 1946.

Honey, Maureen. "Images of Women in the *Saturday Evening Post*, 1931–36." *Journal of Popular Culture* (Fall 1976), 352–58.

Hook, Sidney. "The Radical Comedians: Inside *Partisan Review*." *American Scholar* (Winter 1984–85), 45–61.

Lewis, David Levering. "Mr. Johnson's Friends: Civil Rights by Copyright During Harlem's Mid-Twenties." *Massachusetts Review* (Autumn 1979), 501–19.

Patterson, Oscar III. "Television's Living-Room War in Print: VietNam in the News Magazines." *Journalism Quarterly* 60 (Spring 1984): 34–39.

Perkins, William Eric. Review of a collected edition of the files of the *Crusader*, edited by Cyril V. Briggs between Sept. 1918 and Feb. 1922. Published by Garland in 1987 in a complete edition, *Journal of American History*, vol 75, no.4 (Mar. 1989). The book was compiled by Robert Hill, who edited it and wrote the introduction.

Pool, Gail. "Magazines." *Wilson Library Bulletin* (June 1984), 748–49.

Roberts, Nancy L. "Journalism for Justice: Dorothy Day and the *Catholic Worker*." *Journalism Quarterly* (Spring–Summer 1983), 2–9.

Rothstein, Arthur. *Photojournalism: Pictures for Magazines and Newspapers.* New York: American Photographic Book Publishing Company, 1956.

Span, Paula. "In the Picture: 'The First Decade of *People* Magazine." Washington Post, Mar. 3, 1984.

Sullivan, Paul W. "G.D. Crain, Jr., and the Founding of *Advertising Age*." *Journalism History* (Autumn 1974), 94–95.

Swanberg, W. A. *Luce and His Empire.* New York: Scribner's, 1972.

Tebbel, John. *George Horace Lorimer and the* Saturday Evening Post. New York: Doubleday, 1948.

Tobin, Richard L., ed. *The Golden Age: The* Saturday Review 50th Anniversary Reader. New York: Bantam, 1974.

Tsang, Kuo-jen. "News Photos in *Time* and *Newsweek*." *Journalism Quarterly* 61 (Autumn 1984), 578–84.

Wainwright, Loudon. *The Great American Magazine: An Inside History of* Life. New York: Ballantine Books, 1986.

Part IV: Magazines Since the Second World War

Abbott, William. "The Editorial Connection." *Folio* (Sept. 1983), 203–5.

"Advertising Meets Editorial." *Columbia Journalism Review* (Fall 1962), 14–22.

"The Affluent Market," *Advertising Age*. (Aug. 23, 1984), 11–47.

Alward, E. "Science Fiction: Magazine Growth Industry." *Serials Review* 9 (Fall 1983): 27–32.

"AMA To Launch Magazine for Marketing Researchers." *Marketing News* (Aug. 1, 1988), 2.

American Express Magazines. Full-page advertisement, *New York Times*, Feb. 27, 1989.

Anderson, A. Donald. "Hollywood's Version of Trade Wars." *New York Times*, Aug. 7, 1988.

Anderson, Susan Heller. "*HG* Magazine Is Not What It Used To Be." *New York Times*, June 8, 1988.

Anderson, Susan Heller. "*Traveler*, A Year Old, Influences But Trails." *New York Times*, Aug. 22, 1988.

Armstrong, David. "Remodeling *Mother*." *Columbia Journalism Review* (Jan.–Feb. 1986), 12, 13.

"An Avalanche of Personal Computer Magazines." *Business Week* (Aug. 22, 1983), 90.

Bagdikian, Ben. "The Wrong Kind of Readers: The Fall and Rise of the *New Yorker*." The *Progressive* (May 1983), 32–34.

Barmash, Isadore. "A Magazine with a Profit at Its Debut." *New York Times*, Sept. 30, 1988.

Barmash, Isadore. "At 60, Editor for Women Looks Ahead." *New York Times*, Oct. 17, 1988.

Barmash, Isadore. "Times Magazine Group to Sell *Southern Travel*." *New York Times*, Aug. 24, 1988.

Barmash, Isadore. "Travel Guide Competition to Increase." *New York Times*, Jan. 25, 1989.

Barron, Neil. *Anatomy of Wonder: Science Fiction*. New York: R.R. Bowker Company, 1976.

Bart, Peter. "Giants on Uneasy Footing." *Columbia Journalism Review* (Spring 1962), 32–33.

Beitler, Stephen. "Sportsmen Go by the Books." *Advertising Age* (Mar. 23, 1981), 56.

Benoit, Ellen. "Our New *Astrology*." *Forbes* (Feb. 28, 1983), 93.

Berling-Manuel, Lynn. "Reaching the Gay Market." *Advertising Age* (Mar. 26, 1984), M-4–M-5.

Bernstein, Lester. "Time Inc. Means Business." *New York Times*, Feb. 26, 1989.

Bernstein, Richard. "65th Birthday Party for a Voice of Liberal Opinion." *New York Times*, Jan. 25, 1988.

Bhatia, Gouri. "Jann Wenner: Publisher of *Rolling Stone*." *Folio* (Nov. 1962), 68–69.

Bishop, Katherine. "Magazine Finds a Niche in Medicine and Health." *New York Times*, May 31, 1988.

Blakken, Renee. "Scholastic Offers a Lesson in Making the Grade." *Advertising Age* (Oct. 17, 1983), M-56.

Blau, Eleanor. "Soap Opera Magazines Fight for Fans' Hearts and Dollars." *New York Times*, Oct. 2, 1988.

Blau, Eleanor. "U.S.-Soviet Magazine Plan." *New York Times*, Mar. 27, 1989.

Bogart, Leo. "Magazines Since the Rise of Television." *Journalism Quarterly* (Spring 1956), 153–66.

Bold, Christine. "Voices of the Fiction Factory in Dime and Pulp Westerns." *Journal of American Studies* (Apr. 1983), 29–46.

"Boom in the Business Press." *Newsweek* (Jan. 7, 1985).

Bourne, Tom. "Whatever Happened to the *Ramparts* Gang?" *Quill* (Mar. 1980), 13, 14.

Boylan, James. "Drifting Left: Four New Magazines." *Nation* (Apr. 22, 1978), 454–56.

Brockway, Brian. "Always on Sunday." *Madison Avenue* (Feb. 1985), 98–100.

Budrys, Algis. "Paradise Charted." *Triquarterly* no.49 (Fall 1980): 5–75.

Buffum, Charles. "Sunday Best: Newspaper Magazines and a Parade of Weekend Reading." *Washington Journalism Review* (Oct. 1983), 32–35.

Byron, Christopher H. *The Fanciest Dive: What Happened When the Media Empire of Time/Life Leaped Without Looking into the Age of High-Tech.* New York: Norton, 1986; NAL, 1987.

Camillieri, John. "Overview of the Science Magazine Category." *Marketing & Media Decisions* (Apr. 1985), 124.

Canape, Charlene. "If You Love Fine Food, Read On." *Advertising Age* (Sept. 12, 1983), M-4–M-5.

Chaney, Lindsay, and Michael Cieply. *The Hearsts: Family and Empire — The Later Years.* New York: Simon and Schuster, 1981.

"Christian Science Monitor Group to Begin News Monthly." *New York Times*, Sept. 16, 1988.

Churcher, Sharon. "Radical Transformations." *New York Times Magazine*, July 16, 1989.

"City Magazines Are the Talk of the Town." *Business Week* (Feb. 18, 1967), 184–88.

Cohen, Gordon L. "*Esquire-Coronet*'s *Ken:* Magazine Everyone Hated." *Media History Digest* (Spring–Summer 1987).

Colford, Steven M. "The Pain of Libel Premiums: Small Magazines Hurt the Most." *Washington Journalism Review* (July 1986), 16–17.

Collins, Glenn. "*Variety*'s New Look for New Readers." *New York Times*, Feb. 21, 1989.

"A Computer Magazine Featuring 3-D Glasses." *New York Times*, Feb. 7, 1989.

"Computer Magazine Overbyte." *Fortune* (Nov. 26, 1984).

Cook, Bruce. "The Fight to Control Harper's." *Washington Journalism Review* (Dec. 1983), 43–46.

Cooper, Nancy. "Feminist Periodicals." *Mass Communications Review* (Summer 1977), 15–22.

Curtis, Wayne. "What's New in Magazine Publishing." *New York Times*, Jan. 22, 1989.

Daley, Suzanne. "*Sassy:* Like, You Know, for Kids." *New York Times*, Apr. 11, 1988.

Deloff, Linda-Marie. "The *Century* in Transition: 1916–1922." *Christian Century* (Mar. 7, 1984), 243–46.

Denniston, Lyle. "The SEC Power-Grab." *Quill* (Sept. 1984), 21–24.

Diamond, Edwin. "Celebrating *Celebrity*." *New York* (May 13, 1985), 22–27.

Diamond, Edwin. "The Fate of the Earth: Hubbub at the *New Yorker*." *New York* (Jan. 26, 1987).

Diamond, Edwin. "The New (Land) Lords of the Press." *New York* (Feb. 27, 1989).

Diamond, Edwin. "*Newsweek*'s Testing Time." *New Republic* (Feb. 28, 1983), 13–15.

Diamond, Edwin. "Next, U.S. Timeweek: The News Magazines Converge." *New York* (Dec. 5, 1988).

Diamond, Edwin. "The Unladylike Battle of the Women's Magazines." *New York* (May 20, 1974), 43–46.

Dinan, John A. *The Pulp Western: A Popular History of the Western Fiction Magazine in America.* San Bernardino: Borgo Press, 1983.

Dixler, Elsa."Rich Eccentrics, Red-Baiting and Jameson's Dog." *Nation* (June 8, 1985), 702–4.

Dougherty, Philip H. "Big Changes for Knapp's Magazines." *New York Times*, Sept. 13, 1988.

Dougherty, Philip H. "Diamandis Group at the Helm." *New York Times*, Oct. 2, 1987.

Dougherty, Philip H. "Financial Officers Get a Magazine." *New York Times*, July 11, 1988.

Dougherty, Philip J. "For the Affluent, Yet Another Magazine." *New York Times*, Sept. 21, 1988.

Dougherty, Philip H. *"Harrowsmith." New York Times*, Jan. 21, 1988.

Dougherty, Philip H. "Health Journal Expands Its HMO Circulation." *New York Times*, July 25, 1988.

Dougherty, Philip H. "Hearst Gives Its Magazine Deals a Push." *New York Times*, Jan. 5, 1988.

Dougherty, Philip H. "Magazine Is Prepared for Hospital Patients." *New York Times*, June 27, 1988.

Dougherty, Philip H. "A Magazine on Florida Real-Estate Turf." *Advertising Age* (Jan. 12, 1981), S-10–S-11.

Dougherty, Philip H. "Model Magazine." *New York Times*, Jan. 21, 1988.

Dougherty, Philip H. "New Data on *Modern Maturity*." *New York Times*, Dec. 1, 1987.

Dougherty, Philip H. *"New England Monthly* Is Bustling." *New York Times*, June 15, 1988.

Dougherty, Philip H. "New Mission for Bunny at Playboy." *New York Times*, Feb. 2, 1988.

Dougherty, Philip H. "Old Is Good at *Memories* Magazine." *New York Times*, Sept. 23, 1988, and Dec. 18, 1987.

Dougherty, Philip H. *"Outside*'s Publishers in New Venture." *New York Times*, July 22, 1988.

Dougherty, Philip H. *"PC* Magazine Sees Buyers As 'Heroes.'" *New York Times*, Aug. 9, 1988.

Dougherty, Philip H. *"Psychology* Magazine's New Look." *New York Times*, Dec. 18, 1987.

Dougherty, Philip H. "Shifts at Top of *National Review*." *New York Times*, June 3, 1988.

Dougherty, Philip H. "Social Analysis from *Good Housekeeping*." *New York Times*, Aug. 11, 1988.

Dougherty, Philip H. "A Spinoff for *Sports Illustrated*." *New York Times*, Aug. 4, 1988.

Dreyfuss, Madeleine. "Home and Technology Magazines Booming." *Marketing & Media Decisions* (Dec. 1981), 68–70.

Ehrenreich, Barbara, and Deirdre English. "The Male Revolt: Feminists Were Not the First to Flee the Family." *Mother Jones* (Apr. 1983), 25–31.

Elienthal, Ira. *"Life* Resurrected, and Other Magazine Turnarounds." *Magazine Age* (Mar. 1985), 6–9.

Elliot, Stuart J. "Magazines in Battle for Real-Estate Turf." *Advertising Age* (Aug. 20, 1984), 21.

Elliot, Stuart J. *"Seventeen* Hits 40: Numbers Tell the Story." *Advertising Age* (Aug. 20, 1984), 21.

Ellis, James. "Now Even Playboy Is Bracing for a Midlife Crisis." *Business Week* (Apr. 15, 1985), 66.

English, Mary McCabe. "Black Publishing: Words in the Future Tense." *Advertising Age* (Nov. 29, 1982), 21.

Erin, Geri. "How Special Interest Publications Capture Specialized Audiences." *Magazine Age* (Sept. 1980), 64–69.

Fabrikant, Geraldine. "And Now, a Magazine for the Over-40 Woman." *New York Times,* Feb. 7, 1988.

Fabrikant, Geraldine. "Bauer Starts Second Magazine for Women." *New York Times,* Feb. 16, 1989.

Fabrikant, Geraldine. "Competition in Fashion Magazines." *New York Times,* Oct. 14, 1988.

Fabrikant, Geraldine. "Hachette to Buy Magazine Publisher." *New York Times,* Aprl 13, 1988.

Fabrikant, Geraldine. "The Jury's Out on the Hipper *Vogue.*" *New York Times,* Apr. 30, 1989.

Fabrikant, Geraldine. "A Magazine Entrepreneur's Rescue Attempt." *New York Times,* Dec. 26, 1988.

Fabrikant, Geraldine. "Matchmakers for Magazine Industry." *New York Times,* July 4, 1988.

Fabrikant, Geraldine. "Meredith's Conservative Approach." *New York Times,* Feb. 16, 1988.

Fabrikant, Geraldine. "Murdoch Sets Deal to Sell *Elle.*" *New York Times,* Sept. 11, 1988.

Fabrikant, Geraldine. "An Odd Complex Fills a Gap in Los Angeles," *New York Times,* Mar. 20, 1989.

Fabrikant, Geraldine. "Reader's Digest Buying 2nd Magazine." *New York Times,* Jan. 5, 1988.

Fabrikant, Geraldine. "Si Newhouse Tests His Magazine Magic." *New York Times,* Sept. 25, 1988.

Fabrikant, Geraldine. "The Strategy of Murdoch's Mirabella." *New York Times,* Nov. 25, 1988.

Fabrikant, Geraldine. "Time and Whittle Form Alliance." *New York Times,* Oct. 21, 1988.

Fabrikant, Geraldine. "Time Inc.'s Magazine Drought May Be Ending." *New York Times,* Feb. 6, 1989.

Fabrikant, Geraldine. "*Time* Magazine to Begin New Editorial Features." *New York Times,* Sept. 27, 1988.

Fabrikant, Geraldine. "Wooing the Wealthy Reader." *New York Times,* Oct. 14, 1987.

"Faces: Phillip Moffitt & Christopher Whittle." *Folio* (Aug. 1981), 60.

Fairlie, Henry. "The Vanity of *Vanity Fair:* A Monument to Status Anxiety." *New Republic* (Mar. 21, 1983), 25–30.

Fannin, Rebecca. "Comeback of the Culture Books." *Marketing & Media Decisions* (July 1983), 40–41.

Fannin, Rebecca. "Image Revision for Magazines." *Marketing & Media Decisions* (Jan. 1984), 54–55.

Fannin, Rebecca. "Regional Magazines Gain National Impact." *Marketing & Media Decisions* (Aug. 1982), 64–65.

"*Farm Journal* Now up to 1,106 Versions—More to Come." *Folio* (Feb. 1983), 19–20.

"Farm Media." *Marketing & Media Decisions* (Jan. 1980), 100.

Feinberg, Andrews. "How Far Will the New Travel Magazines Go?" *New York Times*, Jan. 10, 1988.

"*50-Plus* Magazine Isn't What It Used To Be." Advertisement in the *New York Times*, Feb. 13, 1989.

Fischer, John. "The Perils of Publishing: How to Tell When You Are Being Corrupted." *Harper's* (May 1968), 13–16.

Fisher, Kathleen. "Sale Won't Put *PT* and Field Asunder." *Monitor*. Publication of the American Psychological Association. (The reference is to *Psychology Today*.)

Fisher, Lawrence M. "Two Giants in PC Magazines." *New York Times*, Apr. 21, 1988.

Fletcher, Marilyn. "Science Fiction Magazines and Annual Anthologies: An Annotated Checklist." *Serials Librarian* vol. 7, 1 (Fall 1982): 65–72.

Folio Magazine. Issues from 1988 and 1989.

"For Magazine Readers, British Royalty Sells." *New York Times*, Nov. 28, 1988.

Ford, James L.C. *Magazines for Millions: The Story of Specialized Publications*. Carbondale: Southern Illinois Univ. Press, 1969.

Francis, David. "The Boom in Sci-tech Magazines: Is It Going to Last?" *Christian Science Monitor*, Feb. 2, 1984.

Freeman, Laurie. "New Magazines Deliver Professional Niche." *Advertising Age* (Nov. 19, 1984), 45.

Freitag, Michael. "*Wigway*, A Folksy Magazine." *New York Times*, July 24, 1989.

Gage, Theodore. "Consumer Books Carve Niche." *Advertising Age* (Mar. 9, 1981), S-4–S-5.

Gamarekian, Barbara. "Many Opinions Flourish in the Political Hothouse." *New York Times*, Feb. 20, 1989.

Gerard, Jeremy. "*National Geographic* Expands Its Television Horizons." *New York Times*, June 20, 1988.

Gerard, Jeremy. "*TV Guide*'s Power Over the Air." *New York Times*, Aug. 11, 1988.

Gerbner, George. "The Social Role of the Confession Magazine." *Social Problems* (Summer 1958), 29–40.

Gerry, Roberts. "Thoughtleaders: Why Advertisers Look Beyond the Numbers." *Magazine Age* (Aug. 1982), 20–28.

"Get Ready for the Video Publishing Explosion." *Marketing & Media Decisions* (Apr. 1981), 59–63.

Griffith, Thomas. "Merchants of Raunchiness." *Time* (July 4, 1977), 69.

Griffith, Thomas. "Stuck with a Magazine's Genes." *Time* (Aug. 13, 1979), 63.

Hall, Carol. "Home to Work & Homes Working for You." *Marketing & Media Decisions* (Apr. 1986), 24.

Hartman, Thomas. "Reporting For Service: The Big Guns of the Military Press." *Washington Journalism Review* (July–Aug. 1984), 28–32.

Hayes, Harold (obituary). *New York Times*, Apr. 7, 1989.

Hayes, John P. "City/Regional Magazines: A Survey/Census." *Journalism Quarterly* (Summer 1981), 294–96.

Heckman, Lucy. "*Harper's* Magazine." *Serials Review* (Jan.–Mar. 1981), 49–56.

Heller, Steven. "Photojournalism's Golden Age." *Print* (Sept.–Oct. 1984), 68–79.

Henry, William A. III. "*Newsweek*'s Outsider Bows Out, and an Insider Steps Up as Its Sixth Top Editor Since 1972." *Time* (Jan. 16, 1984), 63.

Hersey, Harold. *Pulpwood Editor*. New York: Frederick A. Stokes Co., 1937.

Hinckle, Warren. "The Adman Who Hated Advertising: The Gospel According to Howard Gossage." *Atlantic* (Mar. 1974), 67–72.

Hinds, Michael deCourcy. "Young Consumers: Perils and Power." *New York Times*, Feb. 11, 1989.

Hirsch, James. "Time Inc. Finds New Ways to Profit on Swimsuit Issue." *New York Times*, Feb. 7, 1989.

Hochschild, Adam. "*Ramparts:* The End of Muckraking Magazines." *Washington Monthly* (June 1974), 33–42.

Hochswender, Woody. "Changes At *Vogue* a Complex Tale of Rumor and Fact." *New York Times*, July 25, 1988.

Hogan, Bill. "The Big Business in Business Magazines." *WashingtonJournalism Review* (July–Aug. 1981), 32–34.

Hogan, Bill. "The Boom in Regional Business Journals." *WashingtonJournalism Review* (July–Aug. 1982).

Hogan, Bill. "No Profit *Insight:* The Washington *Times'* Conservative Newsweekly." *Washington Journalism Review* (Mar. 1986), 34–36.

Holder, Stephen. "The Family Magazine and the American People." *Journal of Popular Culture* (Fall 1973), 264–79.

Honan, William. "The Lessons of War Sell in Peacetime." *New York Times*, Dec. 19, 1988.

Hoopes, Roy. "Birth of a Great Magazine." *American History Illustrated* (Sept. 1985), 34–41.

Hopkins, Ellen. "Star Search." *New York* (Oct. 31, 1988), 31.

Horn, Maurice. *Comics of the American West.* New York: Winchester Press, 1977.

"The Importance of Being Rockwell," *Columbia Journalism Review* (Nov.–Dec. 1979), 40–42.

"In-flight Magazines Offering Bonus Pages." *New York Times*, Aug. 5, 1988.

"*Interview* Magazine Is Sold." *New York Times*, May 9, 1989.

Isaacs, Stephen. "America's Magazines: High Stakes, Cover-to-Cover Changes." Washington *Post*, Jan. 2, 1972.

James, Caryn. "Big Little Magazines: A Reader's Guide." *New York Times*, Apr. 30, 1989.

Jones, Alex S. "For News Magazines, Growing Identity Crisis." *New York Times*, June 29, 1988.

Jones, Alex S. "Times Co. to Buy *McCall's.*" *New York Times*, May 30, 1989.

Johns-Heine, Patricke, and Hans H. Garth. "Values in Mass-Periodical Fiction, 1921–1940." *Public Opinion Quarterly* (Spring 1949), 105–13.

Kanner, Bernice. "Special Effects." *New York* (Sept. 19, 1988).

Kanner, Bernice. "Starting Over: Magazine Facelifting." *New York* (Sept. 20, 1988), 22–27.

Keller, Rip. "How the Regional Books Help Advertisers Fill In the Gaps." *Magazine Age* (July 1983), 30–32, 34–41.

Kelley, Pat. "City and State Magazines." *Marketing & Media Decisions* (Mar. 1982), 175–76.

Kirkhorn, Michael J. "Nuclear Arms Reporting: Not With a Bang, but a Whimper." *Quill* (July–Aug. 1983), 11–16.

Kleinfield, N. R. "Grace Mirabella, At 59, Starts Over Again." *New York Times*, Apr. 30, 1989.

Kleinfield, N. R. "In Search of the Next Medium." *New York Times*, Mar. 20, 1989.

Kleinfield, N. R. "Inside *New Yorker*, A Splash of Color." *New York Times*, Feb. 15, 1989.

Knowles, Horace, ed. *Gentlemen, Scholars and Scoundrels: A Treasury of the Best of Harper's Magazine, From 1850 to the Present.* New York: Harper & Bros., 1959.

Kobak, James E. "A Billion-Dollar Year for Acquisitions." *Folio* (Apr. 1985), 82–95.

Kurtzer, Steve. "Magazines: The Next Decade," *Marketing & Media Decisions* (June 1985), 142, 144.

Lacob, Miriam. "Drama in Real Life at Reader's Digest." *Folio* (Feb. 1985), 88–92.

"Legal Publications: A New Growth Industry." *New York Times,* Aug. 19, 1988.

"Liberal Magazines Are Adapting to the 80's." *Folio* (May 1986), 53–54.

Liff, Marc. "Spanish-Language Magazines Begin to Soar." *Advertising Age* (Mar. 19, 1984), M-32–M-33.

"Literary Cross-Pollination: Small Magazines and Book Publishers Are Learning to Draw Upon Each Other's Readers." *Publishers Weekly* (Mar. 31, 1989).

Little, Stuart. "What Happened at Harper's. *Saturday Review* (Apr. 10, 1971), 43–47.

Love, Barbara. "The Past Ten Years: Tracking the Industry's Performance," *Folio* (Sept. 1982), 60–64.

Luce, Robert B. *The Faces of Five Decades: Selections from Fifty Years of the* New Republic, *1914–1964.* New York: Simon and Schuster, 1964.

Lukovitz, Karlene. "Publishing Management Structures." *Folio* (May 1983), 57–69.

"The Macfadden Empire." *New York Times,* May 7, 1989.

Mackenzie, Angus. "Sabotaging the Dissident Press." *Columbia Journalism Review* (Mar.–Apr. 1981), 57–63.

Mackenzie, Angus. "When Auditors Turn Editors: The IRS and the Nonprofit Press." *Columbia Journalism Review* (Nov.–Dec. 1981), 29–34.

"Magazine Industry Campaign." *New York Times,* June 2, 1988.

"Magazine on Addictions." *New York Times,* Nov. 18, 1987.

"Magazines Find Video an Attractive Medium." *New York Times,* July 11, 1988.

"Magazines for Children Address Serious Topics." *New York Times,* June 11, 1985.

"Magazines Raise Reliance on Circulation." *New York Times,* May 8, 1989.

"Magazines Troubled by Campus Editions." *New York Times,* April 20, 1988.

"Magazines Win Older Readers and Eager Advertisers." *New York Times,* Sept. 11, 1988.

"Magazine Takes Aim at Hispanic Market." *New York Times,* Apr. 20, 1989.

"Magazine Targets Consumers in 'Full-Life' Segment." *Marketing News* (June 6, 1988).

Margolick, David. "After 10 Years, Upstart Publisher and His Iconoclastic Magazine Have Both Mellowed." *New York Times,* Mar. 24, 1989.

Mayer, Milton. "Only Seventy-five: Still Time Enough to Change the World." The *Progressive* (July 1984), 82.

McArdle, Kenneth, ed. *A Cavalcade of* Collier's. New York: A. S. Barnes, 1959.

McFadden, Maureen. "Why Some Magazines Are Replacing Fiction with Fact." *Magazine Age* (July 1983), 22–26.

"Men's Entertainment Magazines Target of 'Vigilante Groups.'" *Folio* (Nov. 1984), 65.

Meyers, Bill. "The Advance of Science." *Washington Journalism Review* (Nov. 1981), 36–37.

Miller, Cyndee. "Fashion Magazine Targets 'Classiest Zip Codes.'" *Marketing News* (June 20, 1988), 13.

Mitgang, Herbert. "American Libraries Are in Crisis Over the Cost of Scholarly Journals." *New York Times,* Sept. 5, 1988.

Moskowitz, Sam. *Explorers of the Infinite: Shapers of Science Fiction.* Westport: Hyperion Press, 1974.

Moskowitz, Sam. *Seekers of Tomorrow: Masters of Modern Science Fiction.* Cleveland: 1961.

Moore, Thomas. "Why Cable TV Is Unraveling *TV Guide.*" *Philadelphia Magazine* (July 1981), 108–11.

"The Moving Target." *Marketing & Media Decisions* (Nov. 1979), 64–65.

"Murdoch Buys Triangle Publications." *New York Times,* Aug. 8, 1988.

"Murdoch to Introduce *Sports Travel* in June," *New York Times,* Jan. 25, 1989.

"Murdoch to Sell *Travel Business.*" *New York Times,* May 7, 1989.

The *Nation:* 100th Anniversary Issue (Sept. 30, 1965).

Neher, Jacques. "Video Boom Spawns Magazines." *Advertising Age* (Jan. 12, 1981), S-10–S-11.

"New Cash for Old Bostonians," *Time* (Mar. 17, 1980), 97–98.

"New Journals Tell the Local Business Story." *Business Week* (Oct. 5, 1981), 126, 130.

"*Newlywed* Magazine to Begin in February." *New York Times,* Sept. 12, 1988.

"New Magazine for Women." *New York Times,* May 24, 1989.

"New Magazine on Publishing." *New York Times,* Apr. 25, 1989.

Novit, Earl. "Modernism and Three Magazines: An Editorial Revolution." *Sewanee Review* (Fall 1985), 540–53.

Owen, David. "The Fifth Estate: Eavesdropping on American Business Talking to Itself." *Atlantic* (July 1985), 80–84.

Patterson, Oscar III. "Television's Living-Room War in Print: Vietnam in the News Magazines." *Journalism Quarterly* (Spring 1984), 34–39.

Peer, Elizabeth. "The High Life of Malcolm the Audacious." *New York,* (Jan. 30, 1984).

Peter, John. "12-30 Corporation." *Folio* (Sept. 1978), 37, 60.

Peter, John. "Women's Magazines: A Survey of the Field." *Folio* (Apr. 1977), 74–79.

"*Philip Morris* Magazine." *New York Times,* June 29, 1988.

Pohl, Frederik. "Astounding Story." *American Heritage* (Sept.–Oct. 1989).

Pool, Gail. "*Atlantic Monthly.*" *Serials Review* (Spring 1982), 29–32.

Pool, Gail. "Magazines." *Wilson Library Bulletin* (Feb. 1964), 445–46.

Pool, Gail. "Magazines." *Wilson Library Bulletin* (Mar. 1982), 538, 539, 558.

Pool, Gail. "Magazines." *Wilson Library Bulletin* (Sept. 1982), 67–68.

Pool, Gail. "Magazines." *Wilson Library Bulletin* (May 1983), 780–81.

Pool, Gail. "Magazines." *Wilson Library Bulletin* (Jan. 1984).

Pool, Gail. "Magazines." *Wilson Library Bulletin* (Dec. 1985), 54–55.

"Premier Deal for Murdoch." *New York Times,* Feb. 17, 1989.

Prial, Frank. "A Wine Magazine That Ages Well." *New York Times,* June 11, 1988.

Pugh, David. "History as an Expedient Accommodation: The Manliness Ethos in Modern America." *Journal of American Culture* (Spring 1980), 53–68.

Rao, S. Sreenvias. "Why *Coronet* Failed." *Journalism Quarterly* (Spring 1965), 271.

Reed, Robert. "Flying Those Not-Always-So-Friendly Skies." *Advertising Age* (July 5, 1982), M-2–M-3.

"Reed Wants Even More Magazines." *New York Times,* Apr. 8, 1988.

Reilling, Patrick. "*Playboy* Sees Happy 39th." *Advertising Age* (Sept. 5, 1988).

Reinhold, Robert. "California Trend: Serious Magazines." *New York Times,* July 30, 1989.

Reuss, Carol. "*Better Homes and Gardens:* Consistent Key to Long Life." *Journalism Quarterly* (Summer 1974), 292–96.

Ross, Michael E. "Environmental Journal Trying to Go It Alone." *New York Times,* June 19, 1989.

Ross, Val. "Paying the Wages of War." *MacLean's* (Apr. 11, 1983), 62–66.

Rossi, Lee D. "The Whore vs. the Girl-Next-Door: Stereotypes of Women in *Playboy*, *Penthouse*, and *Oui*." *Journal of Popular Culture* (Summer 1975), 9–94.

Roth, Eleanor. "Writing for Today's Confession Magazines." *Writer's Digest* (Aug. 1972), 27–29.

Rothenberg, Randall. "Condé Nast Changing Style of *Self*." *New York Times*, Nov. 15, 1988.

Rothenberg, Randall. "Condé Nast to Buy Woman Magazine." *New York Times*, Nov. 16, 1988.

Rothenberg, Randall. "First Glimpse of Murdoch's *Mirabella*." *New York Times*, Feb. 7, 1989.

Rothenberg, Randall. "*Mother Jones* Settles It: 60s Are Over." *New York Times*, Jan. 25, 1989.

Rothenberg, Randall. "*New England Monthly* to Be Sold." *New York Times*, Mar. 1, 1989.

Rothenberg, Randall. "Resignation and Boycott at *Sassy*." *New York Times*, Apr. 11, 1988.

Rothenberg, Randall. "A Rosy Year for a Travel Magazine." *New York Times*, Nov. 1, 1988.

Rothenberg, Randall. "With Media Losing Mass, What's Left?" *New York Times*, July 11, 1989.

Rothenberg, Randall. "A Young Elite's Power Over Ads." *New York Times*, Feb. 1, 1989.

"Saul Bellow on Magazines." Interview in *Publishers Weekly* (Mar. 3, 1989).

Scardino, Albert. "Big Spender at *Vanity Fair* Raises the Ante For Writers." *New York Times*, Apr. 17, 1989.

Scardino, Albert. "Former Editor of *Vogue* to Head New Magazine." *New York Times*, Nov. 16, 1988.

Scardino, Albert. "Magazines: Glimpsing a Day When No 2 Copies Will Be Alike." *New York Times*, Apr. 24, 1989.

Scardino, Albert. "The Magazine That Lost Its Way." *New York Times*, June 18, 1989.

Scardino, Albert. "The New Baby Boom Spurs Local Magazines for Parents." *New York Times*, June 26, 1989.

Scardino, Albert. "Reader's Digest Sets Stock Offer." *New York Times*, May 26, 1989.

Scardino, Albert. "Shakeout Raises Concerns on Distribution of Magazines." *New York Times*, May 15, 1989.

Schmidt, Dorothy. "Magazines, Technology and American Culture." *Journal of American Culture* (Spring 1980), 3–16.

Schrambling, Regina. "A *Sassy* Approach." *Profiles, Inc.* (June 1988).

Scott, Joseph E., and Jack L. Franklin. "The Changing Nature of Sex Reference in Mass-Circulation Magazines." *Public Opinion Quarterly* (Spring 1972), 80–86.

Shelley, Gerald U., and William J. Lundstrom. "Male Sex Roles in Magazine Advertising, 1959–1979." *Journal of Communication* (Autumn 1981), 52–57.

Simonds, C. H. "A 25-year Frolic." *National Review* (Dec. 31, 1980), 1606–34.

Sitcomer, Curtis. "An Outspoken Magazine Enters Its Second Century." *Christian Science Monitor*, Sept. 24, 1984.

"Six New Magazines for Doctors' Offices Promise Exclusive Market for Advertisers." *Marketing Now* (n.d.).

Smith, Ron F., and Linda Decker-Amos. "Of Lasting Interest: A Case Study of Change in the Content of the *Reader's Digest.*" *Journalism Quarterly* (Spring 1985), 127–31.

"A Soft Approach to Hardware." *New York Times,* Aug. 1, 1988.

Southard, Bruce. "Language of Science-Fiction Fan Magazines." *American Speech* 57 (Spring 1982): 19–31.

Sprague, Peter F. "What Makes a Successful Publishing Company?" *Folio* (Nov. 1988), 256.

Stein, M. L. "Opinion Journals Meet." *Editor & Publisher* (Apr. 27, 1985), 18, 20.

Stepp, Carl Sessions. "Looking Out for #1—Magazines That Celebrate Success." *Washington Journalism Review* (Nov. 1985), 41–44.

Stevenson, Richard. "Fitness Magazine Explosion." *New York Times,* June 15, 1985.

Stevenson, Richard W. "A Sports Magazine from ESPN." *New York Times,* Oct. 13, 1988.

Stewart, Penni. "He Admits . . . But She Confesses." *Women's Studies International Quarterly* no. 1 (1980): 105–14.

Tan, Alexis S., and Kermit Joseph Scruggs. "Does Exposure to Comic-Book Violence Lead to Aggression in Children?" *Journalism Quarterly* 57 (Winter 1980): 579–83.

"Time Inc. and Warner to Merge." *New York Times,* Mar. 5, 1989.

"Time Inc. Stake in *Hippocrates.*" *New York Times,* July 28, 1988.

"Times Buys *Golf* Magazine." *New York Times,* Dec. 8, 1988.

Toth, Debora. "What's New in Magazine Redesign." *New York Times,* Jan. 1, 1989.

"Update on *Reader's Digest.*" *New York Times,* Aug. 3, 1988.

Vanden Bergh, Bruce G., Alan D. Fletcher, and Mary A. Adrian. "Local Business Press: New Phenomenon in the News Marketplace." *Journalism Quarterly* (Autumn 1984).

Wainwright, Loudon. Life: *The Great American Magazine.* New York: Ballantine Books, 1986.

Weeks, Edward, and Emily Flint, eds. *Jubilee: One Hundred Years of the* Atlantic. Boston: Atlantic Monthly Press, 1957.

Whitney, Catherine. "Magazines in the Bookstore." *Publishers Weekly* (Apr. 1, 1988).

Yarrow, Andrew L. "*Business Week's* Work of Art." *New York Times,* Dec. 5, 1988.

"A Year Old, *7 Days* Faces New Tests. *New York Times,* Mar. 22, 1989.

"Your Family Debut." *New York Times,* Apr. 5, 1989.

"Youth Media Market Diverse." *Advertising Age* (Apr. 1980), S-2.

Yovovich, B.G. "Leisure Reading for Sports." *Advertising Age* (Dec. 1, 1980), 52.

Index